Colloidal Drug Delivery Systems

DRUGS AND THE PHARMACEUTICAL SCIENCES

A Series of Textbooks and Monographs

edited by

James Swarbrick
Applied Analytical Industries, Inc.
Wilmington, North Carolina

1. Pharmacokinetics, *Milo Gibaldi and Donald Perrier*
2. Good Manufacturing Practices for Pharmaceuticals: A Plan for Total Quality Control, *Sidney H. Willig, Murray M. Tuckerman, and William S. Hitchings IV*
3. Microencapsulation, *edited by J. R. Nixon*
4. Drug Metabolism: Chemical and Biochemical Aspects, *Bernard Testa and Peter Jenner*
5. New Drugs: Discovery and Development, *edited by Alan A. Rubin*
6. Sustained and Controlled Release Drug Delivery Systems, *edited by Joseph R. Robinson*
7. Modern Pharmaceutics, *edited by Gilbert S. Banker and Christopher T. Rhodes*
8. Prescription Drugs in Short Supply: Case Histories, *Michael A. Schwartz*
9. Activated Charcoal: Antidotal and Other Medical Uses, *David O. Cooney*
10. Concepts in Drug Metabolism (in two parts), *edited by Peter Jenner and Bernard Testa*
11. Pharmaceutical Analysis: Modern Methods (in two parts), *edited by James W. Munson*
12. Techniques of Solubilization of Drugs, *edited by Samuel H. Yalkowsky*
13. Orphan Drugs, *edited by Fred E. Karch*
14. Novel Drug Delivery Systems: Fundamentals, Developmental Concepts, Biomedical Assessments, *Yie W. Chien*
15. Pharmacokinetics: Second Edition, Revised and Expanded, *Milo Gibaldi and Donald Perrier*
16. Good Manufacturing Practices for Pharmaceuticals: A Plan for Total Quality Control, Second Edition, Revised and Expanded, *Sidney H. Willig, Murray M. Tuckerman, and William S. Hitchings IV*
17. Formulation of Veterinary Dosage Forms, *edited by Jack Blodinger*
18. Dermatological Formulations: Percutaneous Absorption, *Brian W. Barry*
19. The Clinical Research Process in the Pharmaceutical Industry, *edited by Gary M. Matoren*

20. Microencapsulation and Related Drug Processes, *Patrick B. Deasy*
21. Drugs and Nutrients: The Interactive Effects, *edited by Daphne A. Roe and T. Colin Campbell*
22. Biotechnology of Industrial Antibiotics, *Erick J. Vandamme*
23. Pharmaceutical Process Validation, *edited by Bernard T. Loftus and Robert A. Nash*
24. Anticancer and Interferon Agents: Synthesis and Properties, *edited by Raphael M. Ottenbrite and George B. Butler*
25. Pharmaceutical Statistics: Practical and Clinical Applications, *Sanford Bolton*
26. Drug Dynamics for Analytical, Clinical, and Biological Chemists, *Benjamin J. Gudzinowicz, Burrows T. Younkin, Jr., and Michael J. Gudzinowicz*
27. Modern Analysis of Antibiotics, *edited by Adjoran Aszalos*
28. Solubility and Related Properties, *Kenneth C. James*
29. Controlled Drug Delivery: Fundamentals and Applications, Second Edition, Revised and Expanded, *edited by Joseph R. Robinson and Vincent H. Lee*
30. New Drug Approval Process: Clinical and Regulatory Management, *edited by Richard A. Guarino*
31. Transdermal Controlled Systemic Medications, *edited by Yie W. Chien*
32. Drug Delivery Devices: Fundamentals and Applications, *edited by Praveen Tyle*
33. Pharmacokinetics: Regulatory · Industrial · Academic Perspectives, *edited by Peter G. Welling and Francis L. S. Tse*
34. Clinical Drug Trials and Tribulations, *edited by Allen E. Cato*
35. Transdermal Drug Delivery: Developmental Issues and Research Initiatives, *edited by Jonathan Hadgraft and Richard H. Guy*
36. Aqueous Polymeric Coatings for Pharmaceutical Dosage Forms, *edited by James W. McGinity*
37. Pharmaceutical Pelletization Technology, *edited by Isaac Ghebre-Sellassie*
38. Good Laboratory Practice Regulations, *edited by Allen F. Hirsch*
39. Nasal Systemic Drug Delivery, *Yie W. Chien, Kenneth S. E. Su, and Shyi-Feu Chang*
40. Modern Pharmaceutics: Second Edition, Revised and Expanded, *edited by Gilbert S. Banker and Christopher T. Rhodes*
41. Specialized Drug Delivery Systems: Manufacturing and Production Technology, *edited by Praveen Tyle*
42. Topical Drug Delivery Formulations, *edited by David W. Osborne and Anton H. Amann*
43. Drug Stability: Principles and Practices, *Jens T. Carstensen*
44. Pharmaceutical Statistics: Practical and Clinical Applications, Second Edition, Revised and Expanded, *Sanford Bolton*
45. Biodegradable Polymers as Drug Delivery Systems, *edited by Mark Chasin and Robert Langer*

46. Preclinical Drug Disposition: A Laboratory Handbook, *Francis L. S. Tse and James J. Jaffe*
47. HPLC in the Pharmaceutical Industry, *edited by Godwin W. Fong and Stanley K. Lam*
48. Pharmaceutical Bioequivalence, *edited by Peter G. Welling, Francis L. S. Tse, and Shrikant V. Dinghe*
49. Pharmaceutical Dissolution Testing, *Umesh V. Banakar*
50. Novel Drug Delivery Systems: Second Edition, Revised and Expanded, *Yie W. Chien*
51. Managing the Clinical Drug Development Process, *David M. Cocchetto and Ronald V. Nardi*
52. Good Manufacturing Practices for Pharmaceuticals: A Plan for Total Quality Control, Third Edition, *edited by Sidney H. Willig and James R. Stoker*
53. Prodrugs: Topical and Ocular Drug Delivery, *edited by Kenneth B. Sloan*
54. Pharmaceutical Inhalation Aerosol Technology, *edited by Anthony J. Hickey*
55. Radiopharmaceuticals: Chemistry and Pharmacology, *edited by Adrian D. Nunn*
56. New Drug Approval Process: Second Edition, Revised and Expanded, *edited by Richard A. Guarino*
57. Pharmaceutical Process Validation: Second Edition, Revised and Expanded, *edited by Ira R. Berry and Robert A. Nash*
58. Ophthalmic Drug Delivery Systems, *edited by Ashim K. Mitra*
59. Pharmaceutical Skin Penetration Enhancement, *edited by Kenneth A. Walters and Jonathan Hadgraft*
60. Colonic Drug Absorption and Metabolism, *edited by Peter R. Bieck*
61. Pharmaceutical Particulate Carriers: Therapeutic Applications, *edited by Alain Rolland*
62. Drug Permeation Enhancement: Theory and Applications, *edited by Dean S. Hsieh*
63. Glycopeptide Antibiotics, *edited by Ramakrishnan Nagarajan*
64. Achieving Sterility in Medical and Pharmaceutical Products, *Nigel A. Halls*
65. Multiparticulate Oral Drug Delivery, *edited by Isaac Ghebre-Sellassie*
66. Colloidal Drug Delivery Systems, *edited by Jörg Kreuter*

ADDITIONAL VOLUMES IN PREPARATION

Colloidal Drug Delivery Systems

edited by

Jörg Kreuter

Institut für Pharmazeutische Technologie
Johann Wolfgang Goethe-Universität
Frankfurt am Main, Germany

CRC Press is an imprint of the
Taylor & Francis Group, an **informa** business

CRC Press
Taylor & Francis Group
6000 Broken Sound Parkway NW, Suite 300
Boca Raton, FL 33487-2742

First issued in paperback 2019

ISBN-13: 978-0-8247-9214-5 (hbk)
ISBN-13: 978-0-367-40201-3 (pbk)

Library of Congress Cataloging-in-Publication Data

Colloidal drug delivery systems / edited by Jörg Kreuter.
p. cm. — (Drugs and the pharmaceutical sciences ; v. 66)
Includes bibliographical references and index.
ISBN 0-8247-9214-9
1. Colloids in medicine. 2. Drug delivery systems. I. Kreuter, Jörg. II. Series.
RS201.C6C65 1994 94-14913
615'.6—dc20 CIP

Visit the Taylor & Francis Web site at
http://www.taylorandfrancis.com

and the CRC Press Web site at
http://www.crcpress.com

Preface

The efficacy of many drugs is often limited by their potential to reach their therapeutical site of action. In most cases, only a small amount of the administered dose of the drug reaches this site, while the major drug amount is distributed to the rest of the body depending on the physicochemical and biochemical properties of the drug. In contrast, a site-specific delivery would not only increase the amount of drug reaching the site but also simultaneously decrease the amount being distributed to other parts of the body, thus reducing unwanted side effects. Site-specific or targeted delivery, therefore, would also enable a reduction in the necessary dose to be administered. By decreasing the side effects, it would also increase the therapeutic index of the drug.

A more specific accumulation at the target site may be achieved by altering the chemical structure toward more suitable physicochemical and/or biochemical properties of the drug. This strategy involves the synthesis of either a totally new compound or a so-called prodrug from which the active drug is produced by the body's own metabolism. Although these approaches may be very successful in some cases, in others the synthesis of a more site-specific drug is not possible. For this reason, delivery of the original drug by specially designed drug delivery systems is the better solution—and sometimes the only feasible one.

A large number of drug delivery systems have been conceived and developed. Among these systems, colloidal drug delivery systems hold great promise for reaching the goal of drug targeting.

The idea of drug targeting by such systems was originally formulated by Paul Ehrlich. After visiting the opera *Der Freischütz* by Carl Maria von Weber, in which so-called *Freikugeln* play a major role, Ehrlich came upon the idea of *Zauberkugeln*—magic bullets (1). The *Freikugeln* in the opera could be fired in any direction yet still reach their goal. Ehrlich imagined that similarly targeted tiny drug-loaded magic bullets would significantly improve therapy.

Because of their small particle size—below 1 μm—colloidal drug carriers come close to Ehrlich's idea. These drug delivery systems offer various advantages for many medical, agricultural, veterinary, and industrial applications.

Over the years, controlled drug delivery as well as site-specific delivery have made considerable advances. One area that contributed significantly to this progress is the rapidly developing field of submicron delivery systems—colloidal drug carriers. Although numerous reviews and books have been published concerning this topic, the rapid developments in the area of colloidal drug carriers make frequent updates a necessity.

This volume focuses mainly on the most widely investigated colloidal drug delivery systems: nanoparticles, liposomes and the related niosomes, and microemulsions. In addition, the book also sheds light on a somewhat forgotten fact, namely that some ointments also represent colloidal drug delivery systems.

The length of individual chapters may appear imbalanced—the chapters on nanoparticles and especially liposomes are much longer than the others. However, this organization was intended because much more work is presently being done and published in the liposome and nanoparticle fields than in the other areas.

The chapters in this book are arranged by increasing physical ordering and complexity, beginning with ointments, then describing the related but more particulate microemulsions, followed by the more solid liposomes and niosomes, and finally by solid system nanoparticles.

The editor thanks all contributors for their valuable work and Marcel Dekker, Inc., for its professional cooperation.

Jörg Kreuter

REFERENCE

1. Greiling, W. (1954). *Paul Ehrlich*. Düsseldorf: Econ Verlag, p. 48.

Contents

Contributors

David Attwood Department of Pharmacy, University of Manchester, Manchester, United Kingdom

J. A. Bouwstra Center for Biopharmaceutical Sciences, Leiden/Amsterdam Center for Drug Research, Leiden University, Leiden, The Netherlands

Daan J. A. Crommelin Department of Pharmaceutics, Utrecht Institute for Pharmaceutical Sciences, Utrecht University, Utrecht, The Netherlands

H. E. J. Hofland Center for Biopharmaceutical Sciences, Leiden/Amster-Center for Drug Research, Leiden University, Leiden, The Netherlands

Hans E. Junginger Department of Pharmaceutical Technology, Leiden/Amsterdam Center for Drug Research, Leiden University, Leiden, The Netherlands

Jörg Kreuter Biozentrum, Institut für Pharmazeutische Technologie, Johann Wolfgang Goethe-Universität, Frankfurt am Main, Germany

H. Schreier Center for Lung Research, Vanderbilt University School of Medicine, Nashville, Tennessee

1

Ointments and Creams as Colloidal Drug Delivery Systems

Hans E. Junginger

Leiden/Amsterdam Center for Drug Research, Leiden, The Netherlands

I. INTRODUCTION

Ointments and creams have been used for many centuries to improve the healing conditions of wounds and to treat in empirical ways skin diseases as well as to retard the aging process of the skin and to preserve natural beauty. Dermatological preparations and cosmetics that represent the modern generations of these systems are also called semisolids due to their unique properties of being in the solid state under ambient conditions and being transformed to the liquid state when mechanically stressed during the application on the skin. These properties allow the systems to spread easily on the surface of the skin. Depending on the formulation, ointments in general remain on the surface of the skin and show skin protection and occlusion, whereas creams may be able to penetrate into the different layers of the skin, especially the stratum corneum (horny layer) and to exert interactions with both the keratines in the horny cells (corneocytes) and the lipid bilayers in which the horny cells are embedded.

Ointments are defined as water-free systems ranging from systems with extreme lipophilic properties (e.g., liquid paraffins, white petrolatum, and paraffin waxes) to systems with good hydrophilic properties (e.g., polyethyleneglycol ointments). Ointments with medium polar character are W/O or O/W absorption bases (mostly lipophilic vehicles) in which O/W or W/O emulsifiers are incorporated. Addition of water to these absorption bases results in the formation of O/W or W/O creams, which are generally defined as water-containing systems. We speak of O/W creams when water forms the outer (continuous) phase and of W/O creams when the lipid phase is the outer phase. Creams may be viewed as simple O/W or W/O emulsions that are predominantly stabilized by a colloidal gel structure. In his pioneering work Münzel (1) described semisolids as plastic gels for cutaneous application.

The dominant colloidal structural elements of semisolid preparations are three-dimensional colloidal solid networks in which a liquid is incorporated. Such a bicoherent (spongelike) structure may be referred to as a gel. For this reason a differentiation in gels between continuous and dispersed phases is impossible because of the complete interpretation of the bicontinuous phases, whereas it is possible for emulsions and suspensions.

It is in fact this bicoherent gel structure of colloidal-sized structures, caused by van der Waals interactions and hydrogen bondings, that makes these systems different from the other colloidal drug carrier systems such as nanoparticles, liposomes, niosomes, etc. They preferentially exist in a dispersed state.

The gel structures of semisolid preparations may be either in a crystalline or in a liquid-crystalline state. The type of the gel structure mainly determines the systems' (1) rheological properties, (2) stability, (3) possible interactions with the skin, and (4) drug release. The knowledge of the colloidal structures of these semisolid systems is of essential importance for the understanding of the behavior of these systems and for their proper application on the skin.

In this chapter the colloidal structures of the most-used dermatological preparations, i.e., creams with crystalline and liquid-crystalline gel structures, will be presented, as well as some aspects of in vitro drug release and its dependence on the colloidal gel structures of these systems.

II. COLLOIDAL GEL STRUCTURES OF CREAMS

A. Colloidal Structures of O/W Creams

1. Water Containing Hydrophilic Ointment*

The formulation of Water Containing Hydrophilic Ointment DAB 9 (German Pharmacopoeia, 9th edition) is as follows:

*The German Pharmacopoeia, 9th edition, is not consistent in its definitions, and water containing hydrophilic ointment DAB 9 is a cream.

Emulsifying wax	9.0	% wt/wt
Liquid paraffins	10.5	% wt/wt
White petrolatum	10.5	% wt/wt
Water	70.0	% wt/wt

X-ray investigations by the Kratky low angle technique (SAXD) and by wide angle x-ray diffraction (WAXD) in combination with quantitative differential scanning calorimetry (DSC) led to the following structure model of Water Containing Hydrophilic Ointment DAB 9 (2–4). It was found that such O/W creams may be regarded as four-phase systems (Fig. 1). The dominant matrices are the hydrophilic and the lipophilic gel phases. Both gel phases form colloidal-sized mixed crystals consisting of surfactant bilayers. The surfactants in the bilayers are oriented in such a way that the hydrocarbon tails are directed toward each other, as are the polar groups (Fig. 1, region a). The hydrophilic gel phase consists of cetostearyl alcohol as well as the whole amount of the ionic sodium-*n*-alkylsulfates, which are randomly distributed within the cetostearyl alcohol molecules. The latter act as lateral spacers for the strong polar sodium-*n*-alkylsulfate molecules. In the crystalline bilayer structure therefore strong hydrophilic moieties and hydrophobic cores alternate with each other (Fig. 1, region a).

A part of the total water amount of the system is interlamellarly inserted between the polar groups of the surfactant molecules (Fig. 1, region b). This

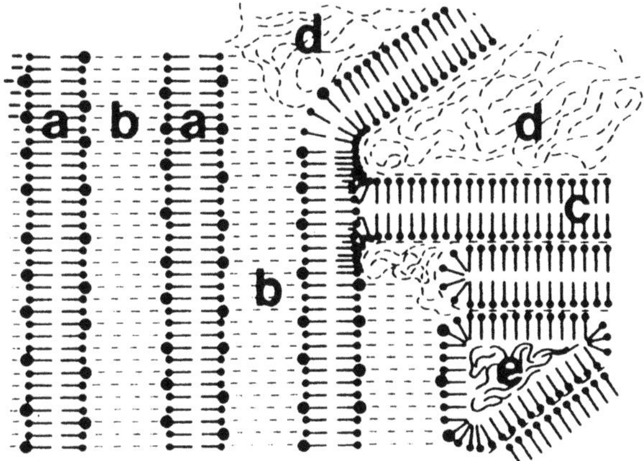

Figure 1 Gel structures of Water Containing Hydrophilic Ointment DAB 9. a, mixed crystal bilayer of cetostearylalcohol and cetostearylalcoholsulfates; b, interlamellarly fixed water layer; a + b, hydrophilic gel phase; c, lipophilic gel phase (cetostearylalcohol semihydrate); d, bulk water phase; e, lipophilic components (dispersed phase).

part of the water is called interlamellarly fixed water. The regions a and b together form the hydrophilic gel phase. The water molecules fixed interlamellarly in the hydrophilic gel phases are in equilibrium with the molecules of the bulk water phase (Fig. 1, region d). The bulk water phase is the liquid component of the gel structure, and the solid phase is the hydrophilic gel phase (although it contains part of the water interlamellarly bound). The bulk water is fixed within the network of the hydrophilic gel phase mainly by capillary attraction forces. Furthermore, it is assumed that the interlamellarly fixed water molecules exhibit other physicochemical properties than those of the bulk water phase.

The surplus of cetostearyl alcohol not incorporated in the hydrophilic gel phase builds up a separate matrix with lipophilic properties (Fig. 1, region c) called the lipophilic gel phase. The inner or dispersed phase (Fig. 1, region e) is mainly immobilized mechanically by this lipophilic gel phase. The lipophilic gel phase consisting of pure cetostearyl alcohol is only able to form a semihydrate with water.

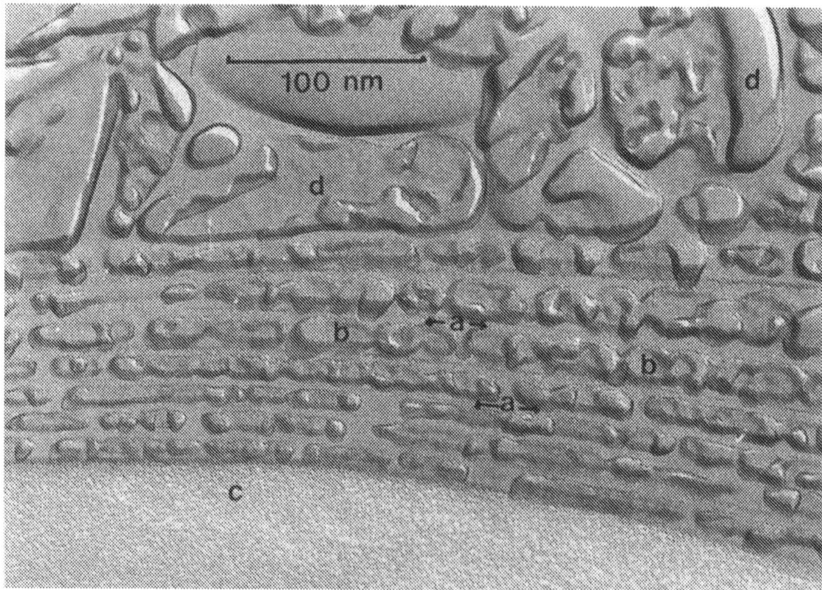

Figure 2 Freeze fracture micrograph of Emulsifying Wax DAB 9 (main structural component of Water Containing Hydrophilic Ointment DAB 9) with 70% (wt/wt) water. a, mixed crystal bilayer; b, interlamellarly fixed water; a + b, hydrophilic gel phase; c, fracture edge of a lipophilic plane; d, bulk water phase.

Freeze fracture electron microscopy (FFEM) has added a new dimension to the studies of colloidal O/W cream organization. This technique allows the visualization of the previously mentioned structural elements (5–7). From Fig. 2 the hydrophilic gel phase can be recognized very clearly. In this photograph the alternating layers of the hydrophilic gel phase are nearly rectangular to the fracture plane. Together with areas of bulk water d entrapped in the hydrophilic gel phase, the interlamellarly bound layers of water b and the bilayers of the surfactant molecules a are visible. Together, a and b form the hydrophilic gel phase.

Investigations about the swelling ability of Emulsifying Wax DAB 9 (main surfactant component of Water Containing Hydrophilic Ointment DAB 9) with water (Fig. 3) show a swelling of the lamellar gel structure when the long spacings as obtained from small angle x-ray diffraction (SAXD) are plotted versus the ratio water/surfactant (wt/wt) (see Fig. 3, where C is the

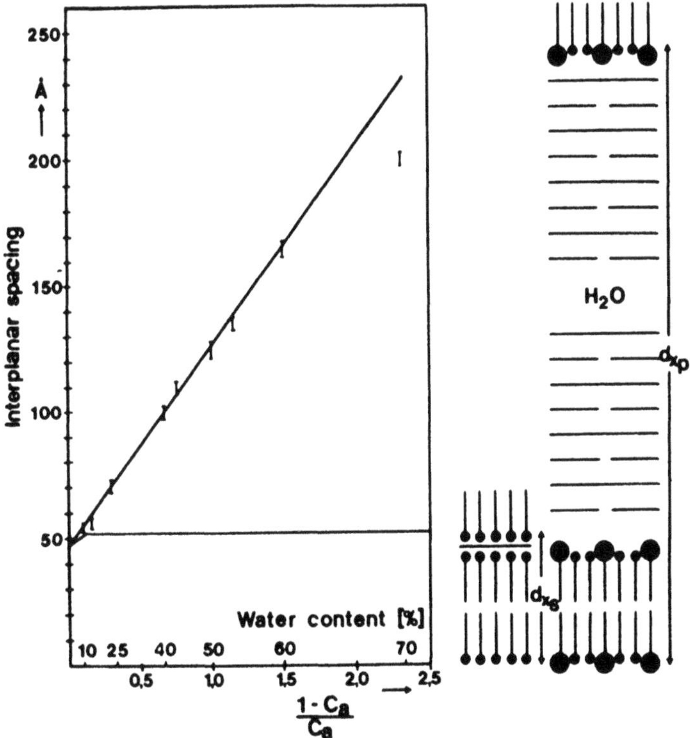

Figure 3 Swelling behavior of Emulsifying Wax DAB 9 with water. C_a, weight fraction of surfactant; $1 - C_a$, weight fraction of water; d_{xs}, interplanar spacings of cetostearylalcohol semihydrate (lipophilic) gel; d_{xp}, interplanar spacings of the hydrophilic gel phase.

weight fraction of surfactants; 1 − C is the weight fraction of water). At a water content of 70% (wt/wt) the thickness of the interlamellar fixed water layer is about 15 nm (long spacing = 20.0 ± 5 nm). For comparison, the molecular sizes for the cetostearylalcohol (lipophilic semihydrate gel phase) and for the hydrophilic gel phase are given on the right side of Fig. 3.

It must be emphasized that the degree of swelling of the hydrophilic gel phase depends on the total water content of the cream. Thus a dynamic equilibrium exists between the bulk water and the interlamellarly fixed water. The bulk water phase forms the continuous phase of the system, but the interlamellar water fraction also contributes to this continuity. The capacity of the hydrophilic gel phase to incorporate interlamellar water is high enough to obtain clearly defined melting and recrystallization peaks measured by DSC, which strongly vary from the water-free systems. At a certain water content of the system a maximum swelling of the interlamellar water layer is reached. Beyond this state the water molecules in the middle of the interlamellar layer possess the same mobility as water molecules of the bulk phase. Consequently, the colloidal structure of the hydrophilic gel phase breaks down and the three-dimensional gel structure is lost. This transition of the cream systems into this state represents the transition of the cream into the (unstable) state of a suspension (emulsion). Undergoing this transition, the plastic flow behavior properties of the cream are lost and the system exhibits the pseudoplastic flow behavior of an emulsion or a suspension.

The lipophilic gel phase (Fig. 1, region c), however, is only able to form a semihydrate, independent of the total water present in the system. After the transition from a cream into an emulsion (suspension) state, the lipophilic gel phase is still surrounding the dispersed inner phase (Fig. 1, region e).

To characterize these O/W creams further, a technique was searched for that allows a more quantitative differentiation between the different types of water existing in these bases. It may be speculated that the swelling capacity of the hydrophilic gel phase may influence not only the water release from the gel to the skin and the diffusion rate of drugs across the vehicle but also the penetration ability of drugs through the hydrated horny layer of the skin. By means of a dynamic thermogravimetric analysis (TGA), a method was developed that enables us to differentiate between interlamellarly fixed water and the bulk water fraction (3,4). The results with the Water Containing Ointment DAB 9 using the TGA are summarized in Fig. 4. The total water contents of the different creams are plotted against the total amount of interlamellarly fixed water fractions. At a total water amount of 70% (wt/wt), about 40% (wt/wt) is present interlamellarly in the hydrophilic gel phase. If the total water amount exceeds 80% (wt/wt), the hydrophilic gel phase reaches a saturation state, in which the water molecules in the middle of the water layer in between the bilayers are supposed to have the same free energy

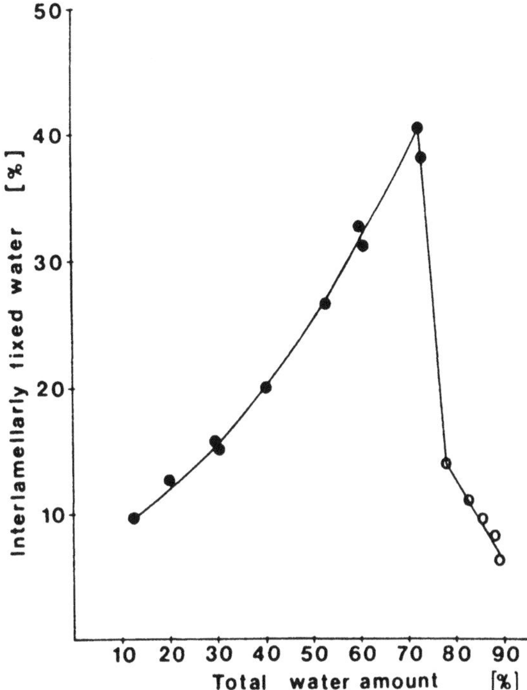

Figure 4 Amount of interlamellarly fixed water in Water Containing Hydrophilic Ointment DAB 9 depending on the total water amount of the system (○ = unstable systems).

as those of the bulk water molecules at the same temperature. This supersaturated state of the hydrophilic gel phase represents the transition from a cream into an emulsion. Simultaneously, the fraction of interlamellarly fixed water markedly decreases in favor of the fraction of the bulk water phase (Fig. 4).

2. Colloidal Structures of Stearate Creams

For stearate creams a formulation proposed by Tronnier was chosen as a model system (8). The stearate creams investigated had the following compositions:

Stearic acid	12	% wt/wt
Palmitic acid	12	% wt/wt
Triethanolamine	1.2	% wt/wt
Glycerol	13.5	% wt/wt
Water	10–61.3	% wt/wt

Addition of water to the water-free system that consists of stearic acid, palmitic acid, and triethanolamine results in lamellar mixed crystals in which the water is incorporated between the hydrophilic moieties of the lamellae. At increasing water concentrations (10–61.3% wt/wt) a swelling capacity of the hydrophilic gel phase could be found (Fig. 5). The thickness of the interlamellar water layer is about 6.5 nm at a water content of 60% (wt/wt) (long spacing = 12.6 ± 0.2 nm) (9).

The results obtained by means of SAXD, DSC, and TGA led to a structure model for stearate creams as given in Fig. 6 (3,9). One part of the lamellar mixed crystals, which consists of free fatty acids and their triethanolamine salts, are able to form the hydrophilic gel phase. Between the polar moieties of the mixed crystals (Fig. 6, region a) water molecules are present (Fig. 6,

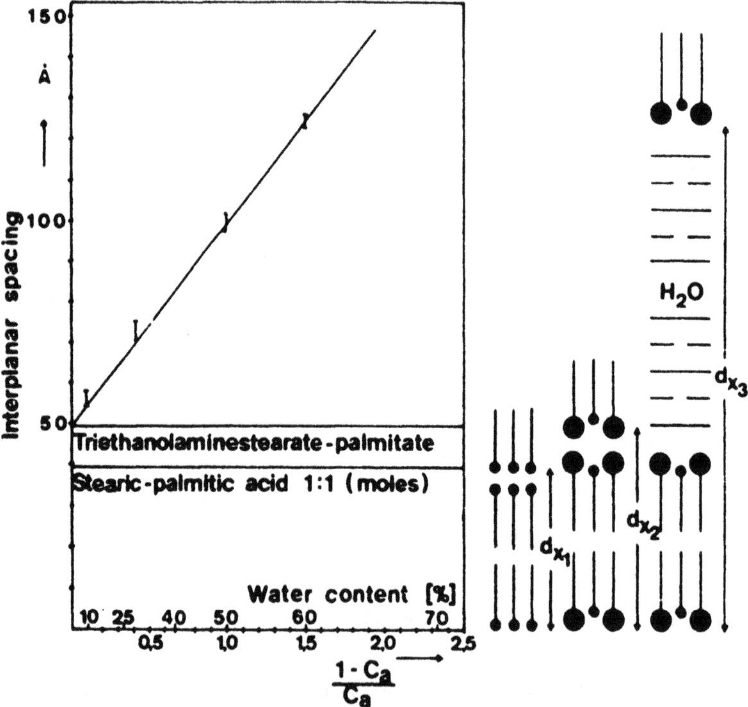

Figure 5 Swelling behavior of the gel forming components (triethanolamine-stearate-palmitate) of a stearate cream with water. C_a, weight fraction of surfactants; $1-C_a$, weight fraction of water; d_{x1}, interplanar spacing of stearic-palmitic acid mixture (1:1 mol); lipophilic gel phase; d_{x2}, interplanar spacing of triethanolamine-stearate-palmitate (water free); d_{x3}, interplanar spacing of triethanolamine-stearate-palmitate swollen with water (hydrophilic gel phase of the stearate cream).

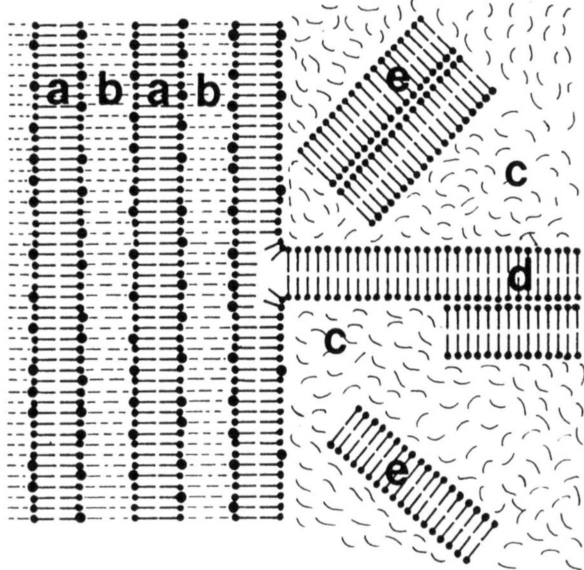

Figure 6 Gel structures of stearate creams. a, mixed crystal bilayer of triethanolamine-palmitate-stearate; b, interlamellarly fixed water; a + b, hydrophilic gel phase; d, bulk water phase; c, lipophilic gel phase ("stearate"); e, isolated "stearate" platelets, dispersed in the bulk water phase.

region b). This water of the hydrophilic gel phase is in equilibrium with the bulk water of the continuous phase (Fig. 6, region d).

The second part of the gel network, consisting of pure mixed crystals of palmitic and stearic acid, which is not able to retain water interlamellarly, forms a lipophilic gel phase (Fig. 6, region c). If a dispersed (lipophilic) phase is present in such a system, it is mainly immobilized by the lipophilic gel phase. Stearate creams show a special pearl effect due to the crystallization of very small isolated platelets (Fig. 6, region e), depending on the amount of added triethanolamine as well as the manufacturing conditions. Platelets are formed preferably in place of a coherent lipophilic gel phase, especially in the absence of a lipophilic phase.

The thermogravimetric results of the stearate cream are depicted in Fig. 7. It becomes quite clear that above all at a high water content only one-third of the water is fixed interlamellarly, but two-thirds of the water is present as a bulk water phase that is directly available for skin hydration. These facts explain the results of Tronnier, who stated that the skin hydration rate is much higher by stearate creams in comparison to other O/W creams (8).

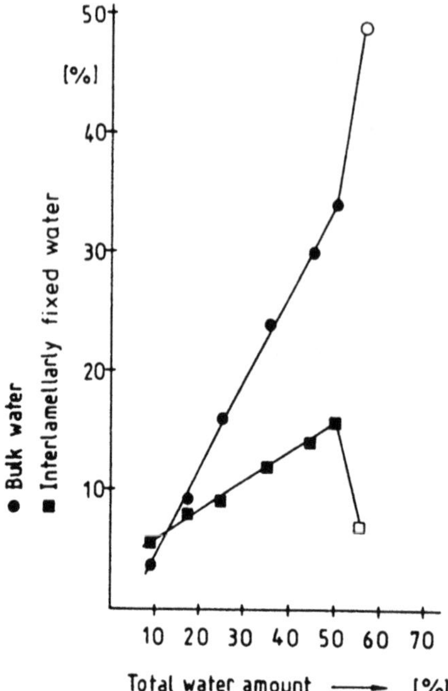

Figure 7 Ratio of bulk water and interlamellarly fixed water of a stearate cream depending on the total water amount of the systems (□, ○ = unstable systems).

These facts also could be confirmed by isothermal TGA, comparing systems with different ratios of interlamellarly fixed water in their hydrophilic gel phases. At ambient temperatures, stearate creams lost a substantial part of the incorporated water much quicker than Water Containing Hydrophilic Ointment DAB 9 with a high amount of interlamellarly fixed water.

At a total water amount higher than 55%, these systems become unstable (Fig. 7), and a transition takes place from a cream with a coherent three-dimensional hydrophilic gel network to an emulsion without these structural elements (4).

3. Colloidal Structures of Nonionic Hydrophilic Creams

Dermatological preparations containing ionic surfactants may show some disadvantages, such as a higher irritation potential of the skin, strong effects on the systems' stability when salts are added, interactions with drugs, etc. For these reasons, preparations containing only nonionic surfactants will be more suitable as a general applicable cream. As one of the most representative formulations of nonionic hydrophilic creams the following formula was studied:

-Poly(oxyethylene) 20 glycerol-monostearate (PGM 20)	7.5	% wt/wt
-Liquid paraffin	7.5	% wt/wt
-Cetylalcohol	5.0	% wt/wt
-Stearylalcohol	5.0	% wt/wt
-Glycerol	8.5	% wt/wt
-White soft paraffin	17.5	% wt/wt
-Water	51.5	% wt/wt

This particular system is known as Unguentum Hydrophilicum Nonionicum Aquosum, DAC (Deutscher Arzneimittel Codex 1979). Similar formulations appeared in the Swiss Pharmacopoeia, 6th edition, and in the Formulary of the Dutch Pharmacists (FNA).

SAXD investigations about the structure of the elementary units of the gel present in the above-mentioned formulation show that PGM 20 and cetostearyl alcohol form mixed crystals if the water free melt recrystallizes (Fig. 8). In the water-free system also the diffraction peaks of pure PGM 20 and cetostearyl alcohol are to be found.

With SAXD no reproducible distances are found if water is added to the water-free system up to 20% (wt/wt) (Fig. 8). The samples remain solidlike, and polarization microscopy shows anisotropic structures. It is assumed that the hydrophilic polyoxyethylene chains together with the water molecules and the hydroxyl groups of cetostearyl alcohol form the hydrophilic layers between the lipophilic sheets. At a water content of 20% (wt/wt) the polyoxyethylene chains are surrounded with the minimum of hydration water needed to build up a homogenous lamellar structure. Thus reduction of the water content leads to formation of mixed crystals containing partially hydrated polyoxyethylene chains as well as cetostearyl alcohol and PGM 20. This picture (Fig. 8) is reinforced by the observation that samples containing less than 20% (wy/wt) water show several endothermal peaks with DTA and no electrical conductivity; consequently, the coherency of the water layer is not reached yet (10,11). The reason for this, therefore, is the insufficient hydration of the polyoxyethylene groups of PGM 20. It has been shown that this insufficient hydration of the polyoxyethylene groups influences the in vitro drug release (12,13; see Sec. IV.A). Between 25 and 60% (wt/wt) water, continuous swelling of the hydrophilic gel phase takes place (Fig. 8).

In addition, in this nonionic cream system polyoxyethylene-glycerolmonostearate (PGM 20) crystallizes together with cetostearylalcohol as mixed crystals (Fig. 9, region a). The swelling degree with water depends on total

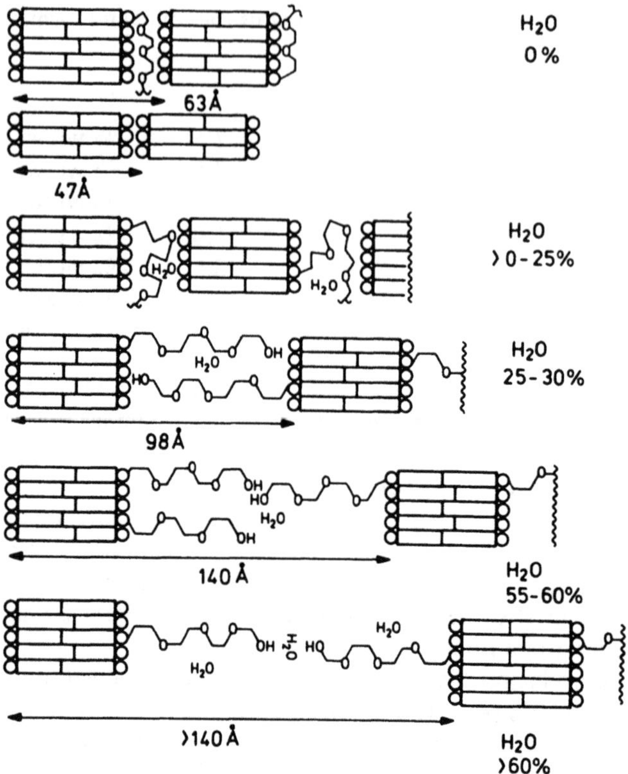

Figure 8 Swelling characteristics of cetostearylalcohol, polyoxyethylene-glycerolmonostearate, and water mixtures. In the water-free state two diffraction peaks are found for cetostearylalcohol (4.7 nm) and cetostearylalcohol-polyoxyethylene-glycerolmonostearate mixed crystals (6.3 nm). Up to 20% (wt/wt) water unreproducible diffraction peaks are found due to insufficient hydration of the polyoxyethylene groups. Between 25 and 60% (wt/wt) water, continuous swelling of the hydrophilic gel phase takes place. At a water content higher than 60% (wt/wt) the hydrophilic gel phase becomes unstable and breaks down.

water content. The length of the polyoxyethylene unit determines the maximum swelling capacity of the systems.

Together with polyoxyethylene bound water the lamellar mixed crystals form the hydrophilic gel phase (Fig. 9, regions a and b) in which partly bulk water is incorporated, too. The hydrophilic gel phase together with (part of) the bulk water (Fig. 9, region d) are the components of the three-dimensional gel-network.

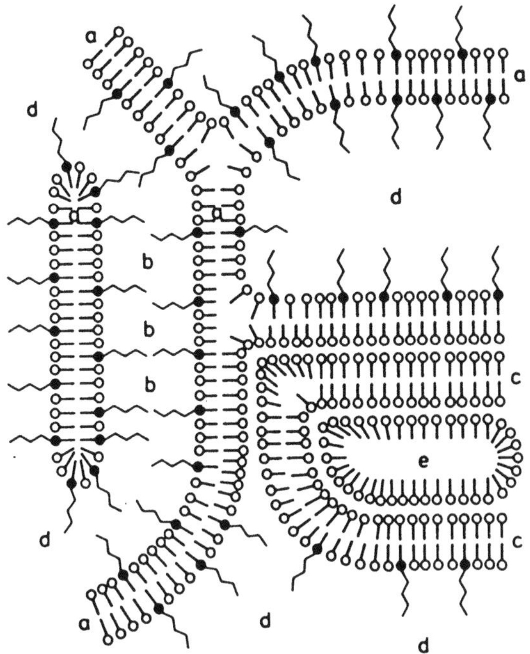

Figure 9 Schematic presentation of the gel structures of nonionic hydrophilic cream (DAC 79). a, mixed crystal bilayer of cetostearylalcohol and poly(oxyethylene) 20 glycerolmonostearate (PGM 20); b, interlamellarly fixed water; a + b, hydrophilic gel phase; c, lipophilic gel phase (cetostearylalcohol-semihydrate); d, bulk water phase; e, lipophilic components (dispersed phase).

The surplus of cetostearylalcohol again forms the lipophilic gel phase (Fig. 9, region c), which immobilizes the lipophilic dispersed phase (Fig. 9, region e) consisting mainly of white petrolatum and liquid paraffins. As a result of these investigations it is concluded that nonionic O/W creams also may be regarded as four-phase systems consisting of the same structural elements as the ionic O/W creams (14). Many nonionic surfactants are obtained with various PEG chain lengths. For this reasons, it becomes feasible to develop nonionic O/W creams with a desired ratio of interlamellarly fixed water and bulk water fraction. It will be clear that knowledge about these gel structures is of fundamental importance for developing a formulation with desired properties such as controlled water release, especially considering the interactions between the vehicles and the skin (12,15).

B. Colloidal Gel Structures of W/O Creams

According to the definition of DAB 9, W/O creams are hydrophobic systems, the continuous phase of which is lipophilic. The general formula of an absorption ointment with incorporated water is as follows (according to DAB 9):

-Anhydrous lanolin (wool fat) (with cholesterol as the most important ingredient)	3.00	% wt/wt
-Cetostearylalcohol	0.25	% wt/wt
-White petrolatum	46.75	% wt/wt
-Water	50.00	% wt/wt

A schematic presentation of the W/O cream gel structures is given in Fig. 10.

The W/O surfactants (cholesterol and other sterols as well as cetostearylalcohol) mainly accumulate at the interface between the water droplets and the oily phase (white petrolatum), forming a monomolecular mixed layer of surfactants at the water/oil interface. Experimental work has proven that the capacity of water strongly increases when mixtures of fatty alcohols and sterols are used. It seems to be important to create a liquid-crystalline monolayer at the oil/water interface. Crystallization of the surfactant film at the interface drastically reduces the water uptake capacity of the system. According to their solubilities the sterols and fatty alcohols are dissolved in the paraffin mixture (white petrolatum), too. A surplus of these O/W emulsifiers may crystallize separately in the lipophilic phase and may additionally enforce the (para)crystalline gel structures of white petrolatum. W/O creams represent simple W/O emulsions stabilized by a high viscous gel of paraffins. The long chain paraffins are able to build up a three-dimensional solid gel network in which the short chain liquid paraffins are mainly immobilized by lyosorption.

C. Colloidal Gel Structures of Amphiphilic Creams

Amphiphilic creams are colloidal systems that transform to O/W creams by addition of an oily phase to a W/O cream and by water addition. They therefore represent a special transition state between the other two cream types. A colloidal state is obligatory to the existence of special gel structures. Prerequisite for the existence of a cream with amphiphilic properties are lamellar mixed crystals that show only a limited swellability upon water addition. Examples for suitable compounds that fulfill these requirements are glycerolmonostearate and esters or ethers of PEG with fatty alcohols or fatty acids, respectively. PEG should have a low polymerization degree n = 2–3. All these compounds are surfactants of the W/O type. The colloidal

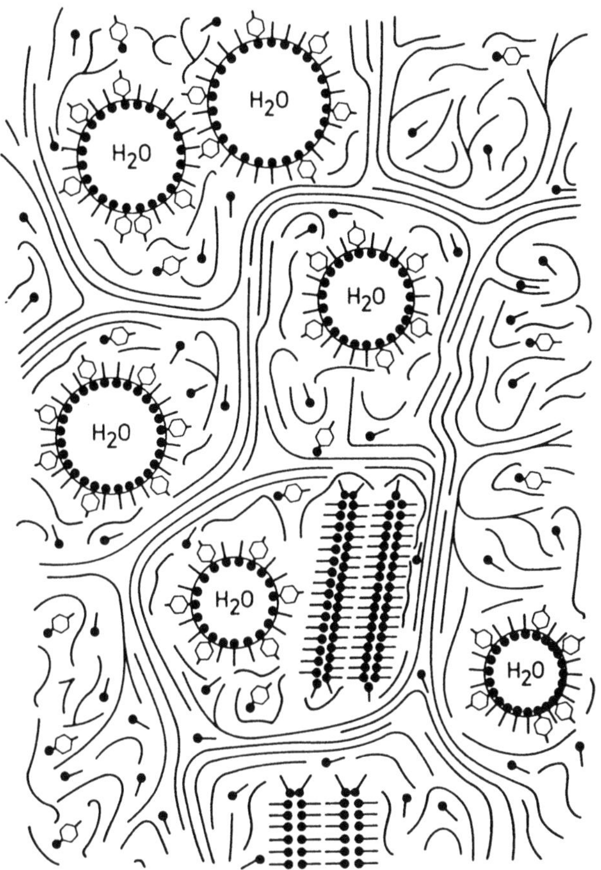

Figure 10 Schematic presentation of the gel structures of a W/O cream. a, long paraffin chains are forming the solid gel in which liquid paraffin chains are immobilized by lyosorption. Both cetostearylalcohol (—●) and cholesterol (derivatives) (—⬡●) are accummulated at the water-paraffin interface, and both are molecularly dispersed in the paraffin gel according to their solubilities. b, a surplus of cetostearylalcohol may crystallize as separate lamellar crystals.

structure of an amphiphilic cream will be illustrated with the "basic cream" of DAC having the following formulation:

-Glycerolmonostearate	4.0	% wt/wt
-Cetylalcohol	6.0	% wt/wt
-Medium chain triglycerides	7.5	% wt/wt
-White petrolatum	25.5	% wt/wt

-Polyoxyethylene 20 glycerol-monostearate (PGM 20)	7.0	% wt/wt
-Propyleneglycol	10.0	% wt/wt
-Water	40.0	% wt/wt

The limited swellability of glycerolmonostearate is well documented in the literature (16,17). Melted glycerolmonostearate together with water continuously swells by interlamellarly incorporating water molecules in between the hydrophilic glycerol rest until a water content of 30% (wt/wt) is reached. At increased water amounts the degree of swelling remains constant and the excess amount of water is incorporated mechanically in the form of droplets (bulk water phase) into the glycerolmonostearate gel structure.

Conductivity measurements of mixtures of glycerolmonostearate, liquid paraffins, and water with a constant amount of glycerolmonostearate of 30% (wt/wt) revealed (Fig. 11) that up to a water content of 20% (wt/wt)

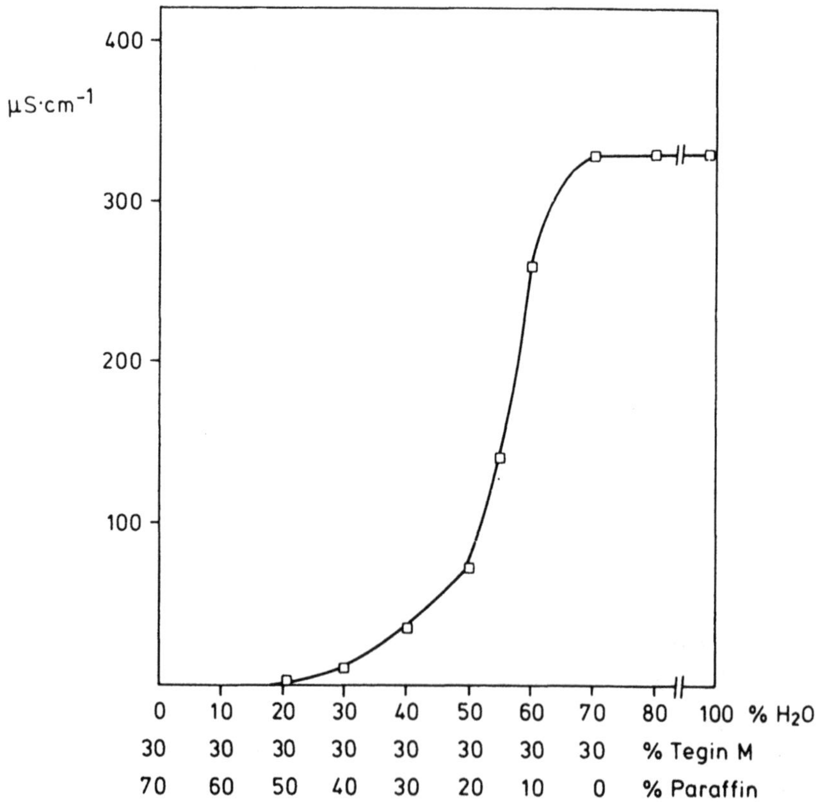

Figure 11 Conductivity of paraffin-water mixtures with a constant amount of glycerolmonostearate (Tegin M®) of 30% (wt/wt).

no conductivity could be measured. Therefore the systems show W/O character up to this water amount.

On the other hand, mixtures with a water content between 60 and 70% (wt/wt) show O/W characteristics. In the range between 20 and 50% (wt/wt) of water the systems behave as amphiphilics, i.e., addition of water results in O/W systems, and additions of liquid paraffins results in W/O systems. However, all these simple systems show no favorable properties with respect to the requirements of well-designed creams for pharmaceutical use. Furthermore, especially the resulting O/W systems are unstable, and phase separation only of the mechanically stabilized bulk water phase occurs. Therefore well-designed formulations of these amphiphilic systems additionally contain nonionic O/W emulsifiers with a high water binding capacity. In the above-mentioned formulation of DAC, poly(oxyethylene) 20 glycerolmonostearate (PGM 20) is added, which is able to form stable O/W creams at high water contents, i.e., exceeding 50% (wt/wt). Amphiphilic creams altogether contain a relatively high amount of O/W and W/O surfactants. Oily and water phases are approximately in the same order of weight range.

Amphiphilic creams are tricoherent systems:

The dominant coherent gel phase is built up by the monoglycerolstearate lamellae.

The continuous water phase is mainly performed by the interlamellary bound water in between the monoglycerolstearate lamellae.

The lipophilic phase is coherent, too.

The liquid paraffins are predominantly mechanically fixed by the monoglycerolstearate lamellae. A schematic representation of the proposed gel structures of an amphiphilic cream and the transitions to an O/W system and a W/O system, respectively, are given in Fig. 12A,B,C.

Originating with the above-described amphiphilic system (Fig. 12B) and adding an oily phase, the water swelling degree of the monoglycerolstearate lamellae is strongly reduced. Due to the favorable phase-volume ratio regarding the lipophilic phase a W/O cream automatically results (Fig. 12A). Increasing the water amount, the amphiphilic basic cream strongly swells—especially the poly(oxyethylene) 20 glycerolmonostearate surfactant (stabilized with cetylalcohol)—by interlamellar incorporation of the added water. It also stabilizes the resulting bulk water phase (Fig. 12C). As a result of water addition a stable O/W cream is obtained.

III. COLLOIDAL STRUCTURES OF CREAMS WITH LIQUID-CRYSTALLINE GEL STRUCTURES

Above the critical micelle concentration (CMC), surfactants in aqueous solutions form micelles. If the surfactant concentration is further increased, the

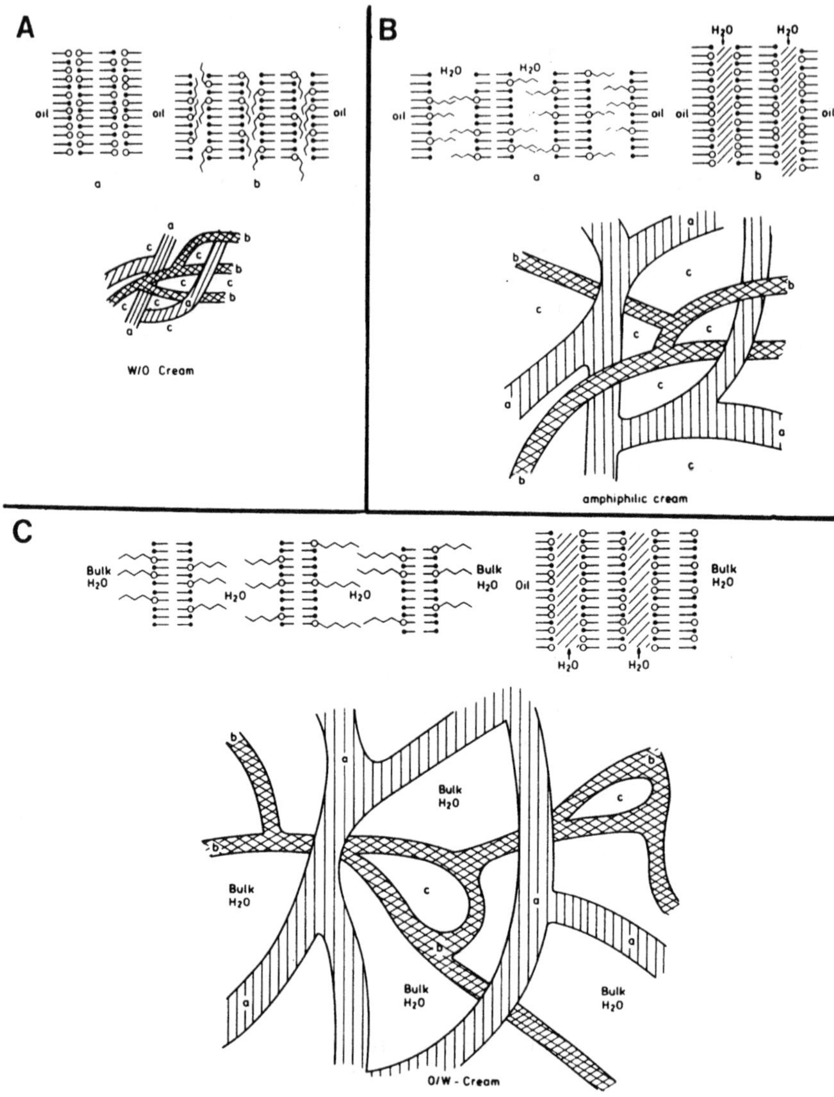

Figure 12 Schematic presentation of gel structures existing during the transition of a O/W cream (A) to an amphiphilic cream (B) and to a W/O cream (C). Poly(oxyethylene) 20 glycerolmonosterate – ; glycerolmonostearate —○; cetostearyl alcohol —●. a, Mixed crystals consisting of poly(oxyethylene) 20 glycerolmonostearate and cetostearyl alcohol (in A in the water-free state, in B in the partly swollen state, in C in the swollen state). b, Mixed crystals of glycerolmonostearate and cetostearyl alcohol with limited swellability (in A in the water-free state, in B and C in the state of limited swelling). c, Lipophilic state (in A as coherent continuous phase, in B as coherent phase, in C as dispersed inner phase).

micelles may grow into cylindrical or rod-shaped structures whose diameter in general is slightly less than twice the fully extended length of the single surfactant molecule. The tendency to form rod-shaped micelles depends upon the nature of the hydrocarbon chain, with such factors as length, branching, and degree of unsaturation, and furthermore on the types of the polar groups. Further increase in surfactant concentration causes them to become closely packed in hexagonal arrays and to separate out as another phase in equilibrium with a more dilute solution containing randomly oriented rods. The new phase, which contains the oriented rods, is called the hexagonal phase (also called the middle phase) due to the hexagonal array of the single rods (Fig. 13H). Depending on the type of surfactant, also a reversed hexagonal phase may occur (reversed middle phase) (Fig. 13RH). With further increase of the surfactant concentration, another phase, a lamellar phase, also called a neat phase, may occur (Fig. 13N).

With some systems a face-centered cubic arrangement of spherical aggregates occurs (Fig. 13C). Such structures are generally observed over only a narrow concentration between the hexagonal and lamellar phases. Furthermore, in some cases a viscous isotropic phase could be identified. The exact structure of the viscous isotropic phase is not yet fully understood, but it is supposed to be continuous in three axial directions, with respect to both the hydrophilic and the hydrophobic regions (see Sec. IV.B) (18–20). These systems are referred to as lyotropic mesophases or lyotropic fluid-crystalline phases (21,22). Lyotropic refers to the property by which the systems assume different structures depending on the addition of a solvent, in particular water, to a solid. Due to the ordered arrays of different micelles, liquid crystals possess both solidlike properties (high viscosity) and liquidlike properties (a relatively high mobility) that on the molecular level often depend on the spatial orientation: the cubic and the viscous isotropic phases show isotropic behavior, whereas the (reversed) hexagonal phase and the lamellar phase show anisotropy, easily detectable under polarizing microscopy. The solidlike and liquidlike behavior of the hexagonal phases and the lamellar (neat) phase are shown in Fig. 13.

It is assumed that in liquid-crystalline phases both the hydrophilic and the lipophilic compartments are separate liquids bridged over an interface. Both compartments are able to swell after addition of water and e.g. liquid paraffins, respectively. This is in contrast to colloidal systems with crystalline gel structure, where only the hydrophilic compartment is able to swell after the addition of water (see Sec. II.A–C). Liquid-crystalline systems for topical application and as drug delivery systems have been described in the literature (22–29) but have not reached the market yet. However, emulsions with a relatively high amount of surfactants may contain highly viscous liquid-crystalline phases between the oil and water phase with an extremely high stabilization potential of the emulsions (19). Liquid-crystalline systems may

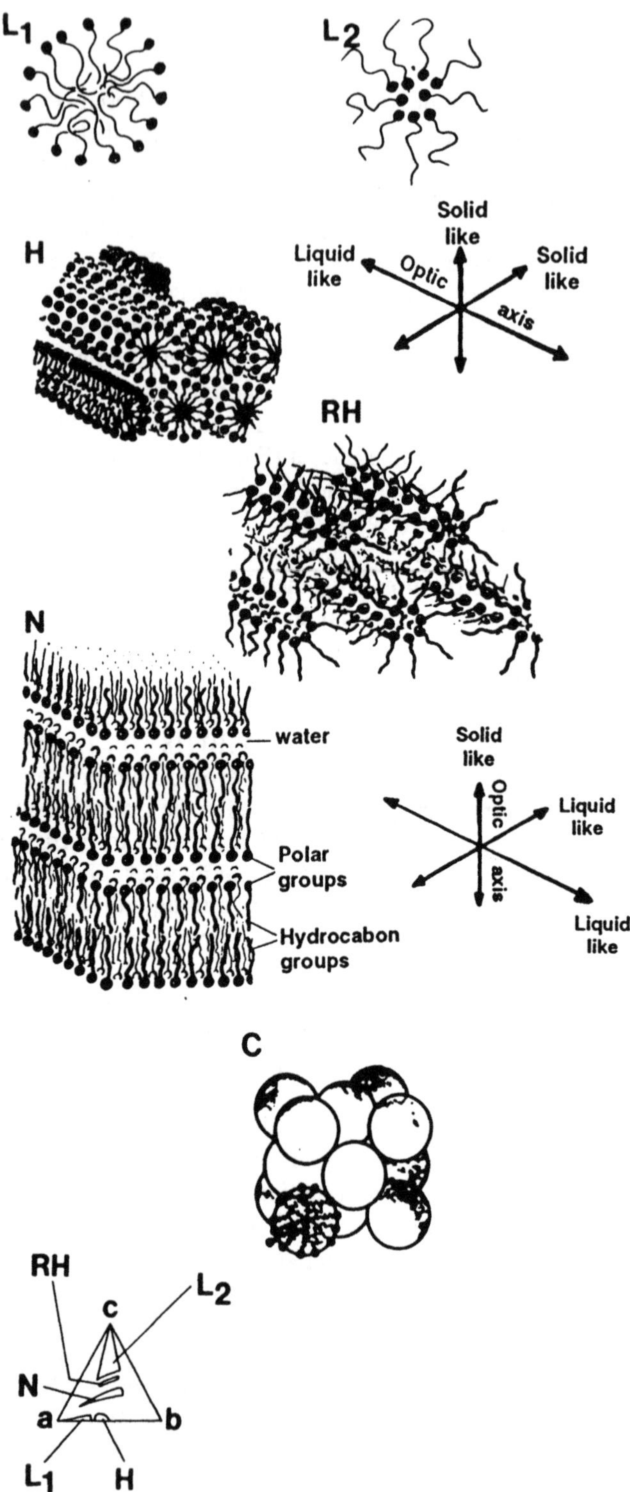
L1
L2
H
Solid like
Liquid like
Optic
Solid like
axis
RH
N
water
Polar groups
Hydrocabon groups
Solid like
Optic
Liquid like
axis
Liquid like
C
RH
L2
c
N
a
b
L1
H

show favorable properties with respect to the solubilization of drugs in the watery, oily, or amphipathic interface and form supersaturated systems. Furthermore, they may interact with the skin lipids of the stratum corneum of the skin and show penetration-enhancing properties. However, after topical application of liquid-crystalline systems, the incorporated water immediately will evaporate, changing the liquid-crystalline character of the system. In most cases, the oil-surfactant residue will form a film on the skin surface. In order to keep their unique liquid-crystalline properties, these systems have to be applied under occlusion. This fact, however, may prevent these novel carrier systems for drug delivery from attaining a broad application to patients.

IV. CREAM STRUCTURE DEPENDENCE ON DRUG RELEASE

A. Drug Release from Creams with Crystalline Gel Structure In Vitro

To study the influence on the colloidal gel structure on drug release, simplified mixtures of the nonionic hydrophilic cream already described in Sec. II.A.3 of this chapter were used with the following compositions:

-Poly(oxyethylene) 20 glycerol-monostearate (PGM 20)	5.0	% wt/wt
-Cetylalcohol	5.0	% wt/wt
-Stearylalcohol	5.0	% wt/wt
-Water	15.6–42.4	% wt/wt

These surfactant/water mixtures may be considered to be simplified versions of the nonionic hydrophilic cream (cf. Sec. II.A.3) with which they share the most important properties of the colloidal hydrophilic gel phase, i.e., over a wide range of water concentrations they adopt the same lamellar crystalline gel structures as described in Fig. 8. In the lamellar structure the surfactants and cosurfactants (fatty alcohols) form platelike lamellar aggregates stacked in a repetitive pattern, whereby lipophilic bilayers alternate with hydrophilic layers that contain interlamellarly inserted water. Generally, a drug permeating through a lamellar gel network may follow an interlamellar or translamellar route, depending on local rates of diffusion and partition. Extremely lipophilic drugs most probably will be trapped inside

Figure 13 Normal (L_1) and reversed (L_2) micelles and different liquid-crystalline structures in a system of (a) water, (b) surface active substances, and (c) cosurfactant. H, hexagonal phase; RH, reversed hexagonal phase; N, neat (lamellar) phase; C, cubic phase.

the lipophilic bilayer (30), or, in case a lipophilic dispersed phase (consisting e.g. of liquid paraffins, white petrolatum, or other lipophilic components) is present, this may act as a drug reservoir for lipophilic drugs (31).

Extremely hydrophilic drugs most likely permeate through the hydrophilic regions between the lamellae, while amphiphilic drugs may move both between and across the lamellae. In order to clarify the role of hydrophilic colloidal gel structures with respect to drug permeation, the release of nicotinamide (pyridine-3-carbamide), an extremely hydrophilic nonionic drug lacking acid-base behavior, was investigated from crystalline systems with lamellar structures. A 5% (wt/wt) aqueous nicotinamide solution was used during sample preparations. Cumulative release of nicotinamide was measured in vitro from a 5 mm thick cream layer (interface layer, cellulose membrane; surface area, 3.14 cm^2) to an aqueous sink at 25°C in a simple two compartment set-up (12,13).

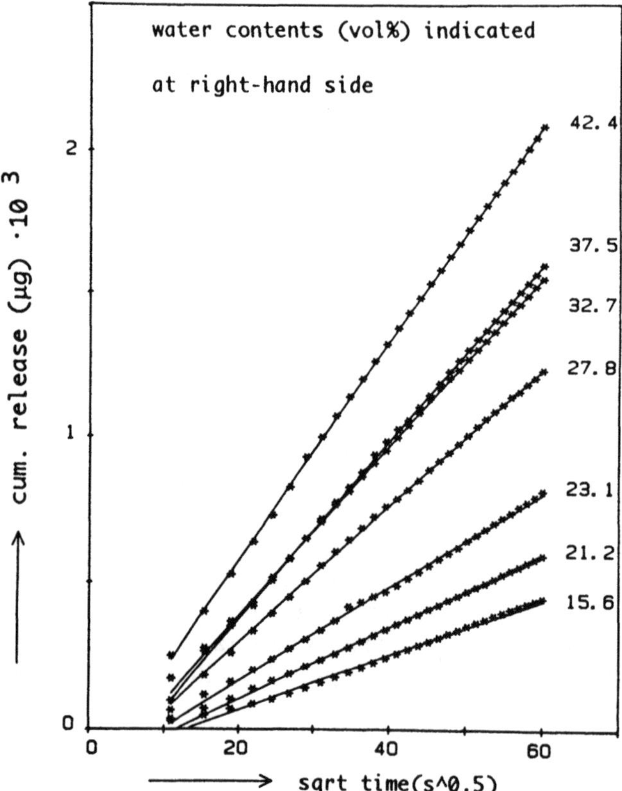

Figure 14 Square root of time plots of the cumulative nicotinamide release from creams with various water contents.

The results of these release experiments are compiled in Fig. 14, which contains plots of the cumulative nicotinamide release versus the square root of time for all systems studied; the straight lines are obtained with linear regression. In all cases there is complete linearity, except for the 15.6 and 21.1 vol % plots. The linéarity of the release plots agrees well with general formulae (32,33) for short term release from semi-infinite source to a perfect sink, based on Fick's diffusion laws.

When the squared slopes of the release profiles of Fig. 14 are plotted against the water contents (Fig. 15), a straight line is obtained after fitting. From

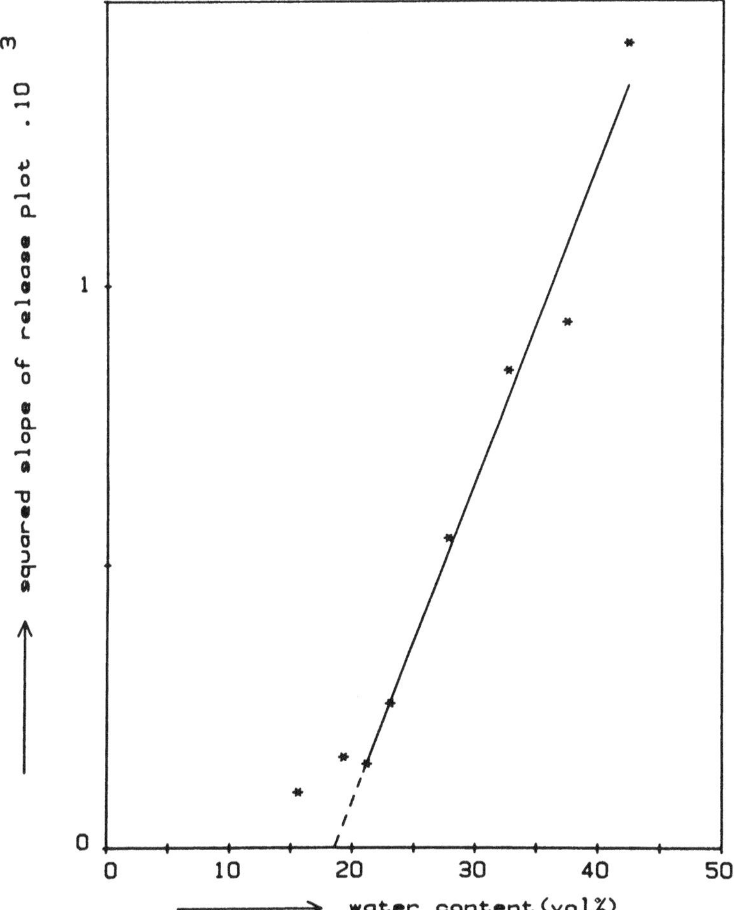

Figure 15 Squared slopes of nicotinamide release versus water contents of the creams.

the slope of this line the intrinsic diffusion coefficient D_0 for nicotinamide can be calculated as $D_0 = 3.6 \pm 0.3\ 10^{-6}\ cm^2\ s^{-1}$. This represents a quite reasonable value for the diffusion of a low molecular weight compound through an aqueous gel phase. This intrinsic diffusion coefficient D_0 is based on the assumptions that the colloidal gel phases possess pores and that the crystalline lamellae have a random orientation resulting in a tortuosity for the pathway of the drug.

Due to the presence of strongly hydrophilic oxyethylene moieties, it is most likely that the water inside the pores only partly contributes to the drug permeation; an "impermeable water fraction" has to be determined, which corresponds with the "nonfreezing water" determined by DTA-measurements (11,13). Linear extrapolation of the fitted straight line of this plot in Fig. 15 toward the X-axis yields a value of 19% (wt/wt) for the "impermeable water fraction," which is in excellent agreement with the value for the nonfreezing water fraction of 18% (wt/wt) by DTA (11,13). Since the "nonfreezing water fraction" is thought to constitute a structured hydration layer, firmly bound to the oxyethylene chains within the hydrophilic gel phase, it is indeed quite likely that it does not contribute to the drug permeation. Therefore for water contents below 19% (wt/wt) the proposed pore model of the colloidal gel structures degenerates and results in an incorrect prediction of the drug diffusion coefficient. At these low water contents, drug diffusion most likely follows a different mechanism and probably takes place along the partly hydrated oxyethylene chains. Finally, it can be concluded that the colloidal hydrophilic gel phases form a porous structure, where the porosity is related to the "free" water content and the tortuosity to the orientation of the lamellae.

Last but not least, it must be emphasized that these release mechanisms presented here only hold for in vitro drug release, but not if these colloidal systems are applied to the skin. The mechanical application of these systems onto the skin surface will change the structure of the colloidal gel phase immediately followed by water evaporation. Therefore these in vitro drug release results may be only of minor importance for the prediction of drug release characteristics from the changed colloidal structures in vivo.

B. Drug Release from Creams with Liquid-Crystalline Gel Structures In Vitro

In this chapter, mixtures of water and the nonionic surfactant polyoxyethylene (10) oleylether (Brij 96™) have been used because they may be considered as simplified versions of creams with colloidal liquid-crystalline gel structures. Henceforth they will be referred to as creams. Depending on the surfactant-water mixture ratio these creams can have various liquid crystalline gel structures (lamellar, hexagonal, and viscous-isotropic) that have been studied extensively (23,24,34). Benzocaine (1% wt/wt) was chosen as a model drug.

Fig. 16 shows the colloidal gel structures determined by SAXD as well as the repeat distances of the cream. Creams having a water content between 15 and 30% (wt/wt) have a lamellar gel structure (L) (neat phase) (see Sec. III). Creams with a relatively high water content have hexagonal structures (H) composed of cylindrical surfactant aggregates arranged in a hexagonal

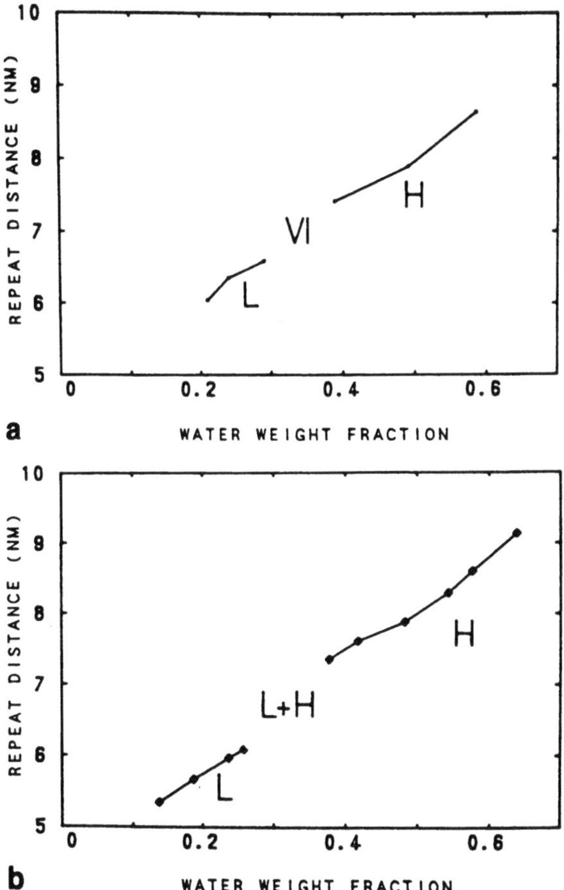

Figure 16 Gel structures and characteristic repeat distances of the creams containing 1% (w/w) benzocaine, as function of the water weight fraction, water/(water + Brij 96™), at 21°C; L, lamellar; H, hexagonal; VI, viscous isotropic. (a) creams containing Brij 96® (EO 9.3); (b) creams containing Brij 96® (EO 8.5).

structure and surrounded by water. Depending on the batch of the surfactant used (see legend to Fig. 16), creams with water contents of about 30% (wt/wt) may exhibit a mixture of lamellar and hexagonal gel structures (Fig. 16b) (24) or a viscous isotropic gel structure (Fig. 16a). The solubilization of 1% (wt/wt) bezocaine does not influence the gel structures compared to the drug-free systems.

Examples of release experiments are given in Fig. 17. The release of the drug was slow; only a fraction of the drug content of the cream was released within the first 10 h. After an initial lag time (25–100 min), complete linearity is obtained when the amount of drug release is plotted versus the square root of time.

The effective diffusion coefficients calculated from the slopes in Fig. 17 are shown in Fig. 18. These differences in the effective diffusion coefficient cannot be explained in terms of (macroscopic) viscosity because the highest effective diffusion coefficients are observed in the hexagonal creams, which

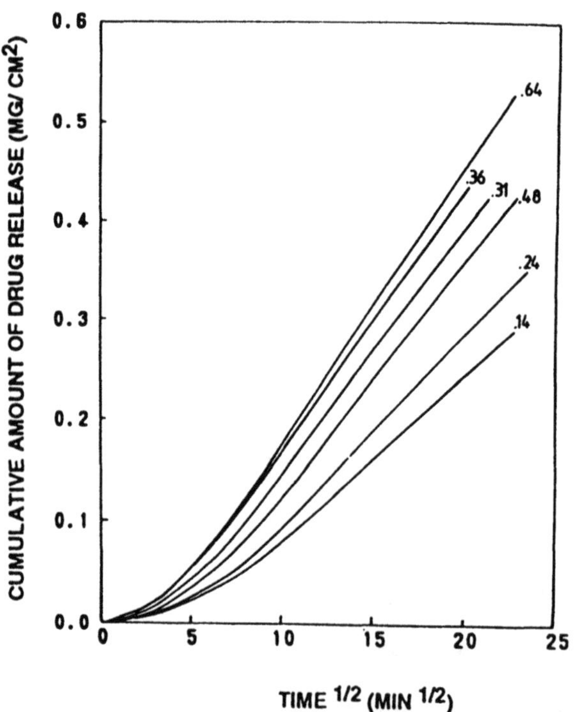

Figure 17 Release of benzocaine from Brij 96[R] (EO 9.3)/water creams at 21°C. The weight fraction water, water/(water + Brij 96[R]) is given at the right-hand side.

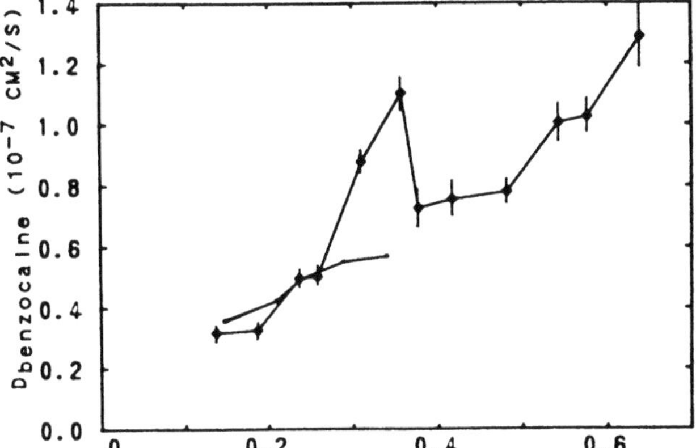

Figure 18 Effective diffusion coefficients of benzocaine versus the water weight fraction, water/(water + Brij 96[R]) in the creams at 21°C. , creams with Brij 96[R] (EO 9.3) (the standard deviation is indicated); •, creams with Brij 96[R] (EO 8.5).

have a viscosity more than 1000 times higher than the lamellar creams. Also the viscosities of the viscous isotropic phases are much higher than those of the lamellar phases. Benzocaine will preferentially partition into the hydrophilic areas of the colloidal gel structures. These hydrophilic areas increase with increasing the total water amount of the systems. Logically, the simultaneous increase of the effective diffusion coefficient from the lamellar (low water content) to the hexagonal phase (high water content) may be explained by this correlation. However, the effective diffusion coefficients of the viscous isotropic phases do not fit with the explanation given for the lamellar and hexagonal creams because their effective diffusion coefficients are relatively high. Effects of tortuosity of the colloidal gel structure and interfacial adsorption may be involved here. Especially the tortuosity factor may count for the high effective diffusion coefficient of benzocaine due to the low tortuosity of the hydrophilic pathway in the bicontinuous gel structures in the viscous isotropic phase.

As already mentioned in Sec. III of this chapter, colloidal drug carrier systems based on liquid crystalline phases for (trans)dermal drug delivery will only retain their properties when applied occlusively, i.e., using suitable application chambers with a waterproof sticking rim. Depending on their composition they may show the following unique properties:

- At low water content (i.e., as lamellar phases) they may be able to take up water from the skin and by doing so prevent skin irritation and show improved skin compatibility.

- Drugs incorporated in these systems show matrix controlled release. A reduced diffusion of a partly exhausted system may be increased when the systems change into other liquid-crystalline structures (i.e., transition from the lamellar phase into the hexagonal phase) that possess higher effective diffusion coefficients.
- Depending on the nature of the surfactant of the colloidal liquid-crystalline phases, a controlled interaction with the skin lipids of the stratum corneum barrier of the skin may occur, resulting in a penetration-enhancing effect (for transdermal drug delivery) or targeting effect into the epidermis (dermal drug delivery).

As a general conclusion, it may be stated that colloidal gel structures with both crystalline and liquid-crystalline structural elements are forming coherent networks with water-filled pores. In both systems drug diffusion depends on the amount of free water, the random distribution of the structural elements, and the related tortuosity (length of the diffusional pathway).

REFERENCES

1. Münzel, K. (1953). Versuch einer Systematik der Salben nach galenischen Gesichtspunkten. *Pharm. Acta Helv. 28*: 320–336.
2. Junginger, H. E., Führer, C., Ziegenmeyer, J., and Friberg, S. E. (1979). Strukturuntersuchungen an der Wasserhaltigen Hydrophilen Salbe DAB 7. *J. Soc. Cosmet. Chem. 30*: 9–23.
3. Junginger, H. E. (1984). Colloidal structures of O/W creams. *Pharm. Weekbl. Sci Ed. 6*: 141–149.
4. Junginger, H. E., Ackermans, A. A. M. D., and Heering, W. (1984). The ratio of interlamellarly bound water to bulk water in O/W creams. *J. Soc. Cosmet. Chem. 35*: 45–47.
5. Junginger, H. E., Heering, W., Führer, C., and Geffers, I. (1981). Elektronenmikroskopische Untersuchungen über den kolloid-chemischen Aufbau von Salben und Cremes. *Coll. Polym. Sci. 259*: 561–567.
6. Junginger, H. E., and Heering, W. (1983). Darstellung kolloider Strukturen von Salben, Cremes, Emulsionen und Mikroemulsionen mittels Gefrierbruch-Ätztechnik und TEM. *Acta Pharm. Technol. 29*: 85–96.
7. Junginger, H. E., and Heering, W. (1990). Gelstrukturen in Cremes. *Dtsch. Apoth. Ztg. 130*: 684–685.
8. Tronnier, H. (1964). *Über die Wirkungsweise indifferenter Salben- und Emulsionssysteme an der Haut in Abhängigkeit von ihrer Zusammensetzung*. Aulendorf: Editio Cantor.
9. Junginger, H. E. (1984). Strukturuntersuchungen an Stearatcremes. *Pharm. Ind. 46*: 758–762.
10. De Vringer, T., Joosten, J. G. H., and Junginger, H. E. (1984). Characterization of the gel structure in a nonionic ointment by small x-ray diffraction. *Coll. Polym. Sci. 262*: 56–60.

11. De Vringer, T., Joosten, J. G. H., and Junginger, H. E. (1986). A study of the gel structure in a nonionic O/W cream by differential scanning calorimetry. *Coll. Polym. Sci. 264*: 691–700.
12. Junginger, H. E., and Boddé, H. E. (1985). Gel structures of dermatological vesicles and their influence on drug release. *Topics in Pharmaceutical Sciences* (D. D. Breimer and P. Speiser, eds.). Amsterdam: Elsevier, pp. 329–343.
13. Boddé, H. E., De Vringer, T., and Junginger, H. E. (1986). Colloidal systems for controlled drug delivery—structure activity relationships. *Progr. Coll. Polym. Sci. 72*: 37–42.
14. De Vringer, T., Joosten, J. G. H., and Junginger, H. E. (1987). A study of the gel structure in a nonionic O/W cream by X-ray diffraction and microscopic methods. *Coll. Polym. Sci. 265*: 167–179.
15. De Vringer, T., Joosten, J. G. H., and Junginger, H. E. (1987). A study of the ageing of the gel structure in a nonionic O/W cream by X-ray diffraction, differential scanning calcorimetry and spin lattice relaxation measurements. *Coll. Polym. Sci. 265*: 448–457.
16. Larsson, K. (1967). The structure of mesomorphic phases and micelles in aqueous glyceride systems. *Z. Phys. Chem. (Neue Folge) 56*: 173–198.
17. Krog, N., and Borup, A. P. (1973). Swelling behaviour of lamellar phases of saturated monoglycerides in aqueous systems. *J. Sci. Food Agric. 24*: 691–701.
18. Rawlins, E. A., ed. (1977). *Bentley's Textbook of Pharmaceutics*. 8th ed. London: Baillière Tindall.
19. Friberg, E., ed. (1976). *Food Emulsions*. New York: Marcel Dekker.
20. Kelker, H., and Hatz, R., eds. (1980). *Handbook of Liquid Crystals*. Verlag Chemie Weinheim.
21. Tiddy, G. J. T. (1980). Surfactant-water liquid crystal phases. *Phys. Rep. 57*: 1–46.
22. Jousma, H., Bouwstra, J. A., and Junginger, H. E. (1988). On the structure of the cubic phase I 1 in mixtures of polyethylene (14) oleylether and water. *Liq. Cryst. Lett. Sect. 5*: 109–115.
23. Jousma, H. (1988). A small angle X-ray scattering and freeze-fracture electron microscopy study of lyotropic mesophases. Thesis, Leiden University.
24. Jousma, H., Joosten, J. G. H., and Junginger, H. E. (1988). Mesophases in mixtures of water and polyoxyethylene surfactant: Variations of repeat spacing with temperature and composition. *Coll. Polym. Sci. 266*: 640–651.
25. Jousma, H., Joosten, J. G. H., Gooris, G. S., and Junginger, H. E. (1989). Changes in mesophase structure of BRIJ 96/water mixtures on addition of liquid paraffin. *Coll. Polym. Sci. 267*: 353–364.
26. Provost, C., and Kinget, R. (1988). Transparent oil-water gels: A study of some physicochemical and biopharmaceutical characteristics. Part I. Formulation of transparent oil-water gels in the 4-component system Emulgin B3, Cetiol HE, isopropyl palmitate, and water. *Int. J. Pharm. 44*: 75–85.
27. Nürnberg, E. (1985). Hydrophile Cremesysteme und transparente Tensidgele: Neuere Erkenntnisse über Struktur und Eigenschaften. *Acta Pharm. Technol. 31*: 123–137.

28. Müller-Goymann, C. C., and Frank, S. G. (1986). Interactions of lidocaine and lidocaine-HCl with the liquid crystal structure of topical preparations. *Int. J. Pharm. 29*: 147–159.
29. Tiemessen, H. L. G. M., Boddé, H. E., van Mourik, C., and Junginger, H. E. (1988). In vitro drug release from liquid crystalline creams: Cream structure dependence. *Progr. Coll. Polym. Sci. 77*: 131–135.
30. Wahlgren, S., Lindstrom, A. L., and Friberg, S. E. (1984). Liquid crystals as a potential ointment vehicle. *J. Pharm. Sci. 73*: 1484–1486.
31. Boddé, H. E., and Joosten, J. G. H. (1985). A mathematical model for drug release from a two-phase system to a perfect sink. *Int. J. Pharm. 26*: 57–76.
32. Crank, J. (1956). *The Mathematics of Diffusion*. London: Academic Press.
33. Higuchi, T. (1963). Mechanism of sustained-action medication. *J. Pharm. Sci. 52*: 1145–1149.
34. Feger, M. (1977). Thesis, Technische Universität Braunschweig.

2

Microemulsions

David Attwood

University of Manchester, Manchester, United Kingdom

I. INTRODUCTION

Since the term "microemulsion" was first introduced by Hoar and Schulman (1) in an article in *Nature* in 1943 there has been much dispute about the relationship of these systems to micellar solubilized systems and to emulsions. The original use of the term by these authors was to describe the transparent systems obtained when normal emulsions were titrated to clarity with medium chain length alcohols. Many investigators have perceived a difference between microemulsions and micellar systems containing solubilized oil or water, and the terms "swollen micellar solutions" or "solubilized micellar solutions" have been applied to the systems examined by Hoar and Schulman. The name "microemulsion" is sometimes restricted to systems in which the droplets are of large enough size that the physical properties of the dispersed oil or water phase are indistinguishable from those of the corresponding oil or water bulk phase. On this basis it is possible, at least in principle, to distinguish oil-in-water microemulsions, in which the droplets have an isotropic oil core, from micellar systems containing small amounts of solubilized oil, since the mixture of the solubilized oil and the lipophilic portions of the surfactant are unlikely to produce an isotropic core. When these micelles are progressively swollen with oil, the properties of the dispersed phase evolve smoothly toward those of a bulk oil phase without any well-defined transition point. However, since there is no simple method of determining the oil content at which the core of the swollen micelle becomes identical in properties to that of a bulk oil phase, the distinction between solubilized micellar systems and microemulsions is not clear-cut, and in this chapter the term "microemulsion" will be used to describe both systems. This is in agreement with the definition proposed by Danielsson and Lindman (2), who considered a microemulsion to be "a system of water, oil and amphiphile which is a single optically isotropic and thermodynamically stable liquid solution." The concept of a microemulsion as described by these authors includes aqueous micellar surfactant solutions containing solubilizate but specifically excludes aqueous solutions of surfactants without added solubilizate, liquid-crystalline systems, and normal emulsions.

The distinction between microemulsions and normal emulsions is very much more pronounced than any perceived difference between microemulsions and solubilized systems. The transparency of microemulsions arises from their small droplet diameter, typically less than 140 nm. Such small droplets produce only weak scattering of visible light when compared with that from the coarse droplets (1–10 μm) of normal emulsions. In this chapter we will examine briefly how the low interfacial tension required to produce the small droplets can be achieved by the addition of cosurfactants to the formulation. We will also examine methods for particle sizing these systems, since the variation of droplet size is a critical test of stability. An essential difference between microemulsions and emulsions is that microemulsions form spontaneously and, unlike emulsions, require no mechanical work in their formation. This is an important distinction between the two systems. Inspection of the patent literature reveals several "microemulsions," particularly those containing phospholipids, that can only be prepared by prolonged sonication or microfluidization of the components. Despite their small droplet size and consequent transparency, these are not microemulsions, at least according to the generally accepted definition of these systems. An important feature of microemulsions from a formulation viewpoint is their stability. Normal emulsions age by coalescence of droplets and by Ostwald ripening (transfer of material from small droplets to larger ones), since these processes lead to a decrease of the interfacial area and hence the free energy of the system. In microemulsions, the interfacial tension is sufficiently low to compensate the dispersion entropy, and the systems are thermodynamically stable.

Research on microemulsions increased in the late 1960s, when it was realized that these systems might have important applications in enhanced oil recovery. More recently, their pharmaceutical applications have been explored, and research in this area is currently very productive. A review of this work is given in this chapter, and the potential of these systems in drug delivery is discussed.

II. FORMULATION

A. Role of the Cosurfactant

In their simplest form, microemulsions are small droplets (diameter 5–140 nm) of one liquid dispersed throughout another by virtue of the presence of a fairly large concentration of a suitable combination of surfactants. They can be dispersions of oil droplets in water (o/w) or water droplets in oil (w/o). An essential requirement for their formation and stability is the attainment of a very low interfacial tension γ. Since microemulsions have a very large interface between oil and water because of the small droplet size, they can only be thermodynamically stable if the interfacial tension is so low that the positive interfacial energy (given by γA, where A is the interfacial area) can be compensated by the negative free energy of mixing ΔG_m. We can calculate a rough measure of the limiting γ value required as follows: ΔG_m is given by $-T\Delta S_m$ (where T is the temperature), and the entropy of mixing ΔS_m is of the order of the Boltzmann constant k_B; hence $k_BT = 4\pi r^2\gamma$, and thus for a droplet radius r of about 10 nm, an interfacial tension of 0.03 mN m^{-1} would be required. The role of the surfactants in the system is thus to reduce the interfacial tension between oil and water (typically about 50 mN m^{-1}) to this low level.

With the possible exception of double alkyl chain surfactants and a few nonionic surfactants, it is generally not possible to achieve the required interfacial area with the use of a single surfactant. If, however, a second amphiphile is added to the system, the effects of the two surfactants can be additive provided that the adsorption of one does not adversely affect that of the other and that mixed micelle formation does not reduce the available concentration of surfactant molecules. The second amphiphile is referred to as the cosurfactant.

The importance of the cosurfactant is illustrated in the following example taken from a review by Overbeek (3). The interfacial tension between cyclohexane and water is approximately 42 mN m^{-1} in the absence of any added surfactant. The addition of the ionic surfactant sodium dodecylsulphate (SDS) in increasing amounts causes a gradual reduction of γ to a value of about 2 mN m^{-1} at a SDS concentration of 10^{-4} g mL^{-1}. Further reduction of interfacial tension does not occur, since the cyclohexane/water interface is now saturated with SDS and any SDS added in excess of this limiting concentration forms micelles in the aqueous solution. Addition of 20% pentanol to the cyclohexane/water system in the absence of SDS reduces the interfacial tension to 10 mN m^{-1}. It is then theoretically possible by the addition of SDS to achieve a negative interfacial tension at SDS concentrations below the level at which it forms micelles (the critical micelle concentration, CMC). The changes in interfacial tension occurring in this system are illustrated in Fig. 1. Although pentanol is not generally regarded as a surfactant, it has the ability to reduce interfacial tension by virtue of its amphiphilic nature (a short hydrophobic chain and a terminal hydrophilic hydroxyl group) and functions as the cosurfactant in this system. Its presence in this system means that the SDS is now required to produce a much smaller lowering of the interfacial tension (10 mN m^{-1} rather than 42 mN m^{-1} in its absence) in order to produce a microemulsion.

B. Microemulsion Structure

The simplest representation of the structure of microemulsions is the droplet model in which microemulsion droplets are surrounded by an interfacial film consisting of both surfactant and cosurfactant molecules, as illustrated in Fig. 2. The orientation of the amphiphiles at the interface will, of course,

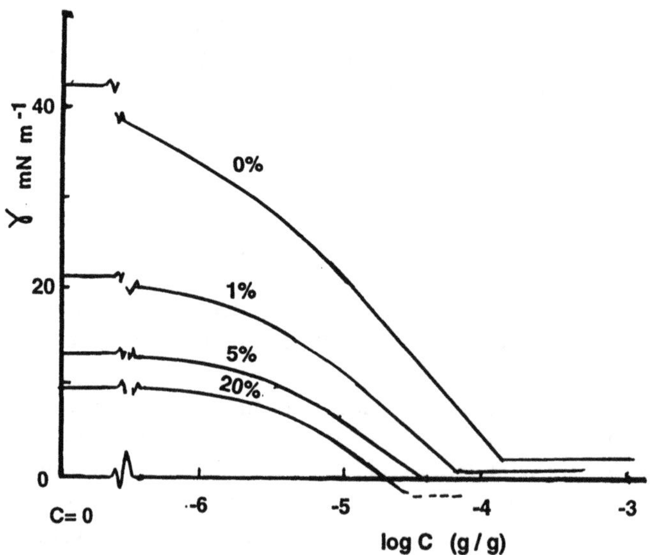

Figure 1 Interfacial tension γ between solutions of sodium dodecylsulphate (SDS) of concentration C in aqueous 0.30 M NaCl and solutions of n-pentanol in cyclohexane with the percentage concentrations indicated. (From Ref. 3, with permission.)

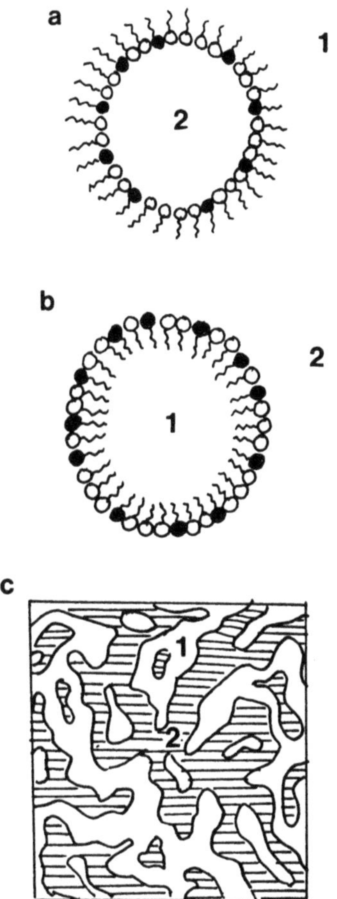

Figure 2 Diagrammatic representation of microemulsion structures. a, Water-in-oil microemulsion droplet; b, oil-in-water microemulsion droplet; c, irregular bicontinuous structure (1 = oil, 2 = water).

differ in o/w and w/o microemulsions. As shown in Fig. 2, the hydrophobic portions of these molecules will reside in the dispersed oil droplets of o/w systems, with the hydrophilic groups protruding in the continuous phase, while the opposite situation will be true of w/o microemulsions. For systems of known composition, an estimation may be made of the droplet radius r using $r = 3\Phi/C_s a_0$, where Φ is the volume fraction of the disperse phase, C_s is the number of surfactant molecules per unit volume, and a_0 is the surface area of a surfactant molecule at the interface. In practice not all of the surfactant can be assumed to be associated with the interface, and C_s is consequently seldom known with any certainty, although it is often assumed that the amount of surfactant in the continuous phase approximates to the surfactant CMC (4).

Whether the systems form o/w or w/o microemulsions is determined to a large extent by the nature of the surfactant. An indication of the type of system that is likely to be favored can be gained by applying an approach proposed by Mitchell and Ninham (5), which considers the geometry of the surfactant molecule. If the volume of the surfactant molecule is V, its head group surface area a, and its length l, then when the critical packing parameter V/al has values of between 0 and 1, o/w systems are likely to form, but when V/al is greater than 1, w/o microemulsions are favored. Values of the parameters V and l can be readily estimated using, for example, equations proposed by Tanford (6). It should, however, be noted that the critical packing parameter is based purely on geometric considerations. Penetration of oil and cosurfactant into the surfactant interface and hydration of the surfactant head groups will also influence the packing of the molecules in the interfacial film around the droplets. In many systems, inversion from w/o to o/w microemulsions can occur as a result of changing the composition or the temperature, as will be discussed in Sec. II.C. In general, o/w microemulsions are favored when small amounts of oil are present and w/o systems form in the presence of small amounts of water. Under such conditions the droplet model is a reasonable representation of the system.

The structures of microemulsions containing almost equal amounts of oil and water or with high surfactant content are not well understood. Similarly, the use of surfactants with a critical packing parameter close to unity gives microemulsions, the microstructure of which does not conform to the droplet model. The extensive recent work that has been carried out to provide models for the bicontinuous structures present in such systems has been reviewed by Chevalier and Zemb (7). The earliest concept of these structures was that of Scriven (8,9), who envisaged areas of water and oil separated by a connected amphiphile-rich interfacial layer as depicted in Fig. 2c. The first quantitative model was that of Talmon and Prager (10–12), which is based on a "random lattice," the cells of which are filled at random with water or oil according to the total volume fraction of these components. The surfactant film is assumed to be adsorbed only on those faces common to two adjacent cells filled with different phases. Figure 3 shows this model applied to systems with increasing water to oil ratio. At low volume fractions of water (Fig. 3a), the system has an oil continuous phase with isolated polygons of water representing a w/o system. The opposite situation occurs at high water content (Fig. 3c), where the isolated polygons of oil represent an o/w system. If the phase volume of water exceeds a critical

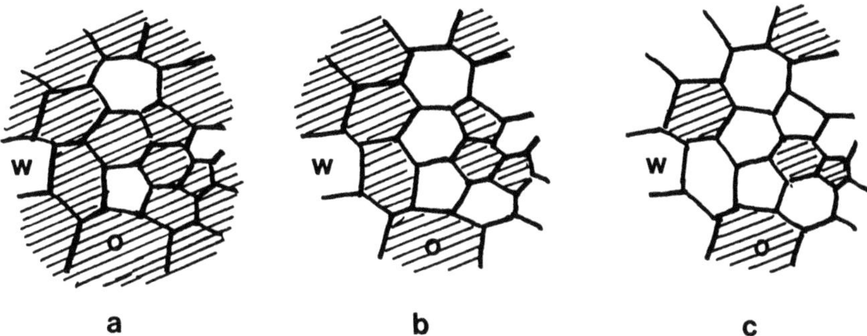

Figure 3 The Talmon-Prager bicontinuous model of microemulsions showing a continuous transition from w/o to o/w systems (a to c) represented by random filling of Voronoi polyhedra.

value, the model predicts a continuous path of water across the sample, representing a continuous microemulsion (Fig. 3b). The Talmon-Prager model was successful in predicting an average size of the oil and water domains that corresponded reasonably to the observed sizes but was unable to account for conductivity behavior or the presence of a peak in the observed small angle x-ray scattering.

The Talmon-Prager model was improved by de Gennes and Taupin (13). For simplicity, the polymers were replaced by cubic cells filled at random with oil or water according to the macroscopic volume fraction determined by the composition of the system. As in the previous model, the interface is located at the edges of the elementary cells containing different contents (see Fig. 4). The model is referred to as the cubic random cell (CRC) model. In their model, de Gennes and Taupin determined the cube edge length ζ_k by the spontaneous curvature. In a modification by Auvray et al. (14), this length is fixed by the specific area covered by surfactant and is given by

$$\zeta_k = \frac{b\Phi_o\Phi_w}{\Sigma} \tag{1}$$

where Φ_o and Φ_w are the volume fractions of oil and water respectively and Σ is the total surfactant interfacial area. Closer agreement with observed scattering behavior was achieved using this modification. Further modifications were made to the model by Widom (15,16) and Andelman et al. (17).

In the disordered open connected (DOC) model (18–20), a network of spherical droplets and connected cylinders is imagined. The average number of cylinders leaving one spherical droplet is called the connectivity Z. In theory, Z can vary between 0 and 13.4. In practice, the observed values

Figure 4 Diagrammatic representation of bicontinuous model of de Gennes and Taupin showing random filling of cubes on a cubic lattice.

are between 0 and 4. Figure 5 shows examples of connectivity of 0 (the hard sphere model) and 2. The DOC model successfully predicts peaks determined by small angle x-ray scattering and also the conductivity behavior of ionic microemulsions (19,21).

C. Phase Studies

Figure 6 shows a hypothetical phase diagram in which S represents the sum of the surfactant and cosurfactant concentration. It is, at least in theory, possible by diluting an oil-rich system with water to cause the transitions shown in Fig. 7. At high oil concentration, the surfactant forms inverse micelles that are capable of solubilizing water molecules in their hydrophilic interior. Continued addition of water to this system can result in the formation of a w/o microemulsion in which the water exists as droplets surrounded and stabilized by an interfacial layer composed of the surfactant/cosurfactant mixture. At a limiting water content, the isotopic clear region changes to a turbid, birefringent one. In hexadecane/hexanol/potassium oleate/water systems, NMR measurements (22) indicate that this liquid-crystalline region is composed of water cylinders dispersed in a continuous oil phase, as shown in Fig. 7, producing a system of high viscosity. On further dilution, a liquid-crystalline system may be formed in which the water

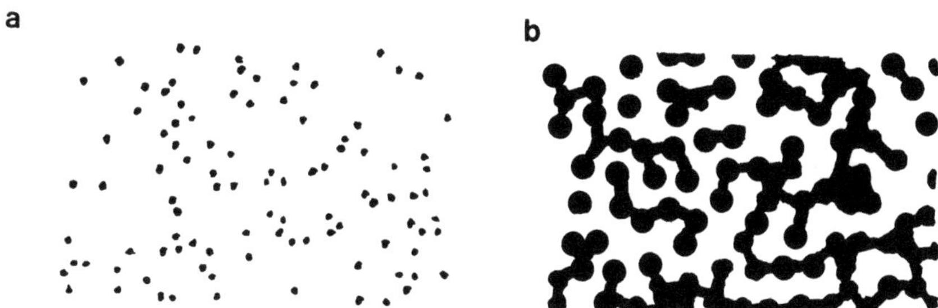

Figure 5 Examples of disordered open connected (DOC) structures built from the same Voronoi lattice. a, Connectivity $Z=0$: collection of spherical droplets. b, DOC cylinder with connectivity $Z=2$: each center is bound to the two nearest neighbors on average.

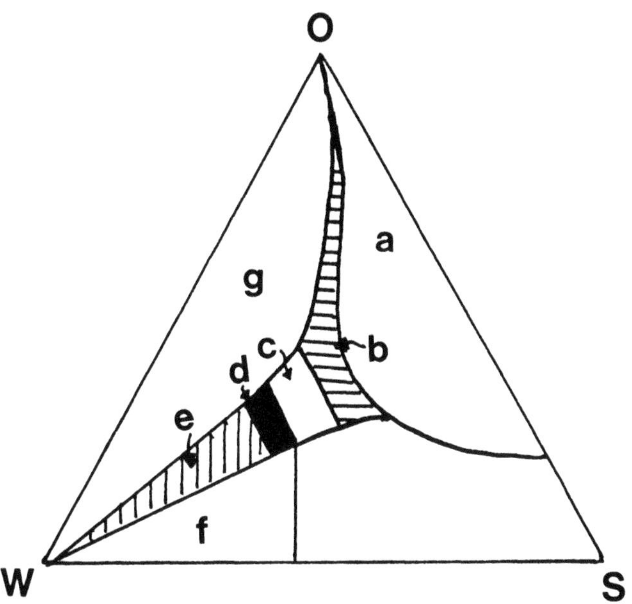

Figure 6 Hypothetical phase regions of microemulsion system composed of oil (O), water (W), and surfactant/cosurfactant (S), showing a, inverted micelles; b, w/o microemulsions; c, cylinders, d, lamellae; e, o/w microemulsion; f, normal micellar; and g, macroemulsion phases.

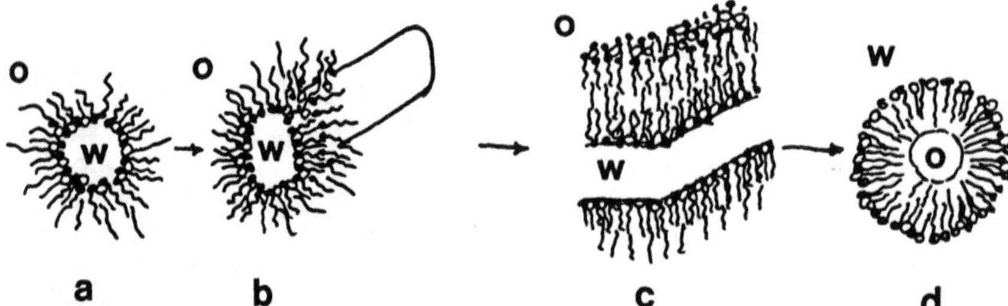

Figure 7 Schematic representation of the mechanism of phase inversion of microemulsions showing a, water spheres in continuous oil phase; b, water cylinders in oil; c, lamellar structures; and d, oil droplets in continuous water phase.

is sandwiched between surfactant double layers. Finally, as the amount of water increases, this lamellar structure will break down and the water will form a continuous phase containing droplets of oil stabilized by a surfactant/cosurfactant interfacial film. Unlike the continuous transition shown between inverted micellar solutions and w/o microemulsions, the normal micellar region in Fig. 6 is not shown to be continuous with the o/w microemulsion region. Dilution of these microemulsions with water does not change the ratio of oil to surfactant, and it is in principle possible to dilute these systems to infinite dilution. In practice, the microemulsion region does not always extend to the water apex of the phase diagram and, as we will see below, this causes complications in the determination of droplet size. As seen from Fig. 6, to convert from o/w microemulsions to micellar solution requires an increase in the surfactant content of the system.

A characteristic feature of microemulsions prepared with nonionic surfactants is the sensitivity of their microstructure to temperature change. Each system is characterized by a narrow temperature range, the HLB temperature or phase inversion temperature (PIT) over which the hydrophilic-lipophilic properties of the surfactant are balanced for a given hydrocarbon-water system (23). Temperature increase can cause a transition from o/w systems at low temperatures and high water concentrations to bicontinuous microemulsions at or near the HLB temperature (24). Further temperature increase beyond the PIT favors the formation of w/o systems (25,26). These possible changes are shown diagrammatically in Fig. 8.

Phase diagrams of the type shown in Fig. 6 are useful in formulation studies as a means of delineation of the area of existence of the microemulsion region. The method used to construct such diagrams depends on the mutual solubilities of the components, but in general it is convenient to use a titration method that allows a large number of compositions to be examined relatively quickly. For example, weighed quantities of surfactant, cosurfac-

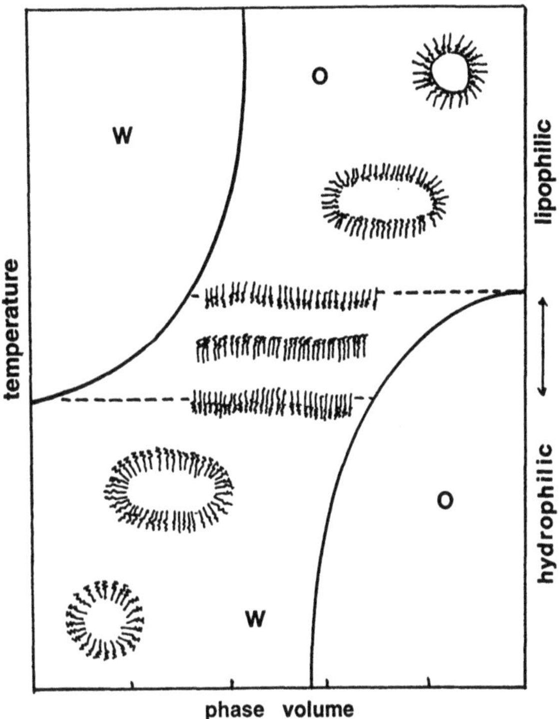

Figure 8 Schematic representation of transition from o/w systems at low temperatures and high water content to bicontinuous microemulsions close to HLB temperature to w/o systems at higher temperatures.

tant, and oil may be mixed to form a monophasic solution and the mixture titrated with water at constant temperature with constant stirring. After each addition of water, the container should be stoppered to minimize loss of any volatile component and the system examined for clarity, birefringence, flow properties, and stability. A more detailed study of the microemulsion region is then required to assess the longer term stability of the systems within this region. Alternatively, surfactant, oil, and water may be mixed together to form a lactecent emulsion and then titrated with the cosurfactant until the mixture becomes clear again (27).

D. Pharmaceutical Microemulsions

1. Choice of Components

The selection of components for microemulsions suitable for pharmaceutical use involves a consideration of their toxicity and, if the systems are to be used topically, their irritancy and sensitizing properties.

Although many nonionic surfactants have suitable properties for topical administration, their potential use in microemulsions for oral or parenteral administration is very limited. For example, peroral administration of microemulsions containing an Arlacel® and Brij® mixture to rats caused side effects, leading to eventual death 30 h after dosage (28). Although polysorbates have been widely reported in pharmaceutical microemulsions, only selected compounds (e.g., polysorbate 80 and 20) have approval for oral ingestion. Phospholipids, particularly the phosphatidylcholines (lecithins) offer a possible nontoxic alternative for parenteral use.

Careful consideration should also be given to the choice of cosurfactant. The inclusion of short or medium chain length alcohols as cosurfactants limits the potential use of the microemulsion due to their toxic and irritant properties. Furthermore, the evaporation of the alcohol can destabilize the system. Inhalation studies (29) of the toxicity of 1-butanol, 2-butanol, and *ter*-butanol in gestating female rats have shown a dose-dependent reduction in fetal weight. Administration of 6.9% butanol, 32% propanol, or 32% ethanol in water was shown (30) to produce elongated mitochondria in rat liver after 1 month and megamitochondria after 2 months. The effect of an oral dose of 2-butanol on hepatic ultrastructure and drug metabolizing enzyme activity was reported by Traiger et al. (31).

The inclusion of a microemulsion in the following review of formulations should be taken to imply that it is of pharmaceutical interest rather than necessarily of use as a dosage form, since several contain ingredients that would limit their application as drug delivery systems. For convenience, microemulsions have been classified according to the surfactant used in their formulation.

2. Nonionic Surfactants

An early study by Jayakrishnan et al. (32) examined the solubilization of hydrocortisone by w/o microemulsions containing a mixture of the nonionic surfactants Brij® 35 (polyoxyethylene 23 lauryl ether) and Arlacel® 186 (glycerolmonooleate-propylene glycol), isopropanol as cosurfactant, water, and *n*-alkane. The influence of the concentration of the oil-soluble surfactant (Arlacel® 186), the chain length of the oil, and the alcohol concentration on the amount of water that could be incorporated in the w/o microemulsion were examined. With fixed quantities of the components, the water solubilization increased with increase of the chain length of the alkane between C_8 and C_{16} and increase of the concentration of Arlacel® 186, reaching a maximum at a 5/1 weight ratio of Arlacel® to Brij®. A formulation containing 10 mL decane, 4 mL isopropanol, 5 g Arlacel® 186, and 1 g of Brij® 35 was capable of incorporating about 8 mL of water. Ritschel et al. (33) have added silicium dioxide (6–10%) to this microemulsion to obtain a gellike consistency and studied its effectiveness as a vehicle for the delivery of cyclosporine A, when filled into hard gelatin capsules, as discussed in Sec. IV.C. Arlacel® 186/Brij® 35 microemulsions containing isopropanol as cosurfactant have also been formulated (34) with several different oil phases including lauric acid hexylester, oleic acid oleylester, glycolyzed ethoxylated glycerides of natural oils, and a mixture of blanched fatty acids with 13 molecules of ethylene oxide. These systems, and also one formed using polyoxylated caprylcaprinic acid glycerides as surfactant, polyglycerol isostearate as cosurfactant, and oleic acid decyl ester as oil phase, were investigated as vehicles for the delivery of peptides. Detailed phase studies were reported for the latter microemulsion.

Oil-in-water microemulsions prepared using blends of the nonionic surfactants polysorbate 60 (Tween® 60) and sorbitan monooleate (Span® 80) together with glycerol as cosurfactant and liquid paraffin as oil were reported by Osipow (35). A more detailed study by Attwood et al. (36) examined the influence on the areas of existence of the microemulsion regions of the polysorbate/sorbitan monooleate ratio and the oil content.

The phase properties of o/w microemulsions prepared using polysorbate/sorbitol/isopropyl myristate/water have been reported in a series of papers by Ktistis (37) and Attwood et al. (38,39). A key factor influencing the phase properties is the surfactant/cosurfactant ratio. Figure 9 shows the change in size and location of the o/w microemulsion region for systems prepared using polysorbate 40, as the polysorbate/sorbitol mass ratio is changed from 1/1 to 1/3.5. Similar studies using polysorbate 80 (37) and 60 (38) have shown a change in the optimum polysorbate/sorbitol ratio (i.e., that producing the largest microemulsion region) from 1/2.5 for polysorbate 80 to 1/2 for polysorbate 60 to 1/1.5 for polysorbate 40. Such effects were attributed to differences in the packing of surfactant and cosurfactant at the oil/water interface. Gasco et al. (40) have used microemulsions of similar composition prepared using polysorbate 60 but with phosphate buffer rather than water, to study the in vitro release of propranolol.

Several workers (41,42) have prepared oil-in-water microemulsions using polysorbate 20 (20.8%) in combination with water/glycerol mixtures (58.3%), decanol/dodecanol (2/1 wt ratio) (14.1%), and 1-butanol (6.8%) for use in the topical delivery of azelaic acid.

Many nonionic surfactants can produce microemulsions without the addition of a cosurfactant. Systems that have been reported have generally been produced from combinations of hydrocarbon, water, and a polyoxyethylene glycol alkyl ether nonionic surfactant (C_mE_n, where m is the hydrocarbon chain length and n is the number of oxyethylene units). Microemulsions formed with $C_{12}E_5$ have been widely studied (43–46).

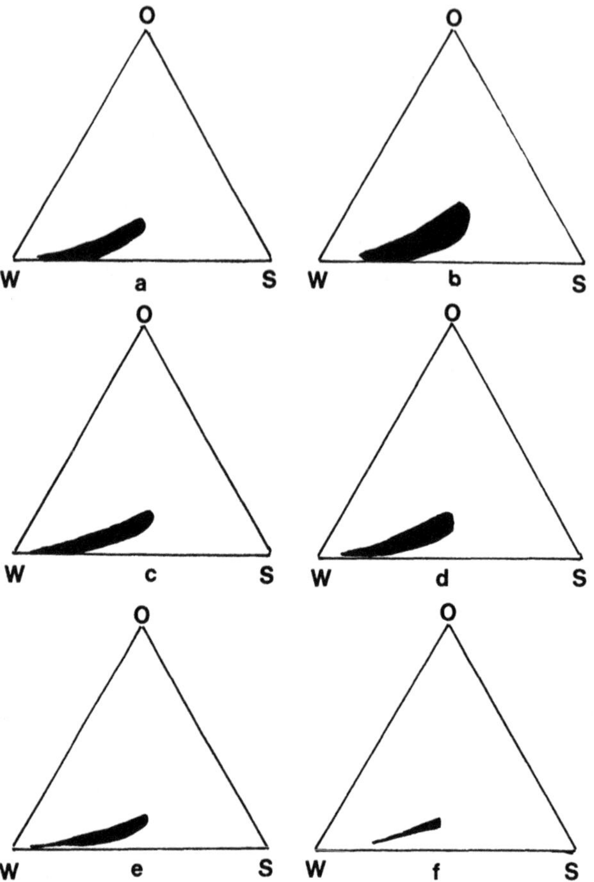

Figure 9 Partial phase diagrams of the system isopropyl myristate (O), polysorbate 40 + sorbitol (S), and water (W) showing areas of existence of o/w microemulsions at 37°C for polysorbate/sorbitol mass ratios of a, 1/1; b, 1/1.5; c, 1/2; d, 1/2.5; e, 1/3; and f, 1/3.5. (From Ref. 39, with permission.)

3. Dioctyl Sodium Sulphosuccinate (Aerosol OT)

Several workers have reported the use of dioctyl sodium sulphosuccinate (Aerosol® OT, AOT) as surfactant in the preparation of microemulsions intended as drug delivery vehicles. Combination of this anionic surfactant with the nonionic surfactant sorbitan monolaurate (Span® 20, Arlacel®20) results (47) in a microemulsion capable of incorporation of large quantities of water. The authors investigated the influence of oil chain length and surfactant weight ratio on the maximum solubilization of water in a clear, isotropic solution. With hexadecane as oil, maximum solubilization occurs when the surfactants are in a 1/1 weight ratio (3/2 molar ratio) of sorbitan monolaurate to AOT. An important feature of these microemulsions is that they are prepared without the addition of alcohol and hence are suitable as topical drug delivery vehicles, free of the irritancy effects normally associated with medium chain length alcohols. In detailed phase studies of these systems, Osborne et al. (48) showed that the sorbitan monolaurate was functioning as a cosurfactant in a similar manner to that of an alcohol in a traditional microemulsion. In the sorbitan monolaurate/AOT systems the ratio of the two surfactants had a significant influence on the area of existence of the w/o microemulsion region, as shown in Fig. 10. A practical limitation in the use of these systems was reported to be the long times required for the waxy semisolid AOT to mix with the viscous liquid sorbitan monolaurate. To overcome this problem, Osborne et al. used a commercially available solution containing 75% AOT, 20% water, and 5% ethanol (AOT-75) instead of AOT. The ethanol in this mixture is acceptable as a pharmaceutical ingredient in such a low concentration since, compared with longer chain alcohols, it shows very low irritation and toxicity. The extent of the microemulsion region for various ratios of AOT-75/sorbitan monolaurate is shown in Fig. 11. Maximum water incorporation occurred with ratios of between 55/45 and 60/40 and was independent of the hexadecane/surfactant mixture

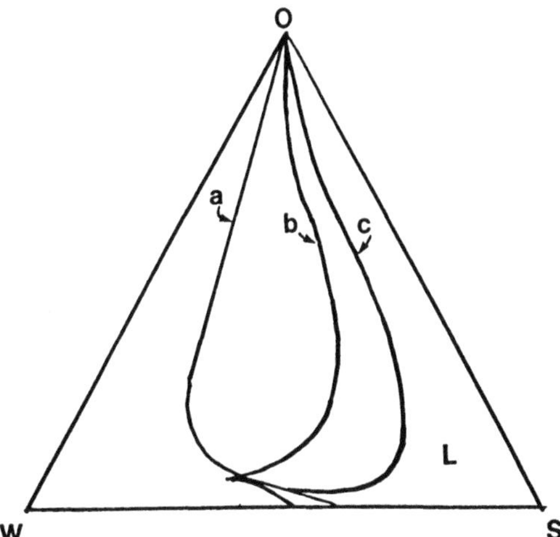

Figure 10 Ternary phase diagram for the water (W)-hexadecane (O)-Arlacel® 20-AOT system in which the Arlacel® 20/AOT weight ratio in the surfactant + cosurfactant mixture (S) is a, 60/40; b, 72.5/27.5; and c, 85/15. The w/o microemulsion side of the diagram is labeled L. (From Ref. 48.)

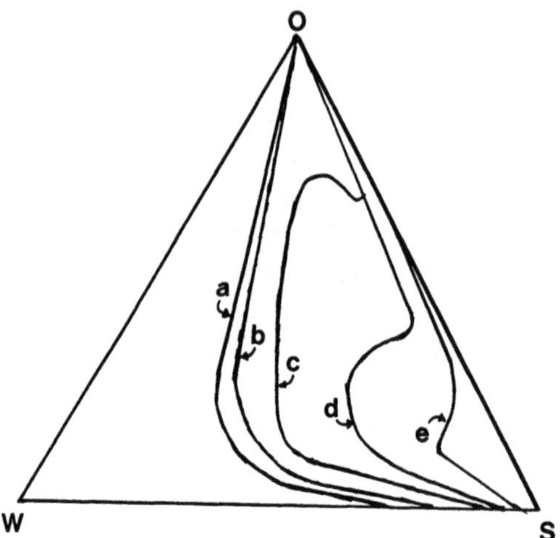

Figure 11 Ternary phase diagram for the water (W)-hexadecane (O)-AOT-75-Arlacel® 20 system. The ratios of AOT-75 to Arlacel® 20 in the surfactant mixture (S) are a, 60/40; b, 70/30; c, 80/20; d, 90/10; and e, 100% AOT-75. (From Ref. 48.)

ratio between 20/80 and 50/50. Although microemulsions prepared from AOT/hydrocarbon/water have been widely studied (49–54), these systems do not tend to incorporate as much water as when the sorbitan monolaurate is used as a cosurfactant. Furthermore, the highest water incorporation for these three-component systems tends to be relatively narrow extensions toward the water apex, making them of less practical use in delivery systems.

The phase properties of ternary systems comprising AOT, water, and a medium chain length alcohol (butanol, hexanol, and octanol) have been reported by Osborne et al. (48,55). Large, isotropic solution regions are observed in phase diagrams for these systems that have been identified as w/o microemulsions in systems containing hexanol or octanol. Figure 12 shows the phase behavior of the water/AOT/octanol system. Addition of an oil (isopropyl myristate) to these ternary systems can result in the formation of an o/w microemulsion. Trotta, Gasco, and coworkers (56–59) have produced o/w microemulsion systems using butanol as cosurfactant. A typical formulation was AOT (14 wt %), isopropyl myristate (19.6 wt %), buffer pH 6 (56 wt %), and butanol (10.4 wt %).

4. Lecithin

The possibility of forming microemulsions using phospholipids is obviously very appealing in view of the low toxicity of these compounds. The problems involved in formulating microemulsions using lecithin have been re-

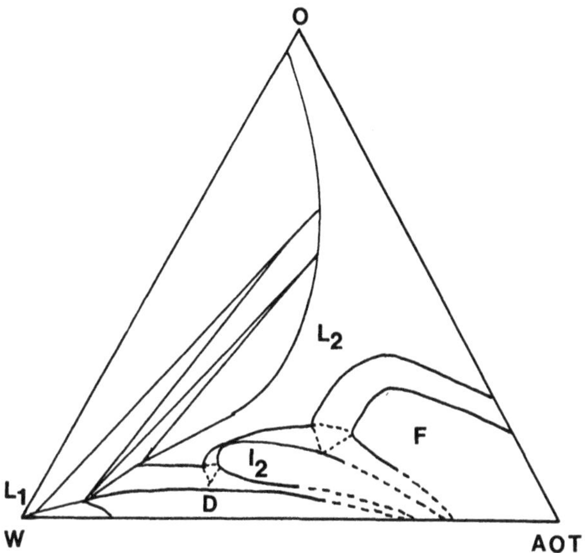

Figure 12 Phase behavior of the water (W)-octanol (O)-AOT system. L is continuous phase, D is the lamellar liquid crystal, l_2 is the cubic liquid crystal, and F is the reversed hexagonal liquid crystal. (From Ref. 48.)

viewed by Shinoda et al. (60). Lecithin is slightly too lipophilic to form spontaneously the zero mean curvature lipid layers needed for balanced microemulsions, and it is necessary both to adjust the HLB and to destabilize the lamellar liquid-crystalline phases, which have a strong tendency to form in these systems. Alteration of the HLB can be achieved by adding short chain alcohols, which make the polar solvent less hydrophilic. In addition, the incorporation of these weakly amphiphilic cosolvents in the polar parts of the lipid layers increases the area of the lipid polar head to produce the required spontaneous curvature of the lipid layers; it also decreases the stability of the lamellar liquid-crystalline phase. These authors have reported the phase properties of the lecithin/propanol/water/*n*-hexadecane systems. About 2.3 wt % lecithin was the minimum amount required to form microemulsions at all mixing ratios. The propanol concentration was in the range 10–15 wt % of the aqueous solvent, the concentration decreasing slightly with increase in oil content. From an examination of the microstructure of the systems, these authors showed a gradual transition from oil droplets in a water-continuous phase through a bicontinuous structure and then to water droplets in oil as the propanol concentration was decreased. This transition is shown in Fig. 13 for lecithin concentrations of 1 and 3 wt % and a weight ratio of (propanol + water) to hexadecane of 1/1.

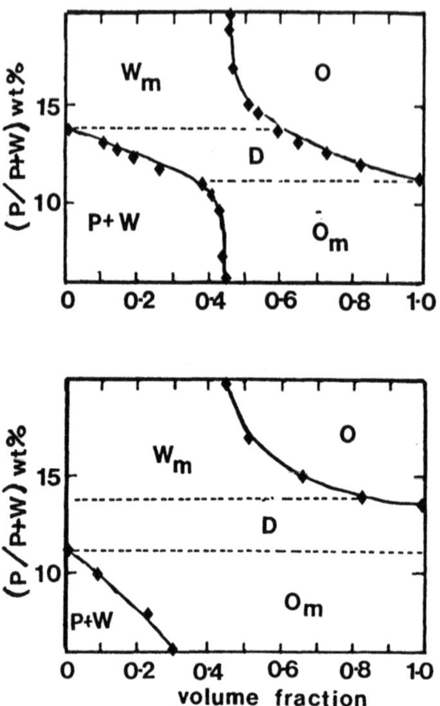

Figure 13 Volume fraction of the different phases for the system of soy lecithin-*n*-hexadecane-1-propanol-water at 25°C at 1 wt % (top) and 3 wt % (bottom) lecithin. Phase notations: O and (P + W) denote oil and aqueous (propanol P + water W) phases containing negligible amounts of lecithin: W_m and O_m denote microemulsions rich in aqueous solvent and oil respectively, and D denotes a microemulsion rich in both solvents. Weight ratio of (propanol + water) and hexadecane is 1/1. (From Ref. 60, with permission.)

Oil-in-water microemulsions prepared using lecithin in combination with *n*-butanol as cosolvent have been recently reported by Attwood and coworkers (61). Figure 14 shows the influence of the lecithin (egg)/butanol ratio on the phase properties in a system containing isopropyl myristate as oil phase. As the lecithin content increased with change of this ratio from 1/0.6 to 1/0.33, the amount of isopropyl myristate incorporated into the microemulsion increased from between 8 and 19 wt % (for the 1/0.6 ratio) to between 24 and 42 wt % (for the 1/0.33 ratio). From a formulation viewpoint, the increased oil content obtained with the 1/0.33 ratio may provide a greater opportunity for the solubilization of poorly water-soluble compounds. Substitution of soya lecithin for the egg lecithin produced differences in the size

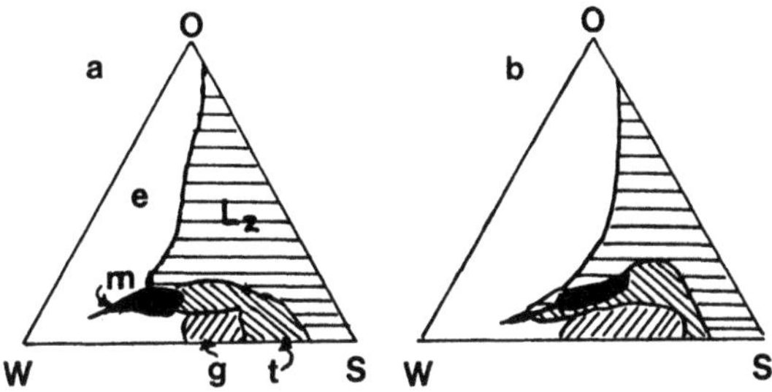

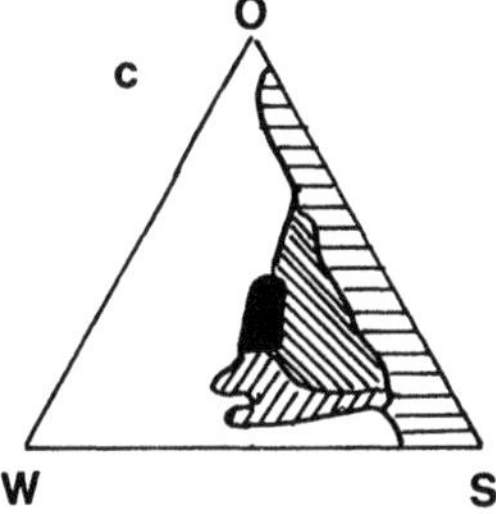

Figure 14 Partial phase diagrams of the system (egg lecithin + butanol) (S)/ isopropyl myristate (O)/water (W) showing stable o/w microemulsion (m), gel (g), monophasic turbid (t), unstable emulsion (e), and isotropic (L_2) regions for lecithin/ butanol weight ratios of a, 1/0.6; b, 1/0.45; and c, 1/0.33. (From Ref. 61, with permission.)

and position of the o/w microemulsion regions that were attributed to differences in the fatty acid impurities in the two types of lecithin, both of which were stated to contain 94–95% of phosphatidylcholine.

These studies were extended by Aboofazeli and Lawrence (62), who recently reported phase properties for systems comprising water/lecithin/ alcohol/isopropyl myristate in which the alcohols were *n*-propanol, isopropanol, *n*-butanol, *sec*-butanol, isobutanol, *ter*-butanol, and *n*-pentanol.

Other workers have used both o/w and w/o lecithin microemulsions as vehicles to deliver a range of drugs but have not reported detailed examinations of phase properties. Fubini et al. (63) and Gasco et al. (64) have prepared microemulsions of egg lecithin, isopropyl myristate, butanol, and water with three lecithin/butanol ratios within the range examined by Attwood et al. (61). A variation of this formulation reported by Gallarate et al. (65) contained octanoic acid to enhance the solubility of timolol in the oil phase and also to facilitate ion-pair formation. The formulations used were egg lecithin (28.7 wt %), 1-butanol (14.9 wt %), isotonic phosphate buffer pH 7.4 (40.0 wt %). The oil phase (16.4 wt %) consisted of solutions of octanoic acid and isopropyl myristate at several weight ratios (10.6/89.4, 15.9/84.1, 21.1/78.9, and 28.4/71.6).

An o/w microemulsion designed for the topical administration of diazepam (66) contained a mixture of egg lecithin (5.6 wt %) and polysorbate 20 (10.1 wt %) as surfactants, in combination with benzyl alcohol (8.9 wt %) as cosurfactant, and a series of mixtures of water and propylene glycol with wt ratios of 10/0, 9/1, 8/2, 7/3, 6/4, and 5/5 that constituted the continuous phase (56.3 wt %).

Water-in-oil microemulsions containing egg lecithin (13.81 wt %), water (11.03 wt %), hexanol (8.6 wt %), and ethyl oleate (66.5 wt %) were shown (67,68) to be suitable reservoirs for doxorubicine and 1-demethoxy-daunorubicine.

III. PARTICLE SIZE ANALYSIS

The size range of the microemulsion droplets and the inability to dilute many microemulsions to infinite dilution makes size determination difficult. Many techniques have been employed with varying success in the size analysis of microemulsions. It is generally true to say that, because of the limitations of each technique, it is preferable to employ a combination of techniques for any system. Some of the more common techniques are reviewed in this section.

A. Electron Microscopy

Although this was one of the earliest methods used to investigate microemulsions (69,70), it is not currently widely used because of the limitations of the technique. In principle, it is possible to measure droplets of the size range of most microemulsions using transmission electron microscopy. Several workers (44,45,71,72) have reported results obtained using freeze fracture electron microscopy (FFEM). In this technique the freezing of the microemulsion must be achieved sufficiently rapidly to avoid phase separation or crystallization. This objective can be achieved by plunging the specimen into a liquid cryogen, by propane jetting, or by spray freezing (44). The sample is subsequently fractured and its visibility is enhanced by deposition of a platinum-carbon layer in vacuo. The microemulsion sample, mounted on a support film and grid, is replicated by breaking apart the film and grid. The replica is then viewed by transmission electron microscopy and assumed to be representative of the bulk microemulsion. Recent developments in the cryofixation techniques have overcome many of the problems associated with artifact formation in earlier studies (73–75), which were due to slow cooling rates. Jahn and Strey (44) and Vinson et al. (45) have developed a mechanical plunging device for rapid freezing and have given detailed experimental procedures. Recent freeze fracture transmission electron microscope studies utilizing rapid cooling rates and adequate environmental control of the samples before freezing (to prevent loss of volatile components) have investigated microemulsions of pentaethylene glycol dodeyl ether ($C_{12}E_5$) water and *n*-octane (44–46,76). However, images produced of the bicontinuous microemulsion structures of identical samples in the studies of Bodet et al. (76) and Jahn and Strey (44) showed two different morphologies. These apparently conflicting results were rationalized in later studies of this system by Vinson et al. (45), who showed that the different images corresponded to adhesive fractures (where the fracture had propagated at or near the grid/microemulsion interface) and cohesive fractures (in regions of the microemulsion where there are no grid bars). The images of adhesive fracture surfaces as presented by Bodet et al. (76) allow visualization of microemulsion structure as it exists near solid surfaces, while the cohesive fractures observed by Jahn and Strey (44) showed evidence of fractured tubulelike or vestibulelike structures that exist in the bulk bicontinuous microemulsion.

A complementary technique is that of direct imaging, in which thin portions of the specimen are directly investigated in the frozen hydrated state by using a cryostage in the transmission electron microscope. The development of glass-forming microemulsions that do not break down during cooling and in which neither dispersed nor matrix phase crystallizes during the cooling process has provided a way for direct studies of the microemulsion structures. The first types of such systems to be reported were o/w microemulsions with a noncrystallizing aqueous matrix obtained by adding propane-1, 2-diol (propylene glycol) to the water in the ratio 1/3 (77,78). It was thought that the propylene glycol functioned as cosurfactant since the usual addition of an alcohol was not necessary. The fact that neither phase crystallized on cooling below 273K meant that possible modifications of the microemulsion structure by crystal growth were avoided. This approach was extended to w/o microemulsions by Green (79), who used glycerol/water mixtures as the dispersed phase, a glass-forming oil (ethylcyclohexane) as the continuous phase, and didodecyldimethylammonium bromide as surfactant. Direct electron microscope imaging of the vitrified microemulsion using freeze fracture techniques revealed clearly separated droplets of 20 nm diameter.

B. Scattering Methods

Scattering methods that have been employed in the study of microemulsions include small angle x-ray scattering (SAXS), small angle neutron scattering (SANS), and static and dynamic light scattering. These techniques have a lower size limit of about 2 nm and an upper limit of about 100 nm for SANS and SAXS and a few microns for light scattering. They are therefore ideally suited to a study of microemulsions.

In the case of monodisperse spheres interacting through hard sphere repulsion, the general expression for scattering intensity I(Q) is

$$I(Q) = nP(Q)S(Q) \tag{2}$$

where n is the number density of the spheres and Q is the scattering vector ($Q = 4\pi \sin \theta/\lambda$ with θ = scattering angle and λ = wavelength). The form factor $P(Q)$ expresses the scattering cross-section of the particle, and the structure factor $S(Q)$ allows for interparticle interference. Analytical expressions may be used to calculate both $P(Q)$ and $S(Q)$ under favorable circumstances.

1. Small Angle X-Ray Techniques

Small angle x-ray techniques have been used by several workers to gain information on droplet size in dilute microemulsion systems. The microstructure of the *n*-dodecyltetraoxyethylene glycol monoether ($C_{12}E_4$)/water/hexadecane system was studied by Shimobouji and coworkers (80) using SAXS. On the basis of the analysis of the surfactant concentration dependence of the peak position, it was concluded that an o/w microemulsion had been formed, oil droplets being identified by a single broad peak on the scattering curves. The radii of the droplets were estimated to be between 3.5 and 140 nm depending on surfactant concentration. Bohlen (81), using SAXS, reported a decrease in droplet size in a poly(oxyethylene) alkyl ether/*n*-alkane/water microemulsion from 94 to 22 nm as the molecular size of the nonionic surfactant decreased in the order $C_{12}E_5$, $C_{10}E_4$, to C_8E_3. The droplet size was also found to decrease with increasing surfactant concentration.

Interpretation of the scattering curves from concentrated microemulsions presents a problem. Zemb et al. (82) and Barnes et al. (20,83) interpreted the SAXS scattering from a w/o system composed of dodecyldimethylammonium bromide and water with various alkanes as the oil, using the disordered open connected model. This model successfully described the structural transitions that occurred in the system as the water content was increased. At low water content the structure resembled a disordered hexagonal phase, which at higher water content changed to isolated water spheres in oil. The approximate regions of the phase boundaries could be determined using the model.

Recent improvements in small angle x-ray methodology have been achieved with the use of synchrotron radiation sources, resulting in an increased performance of SAXS (84,85). By this means, the sample-to-detector distances may be up to 4 m as compared to 30–50 cm with laboratory based x-ray sources. These large sample-to-detector distances result in less diffuse profiles allowing better interpretation of the curves. Another advantage of synchrotron radiation is that it can be used even when the amphiphiles in the microemulsion are poor x-ray scatterers, allowing a wider range of microemulsions to be examined. North et al. (84) used synchrotron SAXS in a study of the three-component w/o microemulsion system AOT, water, dodecane and found that analysis of the low Q value data gave more reliable results than analysis of the high Q value data, which were affected by polydispersity. All microemulsions were close to a critical phase transition at 25°C. A significant change in the small angle scattering profile at low Q values was observed in all samples due to critical scattering on approaching the phase transition within the microemulsion. Hilfiker et al. (85) used synchrotron SAXS to look at spherical deformity in the AOT, water, and *n*-hexane, w/o microemulsion system. The spheres were thought to deform into prolate and oblate ellipsoids, the deformity being more likely when the temperature was increased.

2. Small Angle Neutron Scattering

Small angle neutron scattering has proved a valuable technique for the study of microemulsions, although it suffers the disadvantage of requiring central facilities. Robinson and coworkers (51–53) have reported an extensive investigation of the AOT, alkane, and water microemulsion system using this technique. These workers achieved good SANS contrast by utilizing the fact that hydrogen and deuterium have a different sign in their neutron scattering lengths (−3.74 and 6.67 fm respectively). The negative scattering for hydrogen indicates a phase shift of 180° between waves scattered by hydrogen and those scattered by deuterium. Contrast was achieved between the water and oil phases by selective H/D isotopic substitution. These workers found an approximately linear relationship between the size of the water core radius and the concentration ratio (D_2O/AOT) with heptane as the oil (52). Cebula and coworkers (86) have used SANS in combination with light scattering techniques to investigate the water, xylene, sodium dodecyl benzene sulphonate, and hexanol system. The information from these experimental techniques was consistent. For the system investigated, the radius of the water core of the droplet was found to increase with increase in the water volume fraction while maintaining a constant water/oil interfacial area. In further studies, Cebula et al. (87,88) reported SANS and light scattering measurements on microemulsions formed with water, potassium oleate, hexanol, and dodecane with a water/potassium oleate volume fraction of 1.44. The results from both techniques indicated very little variation in the droplet diameters within a wide region of the microemulsion domain.

There are a number of important differences between SANS and SAXS techniques. Neutrons are absorbed less readily than x-rays, hence the sample is more stable in neutron beams than in x-ray beams. The source-to-detector distances used in SANS are much larger (10 to 30 m) (89) compared to those of SAXS (4 m) (85). A powerful characteristic of neutron scattering is the possibility to enhance selectively the scattering power of various parts of the microemulsion by using protonated or deuterated molecules, while in x-ray scattering, materials must be selected as good x-ray scatterers. The x-ray contrast between oil and water is small, while selective deuteration improves contrast (signal-to-background ratio) in the SANS method. The rate of data collection is faster with SANS due to the longer wavelengths used (1 nm compared with 0.1 nm for x-rays) and larger sample size. SANS profiles suffer from incoherent scattering at high Q values, making interpretation of small signals difficult, whereas with SAXS there is little background scattering at high Q values and consequently the effective Q range for SAXS experiments extends to higher values. In view of the limitations of the two techniques it is preferable to use SANS or SAXS in combination with other techniques, i.e., complementary use of several techniques can help reveal microemulsion structure.

3. Static Light Scattering

Static (total intensity) light scattering techniques have been widely used for the particle sizing of microemulsions. In some systems the area of existence has been such as to allow measurement at sufficiently low concentrations to apply the Rayleigh scattering equation. In cases where this is not possible, the correction must be made for interparticle interference. For a volume fraction Φ of dispersed particles, the Rayleigh ratio at a scattering angle of 90°, $R(90)$, is related to the droplet radius r by

$$R(90) = K\Phi r^3 S(Q)P(Q) \tag{3}$$

where K is an optical constant. An iterative method may be applied in the matching of predicted and experimental $R(90)$ values using the Percus-Yevick expression (90–94) for $S(Q)$. Cebula et al. (86,87) have applied the Percus-Yevick hard sphere model in the interpretation of the light scattering data for w/o microemulsions. According to the droplet model proposed by these workers, the water core of the dispersed droplets may be assumed to constitute the scattering unit since the refractive index of the adsorbed surfactant layer is very similar to that of the dispersion medium. The volume fraction of the scattering unit can thus be equated to that of the water in the system. A complication arises with o/w systems since now the surfactant layer must be considered to be part of the scattering unit that it surrounds, giving rise to uncertainty in the calculation of the volume fraction of the droplets. A method of estimation of the amount of surfactant associated with the droplets has been proposed by Attwood and Ktistis (38), who applied this model to an o/w system composed of polysorbate 60, isopropyl myristate, sorbitol, and water. They reported an increased droplet radius with increasing volume fraction of oil at fixed polysorbate/sorbitol concentration. Tadros et al. (95) investigated the w/o system *n*-tetradecane, water, *n*-butanol with an A-B-A block copolymer, where A is poly(12-hydroxystearic acid) and B is poly(ethylene oxide), as surfactant. By application of the Percus-Yevick model, the workers noticed an increased droplet ratius of the microemulsion with increasing volume fraction of water at constant polymer concentration and a 50% increase in size of the droplets near the phase boundary. Baker et al. (96,97) measured the water core radius as a function of surfactant concentration in the water/xylene/sodium alkyl benzene sulphonate/hexanol system. These workers showed a gradual increase in droplet size as the water concentration was gradually increased at each surfactant concentration. At a given volume fraction of water, the water core radius decreased as the surfactant concentration increased from 5 to 20 wt %.

4. Photon Correlation Spectroscopy

Dynamic light scattering or photon correlation spectroscopy (PCS) analyzes the fluctuations in scattering intensity that occur over very short time intervals due to the Brownian motion of the particles. The diffusion coefficient of the scattering centers may be calculated from the decay of the correlation function, and, in the absence of interparticle interference, the hydrodynamic radius of the particles r_H can be determined from the diffusion coefficient D using the Stokes-Einstein relationship

$$D = \frac{k_B T}{6\pi\eta r_H} \tag{4}$$

where k_B is the Boltzmann constant, T is the absolute temperature, and η is the viscosity of the solvent. With most microemulsions, extrapolation to infinite dilution is not possible and hence errors due to interparticle interference will affect r_H values for these systems and should be corrected (86).

Cebula et al. (87) used PCS to confirm their light scattering and SANS results in the system water, xylene, sodium dodecyl benzene sulphonate, and hexanol, as discussed above. The stability of the microemulsions was demonstrated by the lack of change in the correlation functions after the samples were subjected to ultrasonic radiation. Bedwell and Gulari (98) reported that addition of small amounts of electrolyte to microemulsions of AOT, water, and heptane resulted in an increase in the diffusion coefficient D, which they interpreted as a decrease in the hydrodynamic radius of the particle. In a further examination of this system (99), these workers reported an increase in droplet radius from 6 to 300 nm with increase in temperature and disperse phase volume. Chang and Kaler (100) have used PCS in an examination of the five-component w/o system sodium 4-(1-heptylnonyl) benzene sulphonate/isobutyl alcohol/water/dodecane and sodium chloride. Droplet sizes increased from 6 to 30 nm with increasing volume fraction of water, up to a limiting value. From 25 to 75% water, the systems were turbid, and D values were found to be constant, indicating the presence of a bicontinuous structure. This was confirmed by light scattering studies on the same system. Rosano et al. (101) investigated the influence of the nature of the surfactant, cosurfactant, and oil on microemulsion formation using PCS and transmittance data. Increasing the amount of surfactant (potassium soaps of straight chain fatty acids) decreased the particle size. They reported an increase in size with increasing amounts of alcohol for the droplets in o/w and w/o microemulsions. They attributed this, in the o/w case, to the excess alcohol dissolving in the oil droplets thus causing their swelling. In the w/o systems, the excess alcohol was thought to extract surfactant from the interface, resulting in decreased stabilization of droplets and hence increased size. Müller and Müller (102,103) have reported investigations on w/o microemulsions prepared with potassium oleate (as surfactant), liquid paraffin (as oil), and various alcohols (as cosurfactants). PCS was used to determine the effect of the alcohols on microemulsion stability by sizing the systems at regular time intervals, an increase in particle size indicating a deviation from the stable system. These workers reported an almost constant droplet size in the middle of the microemulsion region over a prolonged period of time. Zulauf and Eicke (50) studied the AOT, water and isooctane system using PCS and presented evidence for a clear distinction between inverted micelles and microemulsion regions according to the degree of hydration or the amount of solubilized water respectively. Cazabat et al. (104) reported a PCS study on a w/o microemulsion composed of sodium dodecyl sulphate and water with cyclohexane or toluene as oil phase and butanol, pentanol, or hexanol as cosurfactant. No attempts were made to interpret the diffusion data at large water volume fractions, but a theoretical treatment was proposed for the variation in the diffusion coefficient with water content at small volume fractions. This treatment assumed the microemulsions to behave as solutions of spherical particles interacting with a potential that was the sum of a hard sphere repulsion force V_{HS} and a small attractive force V_A. In a recent PCS study by Trotta et al. (58), the influence of increasing amounts of alcohol on the diffusion coefficient of microemulsion droplets was investigated. The o/w system studied was AOT, isopropyl myristate, and water with ten alcohols of varying water solubilities, ranging from butanol to pentan-1-ol [although those alcohols with low water solubilities (isoamyl alcohol, pentan-1-ol) were not able to form transparent o/w microemulsions]. All experiments, performed by adding increasing amounts of cosurfactant to the system, showed the presence of a maximum diffusion coefficient corresponding to an optimum amount of cosurfactant for microemulsion formation.

IV. MICROEMULSIONS AS DRUG DELIVERY SYSTEMS

A. Influence of Formulation on Drug Release Characteristics

Drugs incorporated in microemulsions will partition between the aqueous and hydrophobic phases depending on their lipophilicity. The influence of the partition coefficient of a drug on its release characteristics was reported by Trotta et al. (59). The release of five drugs with different lipophilicities from o/w microemulsions composed of isopropyl myristate, butanol, AOT, and buffer (pH 7) was studied by determining mass transfer constants of the drugs through a hydrophilic membrane separating the o/w microemulsion from the receiving aqueous phase. A linear relationship between the release rate of the drug and its isopropyl myristate/water partition coefficient was noted.

Several workers have reported studies in which the lipophilicity of the drug has been increased to enhance its solubility in the dispersed oil droplets. In this way a reservoir of the drug is produced and a sustained release effect is achieved as the drug continuously transfers from the oil droplets to the continuous phase to replace drug released from the microemulsion.

Gallarate et al. (65) have increased the lipophilicity of timolol by ion-pair formation with octanoic acid. The partition of this drug between oil and buffer phases in o/w microemulsions prepared using egg lecithin, butanol, buffer pH 7.4, and mixtures of isopropyl myristate and octanoic acid was shown to increase with increase in the percentage of octanoic acid in the oil phase due to ion-pair formation in this phase. In addition, it was shown that the in vitro permeability constants of timolol through a lipophilic membrane at pH 7.4 were about seven times higher when the timolol was present as an ion pair, suggesting that an improved corneal absorption of this drug might be achieved by topical administration in this microemulsion vehicle. Similarly, the lipophilicity of propranolol has also been enhanced by the formation of lipophilic ion pairs with octanoic acid in o/w microemulsions containing isopropyl myristate, polysorbate 60, butanol, and buffer pH 6.5 (40). The partitioning of the propranolol into the dispersed oil phase was increased by the addition of octanoic acid. As a consequence, its release rate through a hydrophilic membrane was decreased due to its decreased concentration in the continuous phase, producing a prolongation of drug release.

Gallarate et al. (41) have increased the solubility of the dicarboxylic acid, azelaic acid, in the dispersed oil phase of polysorbate 20/butanol/decanol: dodecanol (2/1)/water microemulsions by lowering the pH of the aqueous phase and by the addition of propylene glycol to reduce further its dissociation. An increased partitioning into the lipophilic phase was noted as the propylene glycol concentration was increased. The microemulsions thus provided a vehicle in which the azelaic acid was dissolved rather than suspended, as when applied as a cream, and moreover the reservoir effect achieved by partitioning into the oil could prolong its release over several hours. The authors showed a tenfold increase (up to 27–30% of the initial amount) in the amount of drug released from the microemulsion when compared to a cream clinically used in the treatment of skin disorders.

Examples have been reported of microemulsion systems in which prolonged release of drugs has been unexpectedly achieved by interactions that occur between the drug and the surfactant. The release rates of doxorubicine (56, 68) from o/w microemulsions prepared with AOT or polysorbate 80 and w/o microemulsions prepared with lecithin were greatly reduced due to the formation of lipophilic complexes between this water-soluble drug and each of the surfactants. A similar modification of drug release was later noted when 1-demethoxy daunorubicine was formulated in the same w/o lecithin microemulsion (67).

In view of the appreciable influence that the cosurfactant exerts on the properties of the microemulsion system, it is not unexpected that it should also significantly affect drug release from the microemulsion. The influence of the cosurfactant concentration on the rate of release of six steroids with a range of lipophilicities, from a series of o/w microemulsions obtained by adding increasing volumes of butanol to fixed amounts of isopropyl myristate, water, and Aerosol® OT, was reported by Trotta et al. (57). For all the drugs, the apparent partition coefficient of drug into the dispersed phase increased with butanol addition, producing a decrease in the release rate of the drug. A similar correlation between this partition coefficient and drug release was noted when these steroids were incorporated into microemulsions of fixed composition prepared with a range of alcohols as cosurfactants.

B. Microemulsions for Topical Drug Delivery

In vitro skin permeation studies have been reported by several workers in the assessment of the suitability of microemulsions for topical drug administration. In a study of the effect of formulation on the percutaneous absorption of water from a w/o microemulsion of Aerosol® OT/octanol/water it was shown (55) that the transdermal flux of the water was highly dependent on the water content of the microemulsion. The flux increased sixfold as the water content of the microemulsion increased from 15 to 68%. Low flux at low water content was attributed to the binding of water to the head groups of the Aerosol® OT. At higher water contents, a greater proportion of the water in the microemulsion was available for transport, leading to a higher flux. An interesting synergistic effect was noted in these systems resulting from the presence of Aerosol® OT and octanol, which in combination served as a penetration enhancer. These microemulsion vehicles were later tested for their ability to transport glucose across human cadaver skin (105). Using a flow-through multisample skin diffusion cell it was shown that microemulsions containing 68% of water caused an approximately 30-fold enhancement of glucose transport, whereas no transport was discernible for the 15% water microemulsion. The differences in percutaneous glucose transport therefore corresponded to differences in water transport, and the authors stressed the importance of careful optimization of the microemulsion formulation to achieve maximum percutaneous transport.

Février et al. (106) have reported in vitro experiments designed to simulate the percutaneous penetration of tyrosine when administered using an o/w microemulsion composed of a betaine derivative (as surfactant), benzyl alcohol, hexadecane, and water. The release of radiolabelled tyrosine from this vehicle was compared to that from a liquid-crystal system and an emulsion, using a diffusion cell equipped with rat skin. Both the microemulsion and the liquid-crystal formulation enhanced the penetration of tyrosine through the epidermis when compared to the emulsion. However, cutaneous irritation studies showed a strongly irritant effect from the liquid-crystal formulation but none from the microemulsion.

In a similar study by Ziegenmeyer and Führer (107), tetracycline hydrochloride was reported to show enhanced percutaneous absorption from a microemulsion compared to conventional systems. Complete diffusion from the microemulsion occurred within 5 to 6 h, compared with complete diffusion after 12 h from a gel and after more than 24 h from a cream.

In an in vivo study, Martini et al. (108) showed that the percutaneous absorption of *D*-α-(5 methyl 3H) tocopherol in rats was more rapid from an o/w microemulsion than from a w/o emulsion or white vaseline. It is interesting to note from this work that there was an accumulation of radioactivity in the liver, kidneys, muscle, and body fat, six h after administration of the radiolabelled tocopherol using the emulsion and vaseline, but no detectable accumulation when the microemulsions were used as vehicles. Hence the microemulsion not only enhances the rate of penetration but also appears to influence its distribution and elimination.

One of the problems associated with the use of microemulsions for topical drug delivery is the difficulty of applying these vehicles to the skin because of their fluidity. Gasco et al. (42) have recently addressed this problem in the development of a microemulsion for the topical administration of azelaic acid, which has reported therapeutic effects on some pigmentary disorders and on acne vulgaris. In an earlier study (41) discussed above, these workers demonstrated the formulation conditions required to reduce the dissociation of this acidic drug. Its consequent enhanced partition into the dispersed phase of an o/w microemulsion produced a reservoir of dissolved drug and a high rate of release. The viscosity of the o/w microemulsions used in this study was increased to make them suitable for topical administration by the inclusion of Carbopol® 934. A comparison of the in vitro release profiles through hairless skin membrane from this "viscosized" microemulsion with those from a previously reported gel containing similar components (propylene glycol, Carbopol® 934) is shown in Fig. 15. Clearly, a pronounced enhancement of penetration is achieved; after 8 h about 35% of the azelaic acid present in the microemulsion had been transported compared with only 1.8% from the gel. The release could be further increased by the addition of a penetration enhancer, dimethyl sulphoxide, to the microemulsion. Trotta et al. (66) have reported a similar enhancement of skin permeation of diazepam from "viscosized" o/w microemulsions prepared using egg lecithin, polysorbate 20, benzyl alcohol, isopropyl myristate, and water/propylene glycol mixtures.

The potential application of microemulsions for the ocular administration of timolol was investigated (109) using o/w lecithin microemulsions in which this drug was present as an ion pair with octanoate. The microemulsion, a solution of the ion pair and a solution of timolol alone were instilled in the conjunctival sac of rabbits and the bioavailability of timolol from each compared (Fig. 16). The areas under the curve for timolol in aqueous humor

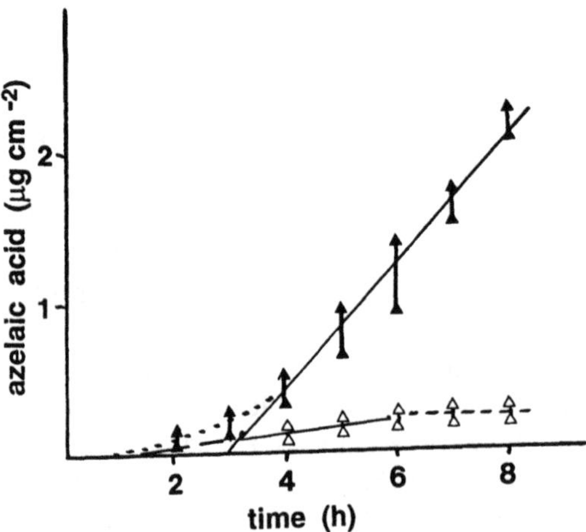

Figure 15 Permeation profiles of azelaic acid from a "viscosized" microemulsion (▲) and from a gel (△). (From Ref. 42, with permission.)

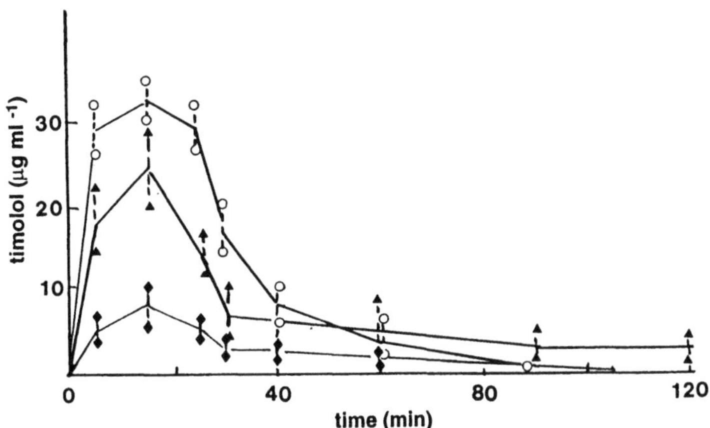

Figure 16 Aqueous humor concentration-time profiles following multiple installation in rabbits' eyes. ♦, timolol alone; ○, timolol as an ion pair in solution; and ▲, timolol as an ion pair in microemulsion. (From Ref. 109, with kind permission of Pergamon Press Ltd.)

after administration of the microemulsion and the ion pair solution were 3.5 and 4.2 times higher respectively than that observed after administration of timolol alone. A prolonged absorption was achieved using the microemulsion, with detectable amounts of the drug still present in the aqueous humor, 120 min after administration. The enhanced corneal absorption of timolol as an ion pair might suggest the possibility of lowering the dose instilled in topical therapy, leading to a possible reduction of the side effects of this drug.

C. Microemulsions as Oral Delivery Vehicles

The assessment of the microemulsions for oral delivery has centered around their potential use for the delivery of peptides, in particular, cyclosporine. Cyclosporine is administered in both oral and intravenous dosage forms. There is some hesitancy for its prolonged use by the intravenous route, and, for long term treatment, the oral route is preferred. However, its oral absorption is slow and incomplete, leading to a wide interindividual variation ranging from 1% to 95% (110). Administration as a coarse dispersion, formed by mixing a vehicle of olive oil, alcohol, and polyoxyethylated oleic glycerides with water may be a contributory factor to its low and erratic absorption. Tarr and Yalkowsky (111) have demonstrated an enhanced intestinal absorption in rats through a reduction of emulsion droplet size by homogenization, which was explained in terms of the greater surface area of the dosage form. The results of this study are supported by a demonstration of significantly higher bioavailabilities of cyclosporine when administered using w/o microemulsions containing a sorbitan ester/polyoxyethylene glycol monoether mixture as surfactant, a low molecular weight alcohol, fatty ester, and water. A gellike consistency was achieved through the addition of pyrogenic silicium dioxide (Cab-o-Sil®), and the microemulsion gel was administered in hard gelatin capsules. The bioavailability from this dosage form was compared to that from a commercially available peroral solution and a solution for intravenous administration using dog (33) and rat (28) models. The higher relative bioavailability of the microemulsion systems compared to the commercial preparation was attributed to their lower droplet size, although differences between two of the microemulsion formulations suggested that an additional mechanism may be involved (see Table 1). In a later, more comprehensive study, Ritschel (34) investigated the gastrointestinal absorption of three peptides using a series of o/w microemulsion formulations. The peptides used were dissolved in the aqueous phase at a suitable pH if water-soluble (insulin, vassopressin) or otherwise added to the microemulsion (cyclosporine) and dispersed by sonication. Three forms of microemulsion were used, a fluid microemulsion for in situ studies on the isolated segment rat model, a microemulsion gel formed by the addition of silicium dioxide (for rectal studies), and an encapsulated microemulsion gel for peroral absorption studies.

Table 1 Absolute (F) and Relative (EBA) Bioavailability in Rats of Cyclosporine A (Average ± SD) after Peroral Solution and 2 Peroral Microemulsion Formulations

Parameters	P.o. solution	P.o. microemulsion A[b]	P.o. microemulsion B[c]
Absolute bioavailability F (%)	11.8 ± 2.8	41.4 ± 18.1	15.0 ± 3.3
Relative bioavailability EBA (%)	100 (standard)	447.1 ± 287.6[a]	147.2 ± 26.5
Rate of bioavailability			
C_{MAX} (μg/mL)	1.95 ± 0.03	4.36 ± 1.65[a]	3.72 ± 1.35
t_{max} (h)	4.35 ± 0.59	9.00 ± 3.47	4.25 ± 0.50

[a]Significantly different from p.o. solutions. $p < 0.05$.

[b]Microemulsion A: w/o microemulsion containing a long chain fatty acid as lipid phase, Arlacel® and Brij® as surfactants, a low molecular weight alcohol as cosurfactant, and distilled water.

[c]Microemulsion B: as A except that branched alkyl fatty esters were used as lipid phase.

Source: Ref. 28, with permission.

Improvements in the bioavailability of these peptides when administered orally were again not solely dependent on droplet size. The author concluded that the systemic peptide uptake from microemulsion in the gastrointestinal tract was dependent additionally on the following formulation factors: type of lipid phase of the microemulsion, digestibility of the lipid used, and types of surfactant in the microemulsion. A detailed hypothesized mechanism of peptide absorption from microemulsions given perorally was proposed.

V. SUMMARY

Microemulsions offer several potential advantages as drug delivery systems arising from their solubilization capacity, transparencies, high stability, and simplicity of manufacture.

Optimization of solubility of poorly water-soluble drugs in pharmaceutical dosage forms presents a challenge, due to the severe restrictions on the choice of solvents suitable for oral, topical, or parenteral use. The finely dispersed oil droplets of o/w microemulsions offer a potential solvent system for such drugs, although as discussed above, the toxicity of the microemulsion components still imposes limitations on their utilization. In an early study (32) of the solubilization capacity of w/o microemulsions prepared using Brij® 35 and Arlacel® 186 as surfactants and isopropanol as the cosurfactant it was shown that the solubility of hydrocortisone was sixfold higher in a microemulsion containing 6 g of total surfactant, 10 mL of oil, and 3.5 mL of isopropanol compared to that in pure isopropanol. A recent comparison has been reported (112) of the solubilization capacity of 2 wt % soybean oil/Brij® 96 (polyoxyethylene-10-oleyl ether) o/w microemulsions and micellar solutions containing identical concentrations (10–20%) of Brij® 96. It was shown that, although there was no significant difference between uptake of testosterone into the microemulsions and micellar solution, there was appreciable enhancement of the solubility of the more lipophilic drugs testosterone propionate and testosterone enanthate, compared with the micellar system.

The transparency of microemulsions enables them to be visually assessed for microorganism growth and also allows inspection for the presence of undissolved drug. Their transparency is also of benefit in topical preparations where clear systems are more aesthetically pleasing. The thermodynamic stability of microemulsions is clearly an important characteristic when compared with kinetically stabilized macroemulsions. The influence of drug incorporation into the microemulsion on its stability was investigated calorimetrically by Fubini et al. (113). It was shown that the addition of prednisone to o/w microemulsions prepared with either lecithin or Aerosol® OT as surfactant/isopropyl myristate/butanol/water had no significant influence on the enthalpy change associated with the microemulsification. However, the addition of menadione to the lecithin microemulsions resulted in less endothermic changes, which were ascribed to the formation of lecithin/menadione complexes or aggregates.

A recent study (114) of the influence of temperature on the physical stability of an o/w microemulsion containing butibufen showed no detectable change in droplet size after storage at 20°C for 1 year and no change in appearance after storage at temperatures between 4 and 40°C for up to 30 days. Slight discoloration was noted after 15 days storage at 75°C. Similarly Gasco et al. (56) have reported no significant changes in the viscosity or turbidity of several o/w and w/o microemulsions after 6 months of repeated shaking and freeze-thaw cycles. Panaggio et al. (115) have shown that 10% soybean o/w microemulsions prepared using a blend of Span® 80 and Tween® 80 showed very little, if any, physical degradation after autoclaving and concluded that this was an effective means of sterilizing these dosage forms.

The formation of microemulsions requires only the most basic mixing equipment. More importantly, their manufacture is not so dependent on the careful control of manufacturing process as, for example, during the preparation of emulsions.

Specific examples of increased bioavailability from microemulsions compared to more conventional dosage forms have been discussed above. Investigations into their use in the delivery of peptides such as cyclosporine may be identified as a particular example of an area that clearly shows their potential as drug delivery vehicles.

The main problem in their commercial exploitation lies in the lack of biological tolerance of the constituents of the microemulsion, particularly the surfactant and cosurfactant. The problems arising from the presence of medium chain length alcohols as cosurfactants in microemulsions intended for topical administration have been addressed by Osborne et al. (48). These cosurfactants tend to be skin or eye irritants, and their use in topical delivery systems is very limited. The alcohol-free w/o microemulsions prepared using Aerosol® OT in combination with Arlacel® 20 (47,48) are clearly an advance toward producing pharmaceutically acceptable topical vehicles. Even greater limitations are imposed on the choice of constituents for oral and parenteral use, and the way forward here may lie with the use of phospholipids as surfactants, providing suitable cosurfactants can be found to replace the medium chain length alcohols currently reported in their formulation. In this respect it should be noted that several groups of workers (116,117) have used combinations of phospholipid and cholesteryl esters to prepare microemulsions following methods similar to that proposed by Ginsberg et al. (118). Such methods often, however, involve prolonged sonication to achieve the clarity normally associated with microemulsions and a subsequent complex size separation technique to ensure uniformity of particle size. Although the final products have the appearance and particle size range of a microemulsion, the lack of spontaneity of their formation differentiates them from the type of microemulsion considered in this chapter, although they are clearly of interest as drug delivery systems.

REFERENCES

1. Hoar, T. P., and Schulman, J. H. (1943). Transparent water-in-oil dispersions: The oleopathic hydro-micelle. *Nature 152*: 102.
2. Danielsson, I., and Lindman, B. (1981). The definition of microemulsion. *Colloids Surf. 3*: 391.
3. Overbeek, J. Th. G. (1986). Microemulsions. *Proc. Roy. Dutch Acad. Sci. Ser. B89 1*: 61.
4. Langevin, D. (1991). Microemulsions—Interfacial aspects. *Adv. Colloid Interface Sci. 34*: 583.
5. Mitchell, D. J., and Ninham, B. (1981). Micelles, vesicles and microemulsions. *J. Chem. Soc. Faraday Trans. 11. 67*: 601.
6. Tanford, C. (1980). Micelles. *The Hydrophobic Effect: Formation of Micelles and Biological Membranes.* New York: John Wiley, p. 42.
7. Chevalier, Y., and Zemb, T. (1990). The structure of micelles and microemulsions. *Rep. Prog. Phys. 53*: 279.
8. Scriven, L. E. (1976). Equilibrium bicontinuous structures. *Nature 263*: 123.
9. Scriven, L. E. (1977). Equilibrium bicontinuous structures. *Micellization, Solubilization and Microemulsions.* Vol. 2 (K. L. Mittal, ed.). New York: Plenum Press, p. 877.
10. Talmon, Y., and Prager, S. (1978). Statistical thermodynamics of phase equilibria in microemulsions. *J. Chem. Phys. 69*: 2984.
11. Talmon, Y., and Prager, S. (1982). The statistical thermodynamics of microemulsions II. The interfacial region. *J. Chem. Phys. 76*: 1535.
12. Kaler, E. W., and Prager, S. (1982). A model of dynamic scattering by microemulsions. *J. Colloid Interface Sci. 86*: 359.
13. De Gennes, P. G., and Taupin, C. (1982). Microemulsions and the flexibility of oil/water interfaces. *J. Phys. Chem. 86*: 2294.
14. Auvray, L., Cotton, J. P., Ober, R., and Taupin, C. (1984). Evidence for zero mean curvature microemulsions. *J. Phys. Chem. 88*: 4586.
15. Widom, B. (1986). Lattice model of microemulsions. *J. Chem. Phys. 84*: 6943.
16. Widom, B., Dawson, K. A., and Lipkin, M. D. (1986). Hamiltonian and phenomenological models of microemulsions. *Physica 140A*: 26.
17. Andelman, D., Cates, M. E., Roux, D., and Safran, S. A. (1987). Structure and phase equilibria of microemulsions. *J. Chem. Phys. 87*: 7229.
18. Hyde, S. T., Ninham, B. W., and Zemb, T. (1989). Phase boundaries for ternary microemulsions. Predictions of a geometric model. *J. Phys. Chem. 93*: 1464.
19. Ninham, B. W., Barnes, I. S., Hyde, S. T., Derian, P.-J., and Zemb, T. N. (1987). Random connected cylinders: A new structure in three-component microemulsions. *Europhys. Lett. 4*: 561.

20. Barnes, I. S., Hyde, S. T., Ninham, B. W., Derian, P.-J., Drifford, M., Warr, G. G., and Zemb, T. N. (1988). The disordered open connected model of microemulsions. *Progr. Colloid Polym. Sci. 76*: 90.
21. Zemb, T. N., Barnes, I. S., Derian, P.-J., and Ninham, B. W. (1990). Scattering as a critical test of microemulsion structural models. *Progr. Colloid Polymer Sci. 81*: 20.
22. Shah, D. O., Tamjeedi, A., Falco, J. W., and Walker, R. D. (1972). Interfacial stability and spontaneous formation of microemulsions. *A.I. Ch. E. Journal 18*: 1116.
23. Shinoda, K. (1967). The correlation between the dissolution state of nonionic surfactant and the type of dispersion stabilized with the surfactant. *J. Colloid Interface Sci. 24*: 4.
24. Olsson, U., Shinoda, K., and Lindman, B. (1986). Change of the structure of microemulsions with the hydrophilic-lipophilic balance of nonionic surfactant as revealed by NMR self-diffusion studies. *J. Phys. Chem. 90*: 4083.
25. Ravey, J. C., and Buzier, M. (1984). Structure of nonionic microemulsions by small angle neutron scattering. *Surfactants in Solution*, Vol. 3 (K. C. Mittal and B. Lindman, eds.). New York: Plenum Press, p. 1759.
26. Kizling, J., and Stenius, P. (1987). Microemulsions formed by water, aliphatic hydrocarbons and pentaethylene glycol dodecyl ether: The temperature dependence of aggregate size. *J. Colloid Interface Sci. 118*: 482.
27. Bhargava, H. N., Narurkar, A., and Lieb, L. M. (1987). Using microemulsions for drug delivery. *Pharm. Technol. 11*: 46.
28. Ritschel, W. A., Adolph, S., Ritschel, G. B., and Schroeder, T. (1990). Improvement of peroral absorption of cyclosporin A by microemulsions. *Meth. Find. Exp. Clin. Pharmacol. 12*: 127.
29. Nelson, B. K., Brightwell, W. S., Khan, A., Burg, J. R., and Goad, P. T. (1989). Lack of selective developmental toxicity of three butanol isomers administered by inhalation to rats. *Fundam. Appl. Toxicol. 12*: 469.
30. Wakabayashi, T., Horiuchi, M., Sakaguchi, M., Onda, H., and Iljima, M. (1984). Induction of megamitochondria in the rat liver by *n*-propyl alcohol and *n*-butyl alcohol. *Acta Pathol. Jpn. 34*: 471.
31. Traiger, G. J., Bruckner, J. V., Jiang, W.-D., Dietz, F. K., and Cooke, P. H. (1989). Effect of 2-butanol and 2-butanone on rat hepatic ultrastructure and drug metabolizing enzyme activity. *J. Toxicol. Environ. Health 28*: 235.
32. Jayakrishnan, A., Kalaiarasi, K., and Shah, D. O. (1983). Microemulsions: Evolving technology for cosmetic application. *J. Soc. Cosmet. Chem. 34*: 335.
33. Ritschel, W. A., Ritschel, G. B., Sabouni, A., Wolochuk, D., and Schroeder, T. (1989). Study on the peroral absorption of the endekapeptide cyclosporin A. *Meth. Find. Exp. Clin. Pharmacol. 11*: 281.
34. Ritschel, W. A. (1991). Microemulsions for improved peptide absorption from the gastrointestinal tract. *Meth. Find. Exp. Clin. Pharmacol. 13*: 205.
35. Osipow, L. I. (1963). Transparent emulsions. *J. Soc. Cosmet. Chem. 14*: 277.
36. Attwood, D., Currie, L. J. R., and Elworthy, P. H. (1974). Studies of solubilized micellar solutions. 1. Phase studies and particle size analysis of solutions formed with nonionic surface active agents. *J. Colloid Interface Sci. 46*: 249.

37. Ktistis, G. (1990). A viscosity study on oil-in-water microemulsions. *Int. J. Pharm. 61*: 213.
38. Attwood, D., and Ktistis, G. (1989). A light scattering study on oil-in-water microemulsions. *Int. J. Pharm. 52*: 165.
39. Attwood, D., Mallon, C., Ktistis, G., and Taylor, C. J. (1992). A study on factors influencing the droplet size in nonionic oil-in-water microemulsions. *Int. J. Pharm. 88*: 417.
40. Gasco, M. R., Carlotti, M. E., and Trotta, M. (1988). In vitro release of propranolol from oil/water microemulsions. *Int. J. Cos. Sci. 10*: 263.
41. Gallarate, M., Gasco, M. R., and Rua, G. (1990). In vitro release of azelaic acid from oil in water microemulsions. *Acta Pharm. Jugoslav. 40*: 533.
42. Gasco, M., Gallarate, M., and Pattarino, F. (1991). In vitro permeation of azelaic acid from viscosized microemulsions. *Int. J. Pharm. 69*: 193.
43. Harusawa, F., Nakamura, S., and Mitsui, T. (1974). Phase equilibria in the water-dodecane-pentaoxyethylene dodecylether system. *Colloid Polymer Sci. 252*: 613.
44. Jahn, W., and Strey, R. (1988). Microstructure of microemulsions by freeze fracture electron microscopy. *J. Phys. Chem. 92*: 2294.
45. Vinson, P. K., Sheeban, J. G., Miller, W. G., Scriven, L. E., and Davis, H. T. (1991). Viewing microemulsions with freeze fracture transmission electron microscopy. *J. Phys. Chem. 95*: 2546.
46. Kahlweit, M., Strey, R., Haase, D., Kuneida, H., Schmeling, T., Faulhaber, B., Borkovec, M., Eicke, H.-F., Busbe, G., Eggers, F., Funck, Th., Richmann, H., Magid, L., Söderman, O., Stilbs, P., Winkler, J., Dittrich, A., and Jahn, W. (1987). How to study microemulsions. *J. Colloid Interface Sci. 118*: 436.
47. Johnson, K. A., and Shah, D. O. (1985). Effect of oil chain length and electrolytes on water solubilization in alcohol-free pharmaceutical microemulsions. *J. Colloid Interface Sci. 107*: 269.
48. Osborne, D. W., Ward, A. J. I., and O'Neill, K. J. (1990). Surfactant association colloids as topical drug delivery vehicles. *Drugs and the Pharmaceutical Sciences, Vol. 42. Topical Drug Delivery Formulation* (D. W. Osborne and A. H. Aman, eds.). New York: Marcel Dekker, p. 349.
49. La Mesa, C., Coppala, L., Ranieri, G. A., Terenzi, M., and Chidichimo, G. (1992). Phase diagram and phase properties of the system water-hexane-aerosol OT. *Langmuir 8*: 2616.
50. Zulauf, M., and Eicke, H.-F. (1979). Inverted micelles and microemulsions in the ternary system H_2O/Aerosol®-OT/isooctane as studied by photon correlation spectroscopy. *J. Phys. Chem. 83*: 480.
51. Robinson, B. H., Toprakcioglu, C., and Dore, J. C. (1984). Small-angle neutron-scattering studies of microemulsions stabilized by Aerosol®-OT. Part 1. Solvent and concentration variation. *J. Chem. Soc., Farad. Trans 1. 80*: 13.
52. Toprakcioglu, C., Dore, J. C., Robinson, B. H., and Howe, A. (1984). Small-angle neutron-scattering studies of microemulsions stabilized by Aerosol®-OT. Part 2. Critical scattering and phase stability. *J. Chem. Soc. Farad. Trans. 1. 80*: 413.

53. Howe, A. M., Toprakcioglu, C., Dore, J. C., and Robinson, B. H. (1986). Small angle neutron scattering studies of microemulsions stabilized by Aerosol®-OT. Part 3. The effect of additives on phase stability and droplet structure. *J. Chem. Soc. Farad. Trans 1. 82*: 2411.
54. Martin, C. A., and Magid, L. J. (1981). Carbon-13 NMR investigations of Aerosol® OT water-in-oil microemulsions. *J. Phys. Chem. 85*: 3938.
55. Osborne, D. W., Ward, A. J. I., and O'Neill, K. J. (1988). Microemulsions as topical drug delivery vehicles 1. Characterisation of a model system. *Drug Dev. Ind. Pharm. 14*: 1203.
56. Gasco, M. R., Pattarino, F., and Voltani, I. (1988). Behaviour of doxorubicine in o/w and w/o microemulsions. *Il Farmaco Ed. Pr. 43*: 3.
57. Trotta, M., Gasco, M. R., and Pattarino, F. (1990). Diffusion of steroid hormones from o/w microemulsions: Influence of the cosurfactant. *Acta Pharm. Tech. 36*: 226.
58. Trotta, M., Gasco, M. R., and Pattarino, F. (1989). Effect of alcohol cosurfactants on the diffusion coefficients of microemulsions by light scattering. *J. Disp. Sci. Tech. 10*: 15.
59. Trotta, M., Gasco, M. R., and Morel, S. (1989). Release of drugs from oil-in-water microemulsions. *J. Contr. Rel. 10*: 237.
60. Shinoda, K., Araki, M., Sadaghiani, A., Khan, A., and Lindman, B. (1991). Lecithin-based microemulsions: Phase behaviour and microstructure. *J. Phys. Chem. 95*: 989.
61. Attwood, D., Mallon, C., and Taylor, C. J. (1992). Phase studies on oil-in-water phospholipid microemulsions. *Int. J. Pharm. 84*: R5.
62. Aboofazeli, R., and Lawrence, M. J. (1993). Investigations into the formation and characterisation of phospholipid microemulsions. 1. Pseudo-ternary phase diagrams of systems containing water-lecithin-alcohol-isopropylmyristate. *Int. J. Pharm. 93*: 161.
63. Fubini, B., Gasco, M. R., and Gallarate, M. (1988). Microcalorimetric study of microemulsions as potential drug delivery systems 1. Evaluation of enthalpy in the absence of any drug. *Int. J. Pharm. 42*: 19.
64. Gasco, M. R., Gallarate, M., and Pattarino, F. (1988). On the release of prednisone from oil-in-water microemulsions. *Il Farmaco Ed. Pr. 43*: 325.
65. Gallarate, M. R., Gasco, M. R., and Trotta, M. (1988). Influence of octanoic acid on membrane permeability of timolol from solutions and from microemulsions. *Acta Pharm. Technol. 34*: 102.
66. Trotta, M., Gasco, M. R., and Pattarino, F. (1991). In-vitro skin permeation of diazepam for microemulsions. *Proceedings of 10th Pharm. Tech. Conf. Bologna, Italy*, p. 388.
67. Pattarino, F., Gasco, M. R., and Trotta, M. (1989). Accumulation of anthracyclines by a w/o microemulsion. *Il Farmaco 44*: 339.
68. Gasco, M. R., Morel, S., and Mazoni, R. (1988). Incorporation of doxorubicine in nanoparticles obtained by polymerisation from non-aqueous microemulsion. *Il Farmaco Ed. Pr. 43*: 373.

69. Schulman, J. H., Stoeckenius, W., and Prince, L. M. (1959). Mechanism of formation and structure of microemulsions by electron microscopy. *J. Phys. Chem. 63*: 1677.
70. Schulman, J. H., Stoeckenius, W., and Prince, L. M. (1960). The structure of myelin figures and microemulsions as observed with the electron microscope. *Kolloid-Z. 169*: 170.
71. Gulik-Krzywicki, T., and Larsson, K. (1984). An electron microscopy study of the L2-phase (microemulsion) in a ternary system: Triglyceride/monoglyceride/ water. *Chem. Phys. Lipids 35*: 127.
72. Gulik-Krzywicki, T., Aggerbeck, L. P., and Larsson, K. (1984). The use of freeze-fracture and freeze-etching electron microscopy for phase analysis and structure determination of lipid systems. *Surfactants in Solution* (K. L. Mittal and B. Lindman, eds.), Vol. 1. New York: Plenum Press, p. 237.
73. Sjoblom, E., and Friberg, S. (1978). Light scattering and electron microscopy determinations of association structures in water in oil microemulsions. *J. Colloid Interface Sci. 67*: 16.
74. Bellocq, A. M., Biais, J., Clin, B., Lalanne, P., and Lemanceau, B. (1979). Study of dynamical and structural properties of microemulsions by chemical physics methods. *J. Colloid Interface Sci. 70*: 524.
75. Biais, J., Mercier, M., Botherel, P., Clin, B., Lalanne, P., and Lemanceau, B. (1981). Microemulsions and electron microscopy. *J. Microsc. 121*: 169.
76. Bodet, J.-F., Bellare, J. R., Davies, H. T., Scriven, L. E., and Miller, W. G. (1988). Fluid microstructure transition from globular to bicontinuous in mid-range microemulsions. *J. Phys. Chem. 92*: 1898.
77. MacFarlane, D. R., and Angell, C. A. (1982). An emulsion technique for the study of marginal glass formation in molecular liquids. *J. Phys. Chem. 86*: 1927.
78. Angell, C. A., Kadiyala, R. K., and MacFarlane, D. R. (1984). Glass-forming microemulsions. *J. Phys. Chem. 88*: 4593.
79. Green, J. L. (1990). Electron microscope study of a glass-forming water/oil pseudo-three-component microemulsion system. *J. Phys. Chem. 94*: 5647.
80. Shimobouji, T., Matsuoka, H., Ise, N., and Oikawa, H. (1989). Small-angle X-ray scattering studies on nonionic microemulsions. *Phys. Rev. A. 39*: 4125.
81. Bohlen, D. S. (1990). A small angle X-ray scattering exploration of nonionic microemulsion structure. *Diss. Abs. Int. Ser B. 51*: 1946.
82. Zemb, T. N., Hyde, S. T., Derian, P.-J., Barnes, I. S., and Ninham, B. W. (1987). Microstructure from X-ray scattering: The disordered open connected model of microemulsions. *J. Phys. Chem. 91*: 3814.
83. Barnes, I. S., Hyde, S. T., Ninham, B. W., Derian, P.-J., Drifford, M., and Zemb, T. N. (1988). Small angle X-ray scattering from ternary microemulsions determines microstructure. *J. Phys. Chem. 92*: 2286.
84. North, A. N., Dore, J. C., Mackie, A. R., Howe, A. M., and Harries, J. (1990). Ultrasmall-angle X-ray scattering studies of heterogeneous systems using synchrotron radiation techniques. *Nucl. Ins. Meth. Phys. Res. B47*: 283.

85. Hilfiker, R., Eicke, H.-F., Sager, W., Hofmeier, U., and Gehrke, R. (1990). Form and structure factors of water/AOT/oil microemulsions from synchrotron SAXS. *Ber. Bunsenges Phys. Chem. 94*: 677.
86. Cebula, D. J., Ottewill, R. H., Ralston, J., and Pusey, P. N. (1981). Investigations of microemulsions by light scattering and neutron scattering. *J. Chem. Soc. Farad. Trans. 1. 77*: 2585.
87. Cebula, D. J., Myers, D. Y., and Ottewill, R. H. (1982). Studies on microemulsions, Part 1. Scattering studies on water-in-oil microemulsions. *Colloid Polym. Sci. 260*: 96.
88. Cebula, D. J., Harding, L., Ottewill, R. H., and Pusey, P. N. (1980). The structure of the microemulsion droplet. *Colloid Polym. Sci. 258*: 973.
89. Caponetti, E., Griffith, W. L., Johnson, J. H., Triolo, R., and Compere, A. L. (1988). Effect of surfactant neutralization on hexadecane/water/1-pentanol/oleic acid/ethanolamine microemulsions, a SANS study. *Langmuir 4*: 606.
90. Percus, K. J., and Yevick, G. J. (1958). Analysis of classical statistical mechanics by means of collective co-ordinates. *Phys. Rev. 110*: 1.
91. Ashcroft, M. W., and Leckner, J. (1966). Structure and resistivity of liquid metals. *Phys. Rev. 145*: 83.
92. Vrij, A., Nieuwenhuis, E. A., Fijnaut, H. M., and Agterof, W. G. M. (1978). Application of modern concepts in liquid state theory to concentrated particle dispersions. *J. Chem. Soc. Farad. Disc. 65*: 101.
93. Agterof, W. G. M., van Zomeren, J. A. J., and Vrij, A. (1976). On the application of hard sphere fluid theory to liquid particle dispersions. *Chem. Phys. Lett. 43*: 363.
94. Vrij, A., and de Kruif, C. G. (1990). Analytical results for the scattering intensity of concentrated dispersions of polydispersed hard-sphere colloids. *Micellar Solutions and Microemulsions* (S. H. Chen and R. Rajagopalan, eds.). New York: Springer-Verlag, Chap. 8, p. 143.
95. Tadros, Th. F., Luckham, P. F., and Yanaranop, C. (1989). Formation and characterisation of w/o microemulsions stabilised by ABA block copolymers. *Structures, Microemulsions and Liquid Crystals.* A.C.S. Symp. Ser. 384, Chap. 2, p. 22.
96. Baker, R. C., Florence, A. T., Tadros, Th. F., and Wood, R. M. (1984). Investigations into the formation and characterisation of microemulsions I. Phase diagrams of the ternary system water-sodium alkyl benzene sulphonate-hexanol and the quaternary system water-xylene-sodium alkyl benzene sulphonate-hexanol. *J. Colloid Interface Sci. 100*: 311.
97. Baker, R. C., Florence, A. T., Ottewill, R. H., and Tadros, Th. F. (1984). Investigations into the formation and characterisation of microemulsions II. Light scattering, conductivity and viscosity studies of microemulsions. *J. Colloid Interface Sci. 100*: 332.
98. Bedwell, B., and Gulari, E. (1984). Electrolyte-moderated interactions in w/o microemulsions. *J. Colloid Interface Sci. 102*: 88.
99. Gulari, E., Bedwell, B., and Alkhafaji, S. (1980). Quasielectric light scattering investigation of microemulsions. *J. Colloid Interface Sci. 77*: 202.
100. Chang, N. J., and Kaler, E. W. (1986). Quasielastic light scattering study of five-component microemulsions. *Langmuir 2*: 184.

101. Rosano, H. L., Lan, T., Weiss, A., Gerbacia, W. E. F., and Whittam, J. H. (1979). Transparent dispersions: An investigation of some of the variables affecting their formation. *J. Colloid Interface Sci. 72*: 233.
102. Müller, B. W., and Müller, R. H. (1984). Particle size analysis of latex suspensions and microemulsions by photon correlation spectroscopy. *J. Pharm. Sci. 73*: 915.
103. Müller, B. W., and Müller, R. H. (1984). Particle size distributions and particle size alterations in microemulsions. *J. Pharm. Sci. 73*: 919.
104. Cazabat, A. M., Langevin, D., and Pouchelon, A. J. (1980). A light scattering study of water-in-oil microemulsions. *J. Colloid Interface Sci. 73*: 1.
105. Osborne, D. W., Ward, A. J. I., and O'Neill, K. J. (1991). Microemulsions as topical delivery vehicles: In vitro transdermal studies of a model hydrophilic drug. *J. Pharm. Pharmacol. 43*: 451.
106. Février, F., Bobin, M. F., Lafforgue, C., and Martini, M. C. (1991). Advances in microemulsions and transepidermal penetration of tyrosine. *S.T.P. Pharma. Sci. 1*: 60.
107. Zeigenmeyer, J., and Führer, C. (1980). Mikroemulsionen als topische arzneiform. *Acta Pharm. Technol. 26*: 273.
108. Martini, M. C., Bobin, M. F., Flandin, H., Caillaud, F., and Cotte, J. (1984). Rôle des microemulsions dans l'absorption percutanée de l'α-tocopherol. *J. Pharm. Belg. 39*: 348.
109. Gasco, M. R., Gallarate, M., Trotta, M., Bauchiero, L., Gremmo, E., and Chiappero, O. (1989). Microemulsions as topical delivery vehicles: Ocular administration of timolol. *J. Pharm. Biomed. Anal. 7*: 433.
110. Kahan, B. D., Ried, M., and Newburger, J. (1983). Pharmacokinetics of cyclosporine in human renal transplantation. *Transplant Proc. 15*: 446.
111. Tarr, B., and Yalkowsky, S. H. (1989). Enhanced intestinal absorption of cyclosporine in rats through the reduction of emulsion droplet size. *Pharm. Res. 6*: 40.
112. Malcolmson, C., and Lawrence, M. J. (1993). A comparison of the incorporation of testosterone, testosterone propionate and testosterone enanthate in nonionic micellar and microemulsion systems. *J. Pharm Pharmacol. 44*: 142.
113. Fubini, B., Gasco, M. R., and Gallarate, M. (1989). Microcalorimetric study of microemulsions as potential drug delivery systems II. Evaluation of enthalpy in the presence of drugs. *Int. J. Pharm. 50*: 213.
114. Gonzalez Tavares, L., Sanz Saiz, P., Perez de la Cruz, M. J., Camacho, M. A., and Martin, J. L. (1991). *S.T.P. Pharma. Sci. 1*: 195.
115. Panaggio, A., Rhodes, C. T., and Worthen, L. R. (1979). The possible use of autoclaving microemulsions for sterilization. *Drug. Dev. Ind. Pharm. 5*: 169.
116. Halbert, G. W., Stuart, J. F. B., and Florence, A. T. (1984). The incorporation of lipid-soluble antineoplastic agents into microemulsions—Protein free analogues of low density lipoprotein. *Int. J. Pharm. 21*: 219.
117. Ekman, S. (1987). ^{3}H cholesterol transfer from microemulsion particles of different sizes to human fibroblasts. *Lipids 22*: 657.
118. Ginsberg, G. S., Small, D. M., and Atkinson, D. (1982). Microemulsions of phospholipids and cholesterol esters. *J. Biol. Chem. 257*: 8216.

3

Liposomes

Daan J. A. Crommelin

Utrecht University, Utrecht, The Netherlands

H. Schreier

Vanderbilt University School of Medicine, Nashville, Tennessee

INTRODUCTION

Liposomes have been intensively studied by scientists from different disciplines since their discovery in the 1960s. Up until the end of 1992, this activity had resulted in the publication of over 15,000 articles with the word "liposome" in the title (source: *Chemical Abstracts*). Although the liposome structure as such has not been patented, almost 1000 patents have been issued or filed dealing with specific aspects of the liposome. Of these, about 500 were filed since 1985. This suggests that the spirit and enthusiasm to explore new research strategies with liposomes are as alive as in the early days of liposome research.

Part of these activities concentrated on establishing liposomes as pharmaceutical entities, and the first topical and parenteral liposome-based drug delivery systems have reached the marketplace already. More liposome-based delivery systems can be expected soon. The rapid progress made in recent years affected established hypotheses and dogmas considerably. This made writing a chapter devoted to the state of the art in the field of liposomes, with special emphasis on pharmaceutical and clinical aspects, a worthwhile exercise in the eyes of the authors.

This chapter is divided into three parts. Part I covers pharmaceutical technological aspects of liposomes; Part II deals with potential therapeutic applications and recent and ongoing clinical trials with liposomal pharmaceutical products. In particular, the therapeutically and clinically oriented Part II focuses on present activities in selected areas of research, instead of providing a full historical perspective. Part III presents conclusions and an outlook to the future. For a full historical picture of progress made in understanding liposome behavior in vivo and in vitro, the reader is referred to standard books and review articles, e.g., (1–14).

I. PART I. TECHNOLOGICAL ASPECTS

A. Classification of Liposomes

Liposomes can be classified either on the basis of their structural properties or on the basis of the preparation method used. These two classification systems are, in principle, independent of each other. In Table 1 the different

Table 1 Liposome Classification

A. Based on Structural Parameters	
MLV	Multilamellar large vesicles, >0.5 μm
OLV	Oligolamellar vesicles, 0.1–1 μm
UV	Unilamellar vesicles (all sizes)
SUV	Small unilamellar vesicles, 20–100 mm
MUV	Medium-sized unilamellar vesicles
LUV	Large unilamellar vesicles, >100 mm
GUV	Giant unilamellar vesicles (vesicles with diameters > 1 μm)
MVV	Multivesicular vesicles (usually large > 1 μm)
B. Based on Method of Liposome Preparation	
REV	Single or oligolamellar vesicles made by reverse-phase evaporation method
MLV-REV	Multilamellar vesicles made by the reverse phase evaporation method
SPLV	Stable plurilamellar vesicles
FATMLV	Frozen and thawed MLV
VET	Vesicles prepared by extrusion methods
FPV	Vesicles prepared by French press
FUV	Vesicles prepared by fusion
DRV	Dehydration-rehydration vesicles
BSV	Bubblesomes[a]

[a]From Ref. 60.
Source: Based on Lichtenberg and Barenholz (33). Adjusted from Barenholz and Crommelin (15).

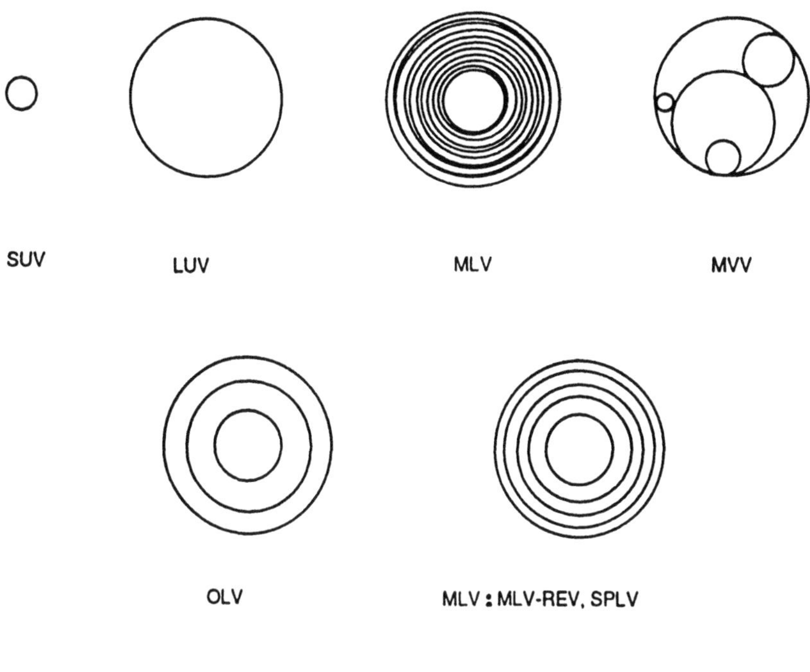

Figure 1 Morphology of different liposome structures. SUV = small unilamellar vesicles. LUV = large unilamellar vesicles. MLV (classical) = multilamellar vesicles. MVV = multivesicular vesicles. OLV = oligolamellar vesicles. MLV (MLV-REV, SPLV) = nonclassical multilamellar vesicles; see Table 1. (From Ref. 15.)

vesicles with the regularly used acronyms are listed. Figure 1 presents a schematic view of the major liposome types.

Dependent on the selection of lipids, the preparation technique, and preparation conditions, liposomes can vary widely in size, number, position of lamellae, e.g., multilamellar vesicles (MLV) versus multivesicular vesicles (MVV), charge, and bilayer rigidity (liquid-crystalline versus gel state). These parameters influence the behavior of liposomes both in vivo and in vitro. The opsonization process, leakage profiles, disposition in the body, and shelf life all depend on the type of liposome involved. Therefore it is important to select liposome constituents and the preparation technique carefully and to characterize the produced liposomes properly. Liposome formation and potential bilayer constituents are dealt with first. Characterization techniques will be discussed in more detail later on (see Sec. II.H).

Unfortunately, liposome nomenclature has not been rigorously standardized yet. For instance, the size range definitions for small unilamellar vesicles (SUV) and large unilamellar vesicles (LUV) can differ. Hope et al. (16) define SUV as "limit size" vesicles, which means liposomes with the smallest possible diameter. This minimum size limit is due to lipid packing constraints preventing smaller liposomes from forming. All other unilamellar liposomes are defined as LUV. On the other hand, Barenholz and Crommelin (15) consider unilamellar liposomes as LUV when there is no detectable effect of the bilayer curvature on the physical properties of the liposomes. This implies that there is no detectable size dependence, neither on the distribution of phospholipids over the inner and outer leaflet of the bilayer (unequal for small liposomes) nor on the gel-to-liquid-crystalline phase transition (which tends to disappear with very small liposomes). On the basis of this criterion, the minimum diameter for LUV is generally taken as 100 nm; unilamellar liposomes with smaller sizes are considered as SUV. In this review LUV are defined as unilamellar liposomes with diameters >100 nm, SUV as unilamellar liposomes with diameters <100 nm.

B. Mechanism of Liposome Formation

Liposomes are vesicular structures consisting of hydrated bilayers. Liposome structures used for pharmaceutical purposes consist of a phospholipid backbone. But other classes of molecules can form bilayer-based vesicular structures as well (e.g., certain ionized cholesterolesters as cholesterolhemisuccinate). On the other hand, not all hydrated phospholipids form bilayer structures. Other forms of self-aggregation such as inverted hexagonal phases (H_{II}, formed by phosphatidylethanolamine under certain conditions) or micelles (lysophospholipids) with completely different properties can occur. The common feature that all bilayer-forming compounds share is their amphiphilicity. They have defined polar and nonpolar regions. In water the hydrophobic regions tend to self-aggregate and the polar regions tend to be in contact with the water phase. Organization in the bilayer structure is one of the possibilities. Israelachvili and coworkers (17,18) made an attempt to identify potential bilayer-forming compounds on the basis of molecular shape (Fig. 2). They defined the critical packing parameter p by

$$p = \frac{v}{a_0 l_c} \qquad (1)$$

where v is the molecular volume of the hydrophobic part, a_0 is the optimum surface area per molecule at the hydrocarbon-water interface, and l_c is the critical half-thickness for the hydrocarbon region, which must be less than the maximum length of the extended lipid chains (18). For $p < 1/3$, spherical micelles are formed. In this category fall single-chain lipids with large head

Lipid	Critical packing parameter $v/a_0 l_c$	Critical packing shape	Structures formed
Single-chained lipids (surfactants) with large head-group areas: *SDS in low salt*	< 1/3	Cone 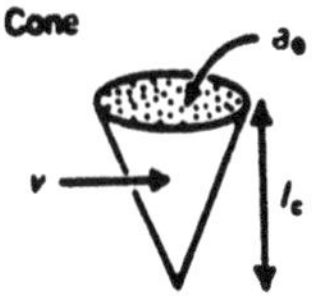 (a_0, v, l_c)	Spherical micelles 
Single-chained lipids with small head-group areas: *SDS and CTAB in high salt, nonionic lipids*	1/3-1/2	Truncated cone 	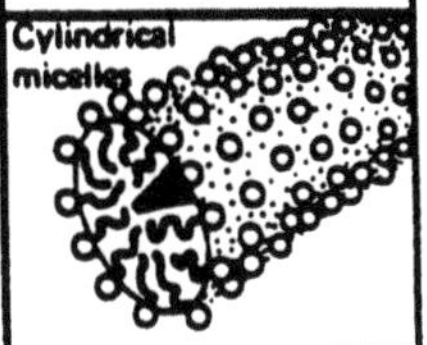 Cylindrical micelles
Double-chained lipids with large head-group areas, fluid chains: *Phosphatidyl choline (lecithin), phosphatidyl serine, phosphatidyl glycerol, phosphatidyl inositol, phosphatidic acid, sphingomyelin, DGDG, dihexadecyl phosphate, dialkyl dimethyl ammonium salts*	1/2-1	Truncated cone 	Flexible bilayers, vesicles 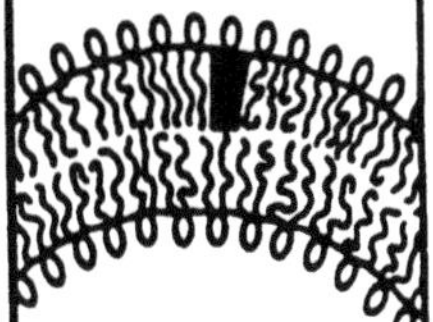
Double-chained lipids with small head-group areas, anionic lipids in high salt, saturated frozen chains: *phosphatidyl ethanolamine, phosphatidyl serine + Ca^{2+}*	~1	Cylinder 	Planar bilayers 
Double-chained lipids with small head-group areas, nonionic lipids, poly *(cis)* unsaturated chains, high *T*: *unsat. phosphatidyl ethanolamine, cardiolipin + Ca^{2+} phosphatidic acid + Ca^{2+} cholesterol MGDG*	> 1	Inverted truncated cone or wedge 	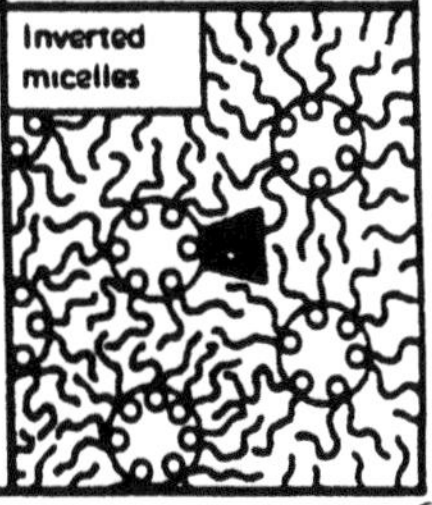 Inverted micelles

Figure 2 Mean dynamic packing shapes of lipids and the structures they form. DGDG, digalactosyl diglyceride, diglucosyl diglyceride; MGDG, monogalactosyl diglyceride, monoglucosyl diglyceride. (From Ref. 18.)

group areas, e.g., lysophosphatidylcholine. For $1/3 < 1/2$ (truncated cone or wedge shape), globular or cylindrical micelles are formed. Double-chain "fluid state" lipids; with large head-group areas ($1/2 < p < 1$; truncated cone) form bilayers and vesicles. This occurs also with double-chain "gel state" lipids with small head groups and $p \sim 1$. For $p > 1$ (inverted truncated cone), inverted structures, such as the inverted hexagonal (H_{II}) phase, can be observed. An additional condition required for bilayer formation is that the compound can be classified as a nonsoluble swelling amphiphile (19). The approach developed by Israelachvili is a simplification. Exceptions are often observed: (1) changing the pH or temperature can change the state of aggregation; (2) the aggregates may tend to interact with each other; and (3) the approach may loose predictive power with lipid mixtures. For example, under certain conditions, lysophosphatidylcholine (micelle former) mixed with fatty acids can form stable bilayers, and phosphatidylethanolamine (a H_{II} hexagonal phase former) can form a bilayer when mixed with cholesterol (a nonsoluble and nonswelling amphiphile). Nevertheless, Israelachvili's approach to look at packing constraints still helps to predict the behavior of a new lipid molecule in water. More recent statistical thermodynamic approaches to predict the state of aggregation of (phospho)lipids were discussed by Seddon (20).

Several groups addressed the question of how vesiculation occurs upon the formation of stacks of hydrated lipid bilayers. For one-component, uncharged systems, the "curvature elastic theory" predicts that the curvature energy of a bilayer in vesicles is higher than in the stacked multilamellar, liquid-crystalline phase; therefore energy is required to generate vesicles (21).

Subsequently, the liposomes are locked into metastable kinetic traps (22). This metastable, trapped situation can last for prolonged periods of time, months or even years. The general rule is, the higher the curvature of the bilayer (the smaller the vesicle), the more unstable the vesicle will be, e.g., (22). However, very small vesicles (size comparable to the lipid molecule size) can theoretically form the most stable configuration of a hydrated bilayer lipid (21). A situation has been described where, by using two-component systems, energetically stable vesicles are formed with a well-defined size. This size is determined by the curvature energy of the interacting system.

In contrast to uncharged phospholipids, under certain conditions charged phospholipids do not form bilayer stacks but swell continuously and spontaneously form uni- and oligolamellar structures (23,24).

Vesiculation of neutral phospholipid bilayers is not a spontaneous process. In Sec. II.D, the abundant literature on liposome preparation techniques will be dealt with. Figure 3 [taken from (22)] gives a classification of the physical and chemical means to produce well-defined liposome dispersions

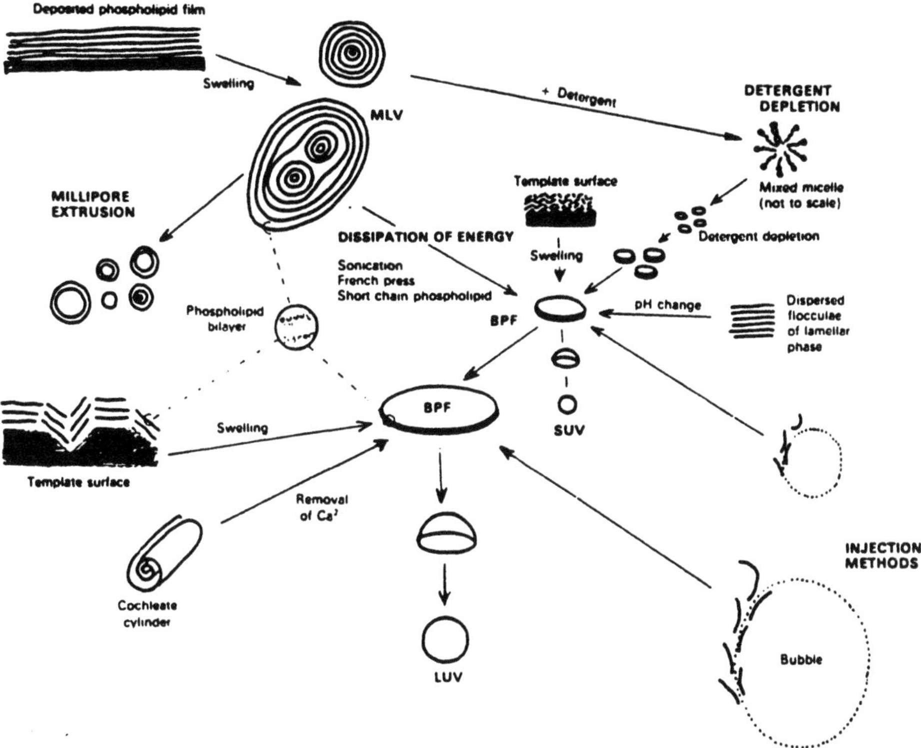

Figure 3 Schematic drawing of the formation of vesicles according to the bilayer phospholipid fragment theory (22). The illustration shows how bilayered phospholipid fragments (BPF) can be formed using different vesicle preparation methods. BPF can be grown by detergent depletion or phospholipid precipitation, or they can be formed from preexisting bilayers. They can also be prepared by using template surfaces. This scheme shows that BPF is an intermediate structure in the vesicle formation process by different preparation procedures. The thermodynamic instability at the edges of the BPF causes bending, and when the BPF closes upon itself a vesicle is formed.

from hydrated lipids. Lasic (22) proposes bilayered phospholipid fragments (BPF) as intermediate structures in the vesicle-formation process. The BPF model is principally different from the alternative "budding off" model, where large sheets of bilayer are supposed to bud off MLV, forming SUV or LUV. Liposomes can be formed by destabilization of mixed micelles consisting of (phospho)lipids, detergents, and, dependent on the situation, an amphipathic drug. The theoretical aspects of the preparation of liposomes through detergent removal were discussed by Lasic (25) in 1982, Fromherz (26) in 1983, and Fromherz and Rüppel (27).

When mixed micelles lose detergent beyond a certain limit, their ability to shield the alkyl chains of the phospholipids effectively from the water phase is lost; this results in bending and, ultimately, vesiculation of the grown micelles to minimize the exposed alkyl chain/water interfacial area. Upon removal of the detergent molecules from the stable mixed micelles, vesicle formation is driven by (1) the unfavorable boundary interaction at the unshielded edges at the circumference of the detergent-depleted mixed micelles, (2) the unelastic curvature energy upon bending of the mixed micellar structures (stretching the area of polar heads in the outer bilayer leaflet and compressing it in the inner leaflet), and (3) the elastic curvature energy.

An important parameter in the formation process of liposomes is the rigidity of the bilayer. Hydrated single-component phospholipid bilayers can be in a liquid-crystalline ("fluid") state or in a "gel" state. By increasing the temperature, the gel state bilayer "melts" at the transition temperature (T_c) and is converted from the gel into a liquid-crystalline state. T_c depends on the acyl chain length, degree of saturation, and polar head group, and can vary between −15°C for egg yolk phosphatidylcholine (high degree of unsaturation of the acyl chains and varying chain length) to over 50°C for fully saturated DSPC (distearoylphosphatidylcholine). Gel state bilayer structures tend to be more rigid and less permeable than the liquid-crystalline structures. For the formation of liposomes usually conditions are chosen where bilayers in the liquid-crystalline state are generated. An interesting phenomenon is the extremely high bilayer permeability observed around the T_c; this high permeability can be utilized for triggered release of liposome encapsulated material (cf. Sec. II.E.e., "Temperature-Induced Destabilization of Liposomes").

Cholesterol is a regular component of liposomal carrier systems. Cholesterol addition to phosphatidylcholine bilayers changes the melting behavior of the bilayer dramatically as cholesterol tends to eliminate the phase transition. Cholesterol inclusion has a condensing effect on "fluid-crystalline" bilayers and strongly reduces bilayer permeability.

C. Raw Materials for Liposome Formation

Liposomes intended as carriers for drugs or diagnostic agents should be prepared from constituents that are safe for use in humans. Although limited experience is available on the safety of liposomes in the clinical setting, a general pattern is now developing concerning the (phospho)lipids that are tolerated, e.g., reviews by (28–30). Phosphatidylcholines and phosphatidylglycerols from natural sources, semisynthetically or fully synthetically produced, and cholesterol and PEG-ylated phosphatidylethanolamine, are frequently encountered in liposomes designed as drug carriers for parenteral administration or for in vivo diagnostic purposes. The structure of commonly used phospholipids in animal experiments and clinical studies is shown in Fig. 4.

These phospholipids display a chiral center at position 2 of the glycerol group. The 1,2-diacyl-sn-glycero-phosphate structure is the natural configuration. As the stereospecificity of the molecule may play a role in the metabolism of the compound upon administration, the correct configuration should be used (see below).

As mentioned above, phosphatidylcholine (PC) is routinely used as a bulk neutral phospholipid. As a negatively charged lipid, phosphatidylglycerol (PG) is often selected. Finally, if it is desirable to reduce the permeability of "fluid-crystalline state" bilayers, cholesterol is added to the bilayer structure. Sometimes lipids with a special affinity for certain target cells in the body are deliberately inserted into the bilayer. This was, for instance, the case when hepatocytic delivery was aimed for and lactosylceramide, a ligand with a specific affinity for hepatocytes, was included in the liposomal bilayer (31,32).

There is no general answer to the question what raw product quality criteria should be selected for liposomal products. The route of administration, dose, and therapeutic indication will play a role. Dermal application of liposomes for local treatment on intact skin will require lipids with different—less stringent—specifications than lipids designed for parenteral administration. Interestingly, emulsions for parenteral nutrition have been used in the clinic for several decades. In these emulsions egg yolk phosphatidylcholine "contaminated" with other phospholipids acts as emulsifier of the oil phase and is used quite satisfactorily.

Several companies (Rhone-Poulenc Rorer Nattermann Phospholipid GmbH, Lipoid KG, Avanti Polar Lipids Inc., Genzyme Corp., Lipid Products) supply reasonably priced lipids in different quality grades. High quality products are available for parenteral use. These products typically contain at least 98% of the claimed phospholipid, less than 1% lysophospholipid, low endotoxin and microbial loads, and only trace levels of heavy metals.

Five groups of phospholipids that can be used for liposome preparation can be discerned (15).

1. Phospholipids from natural sources
2. Modified natural phospholipids
3. Semisynthetic phospholipids
4. Fully synthetic phospholipids
5. Phospholipids with nonnatural head groups

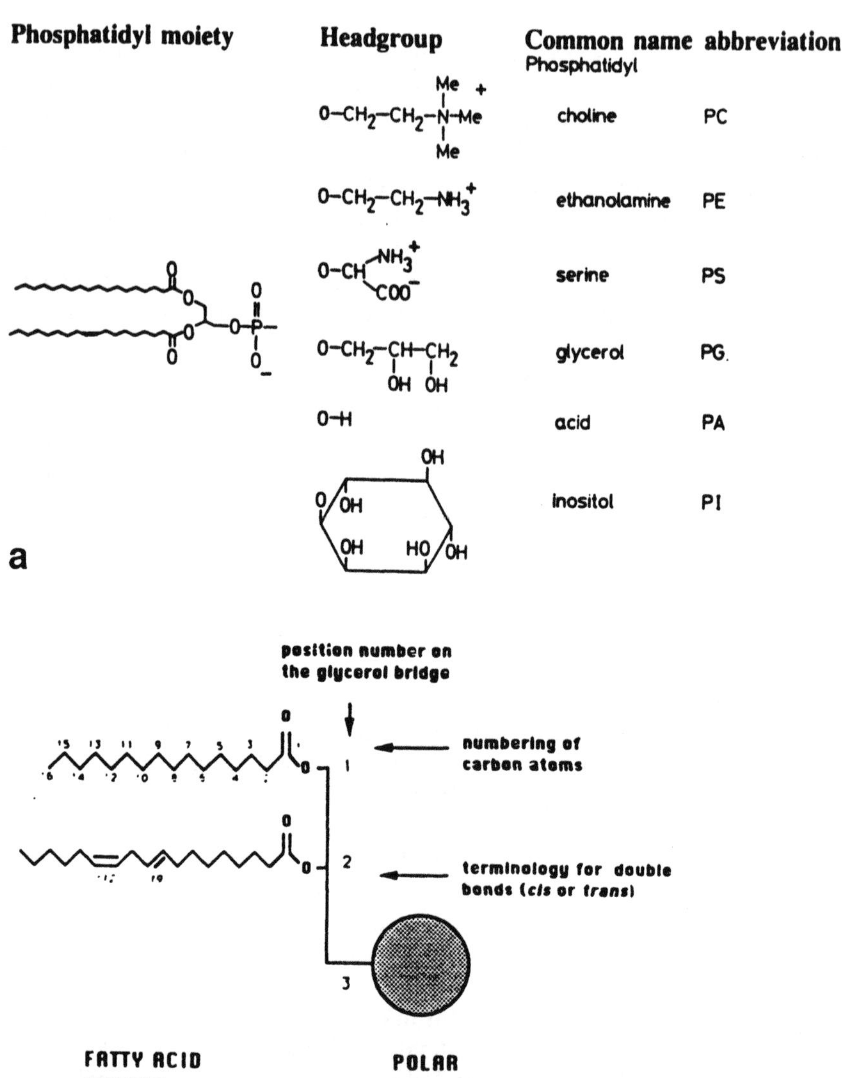

Figure 4 (a) Some common naturally occurring phosphatidyl phospholipids. (b) Numbering of atoms and bonds in phosphatidylglycerides. c = cis, t = trans (very unusual in natural fatty acids of mammalian or plant origin). (From. Ref. 11.)

In the following section some aspects of the availability, preparation, and supply of these phospholipid classes will be discussed.

1. Phospholipids from Natural Sources

The sources for natural phospholipids, mainly PC, but also phosphatidylethanolamine (PE), phosphatidylinositol (PI), and sphingomyelin (SPM), are egg yolks and soybeans. These PCs are mixed acyl ester phospholipids. Apart from source-dependent differences in acyl chain type, considerable interbatch variation has been observed for egg PC (33). The esterified acyl chains of egg PC are different from those of soybean PC. In egg PC the acyl chains in position 1 are almost completely saturated; those in position 2 are mainly unsaturated (11,33,34). In soybean PC unsaturated acyl chains (linoleic acid being a major component) can be found in both positions 1 and 2.

2. Modified Natural Phospholipids

Natural phospholipids can be modified. Because of their high degree of unsaturation, which makes them sensitive to oxydation, PC from natural sources can be catalytically hydrogenated. Partially or fully hydrogenated natural PCs are readily available. The iodine value (measure of the degree of unsaturation) of these lipids is reduced as the number of unsaturated C=C bonds drops. Dependent on the degree of unsaturation left after the hydrogenation process, phase transition temperatures can be identified for liposomal dispersions of the (partially) hydrogenated PCs, e.g., (34).

In natural PCs all double bonds in the acyl chains are in the cis isomeric form. During hydrogenation, isomerization to the trans form occurs in partially hydrogenated PCs. The conversion of the cis form to the trans form will influence the bilayer structure. Presumably, the occurrence of the trans form has an impact on the physical properties of the bilayer (rigidity, permeability). It is not clear to what extent it influences the biological properties. As heavy metals are used as catalysts in the hydrogenation process, the heavy metal content in hydrogenated PCs should be monitored and meet the (strict) specifications.

Head group modification can be performed by using phospholipase D. With this enzyme one can convert PC into PE, PG, or phosphatidylserine (PS). More details can be found elsewhere (35).

3. Semisynthetic Phospholipids

The acyl chains that are attached to phospholipids from natural sources are often unsaturated. This makes them liable to oxidation reactions, which may limit liposome shelf life. Moreover, as mentioned above, reproducibility of the quality of the batches in terms of acyl chains may be poor as well, which may cause variation in stability or liposome properties (rigidity, phase transition temperature of the bilayers). Removal of the original acyl chain and, within certain limits, replacement by a chosen acyl chain is possible. Phospholipase A_2, which cuts the acyl chain at the C2 position of glycerol, can be used, if only replacement of the C2 acyl chain is required. Otherwise, one can use fully synthetic chemical conversion schemes resulting in the requested enantiomeric form (see below).

4. Fully Synthetic Phospholipids

Eibl reviewed different completely chemical pathways for phospholipid synthesis (35,36). A partially enzymatic synthesis with a stereospecific phosphorylation of glycerol as a first step was recently commercialized.

5. Phospholipids with Nonnatural (Head) Groups

The idea of manipulating the fate of liposomes in the body by selecting the appropriate bilayer characteristics has led to modified phospholipids. The circulation time of liposomes in the blood compartment can be considerably prolonged when polyethyleneglycol chains are attached to bilayer constituents. Alternatively, for active targeting purposes ligands for cell surface receptors can be attached. These ligands can be chemically and physically widely different structures, such as monoclonal antibodies (150 kD) or just a simple peptide.

PEG has been linked to PE for the preparation of long circulating liposomes (cf. Part II, Sec. II.A.2). Various reaction schemes have been developed. Molecular weight fractions for maximum prolongation of circulation times for PEG vary between 1900 and 5000. Allen and coworkers (37) described the synthesis of a PEG-carbamate derivative of PE; Klibanov et al. (38) used a succinidyl conjugation method, while Blume and Cevc (39) adopted the procedure that Abuchowski and coworkers (40) described for the preparation of PEG-albumin conjugates (via cyanuric chloride).

The issue of covalent attachment of proteins to lipid bilayer constituents has gained considerable interest in the past decade, and the reader is referred to review articles (41,42,43). The binding reaction should be performed under mild conditions and should have a high yield; no loss of activity and no aggregation of the proteoliposomes through cross-linking should occur. Table 2 provides an overview of frequently used chemical procedures for binding proteins to liposomes (44). The functional groups regularly used for binding are $-COOH$, $-NH_2$, vicinal $-OH$ (e.g., in gangliosides), and $-SH$. By using heterobifunctional cross-linking agents, undesired protein-protein interactions, or liposome-liposome cross-linking, can be minimized. This approach of limiting unwanted cross-linking or aggregation is exemplified by the procedure for the formation of immunoliposomes (Fab′ fragments coupled to

Table 2 Frequently Used Techniques for Binding Drugs or Homing Devices to Carriers

Functional groups involved	Binding agent	Product
$RCO_2H + R'NH_2$	Carbodiimides	$RCONHR'$[a]
$RNH_2 + R'NH_2$	Glutaraldehyde	$RN{=}CH(CH_2)_3CH{=}NR'$[b]
$R(OH)_2 + R'NH_2$	Cyanogen bromide	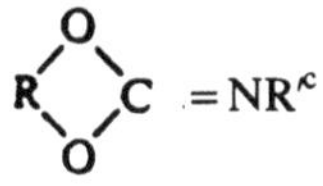 R(–O–)(–O–)C =NR'[c]
$RCO_2H + R'NH_2$	Mixed carbonic anhydride	$RCONHR'$
$RNH_2 + R'NH_2$	SPDP	$RNHCO(CH_2)_2SS(CH_2)_2\ CONHR'$[d]
$RNHB_2 + R'SH$	SMPB	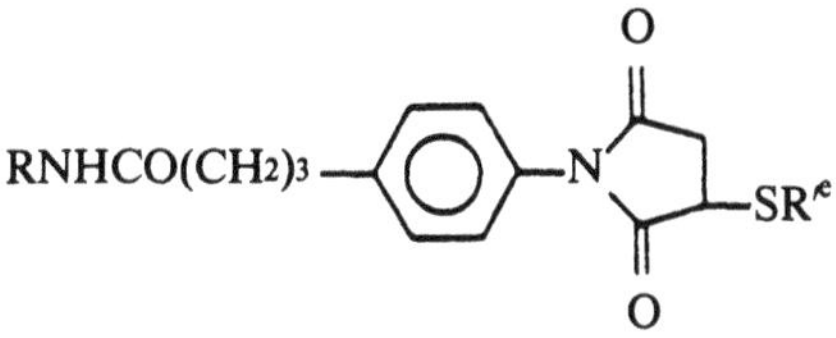 $RNHCO(CH_2)_3$–(phenylene)–N(succinimide, 2 C=O)–SR'[e]

[a]Aggregate formation has been reported.
[b]Changes in antibody binding specificity have been described.
[c]Cyanogen bromide is very toxic and coupling reaction is highly pH dependent.
[d]Bond is unstable in serum.
[e]Bond is stable in serum; nonreducible sulfide bridge.
Source: From Ref. 44.

liposomes) described by Martin and Papahadjopoulos (45) (Fig. 5). Martin, Heath, and New have presented an excellent review with experimental details on the currently most successful liposome-conjugation techniques including advice for trouble-shooting (46).

D. Techniques for Liposome Preparation

Liposome preparation techniques have been described extensively in a number of review articles and books (6,15,33,47). In particular, the book edited by New (11) can be recommended, if information on experimental details for lab scale production of liposomes is required. Here, a compilation of the different approaches will be presented, with special emphasis on new developments and typical pharmaceutical aspects (safety, scale-up potential, simplicity). In the different preparation procedures a general pattern can be discerned: (1) the lipid must be hydrated, then (2) liposomes have to be sized, and finally (3) nonencapsulated drug has to be removed. In some preparation schemes the hydration and sizing steps are combined. Sometimes all drugs are liposome associated and no free drug can be found after stage 2; then stage 3 is lacking.

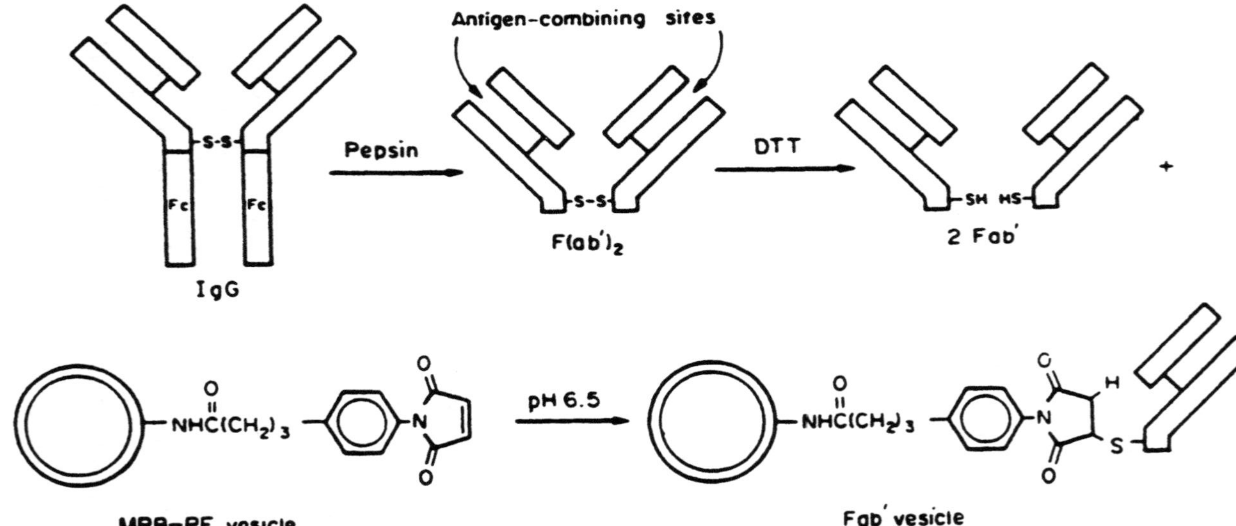

Figure 5 Covalent coupling of Fab′ fragments to *N*-[4-(*p*-maleimidophenyl)butyryl]-phosphatidylethanolamine (MPB-PE) vesicles. $F(ab')_2$ dimers are prepared by pepsin digestion of IgG molecules. Fab′ monomers are generated from these by reduction with dithiothreitol (DTT) at low pH. Immediately following the removal of DTT, Fab′ fragments are mixed with MPB-PE-containing vesicles and the pH is adjusted to 6.5. Addition of the Fab′–SH to the double bond of the maleimide moiety of MPB-PE molecules present in the vesicle membranes results in a stable thioether cross-linkage. (From Ref. 44.)

The major routes for production of pharmaceutical liposomal products are listed in Table 3. Figure 6 schematically depicts the type of vesicles formed.

The different items listed in Table 3 will be systematically discussed below.

1. Hydration Stage

a. Mechanical Methods (see Table 3, item A and item B.a). MLVs were "traditionally" produced by hydrating thin lipid films deposited from an organic solution on a glass wall by shaking at temperatures above the phase transition temperature (if detectable) of the phospholipid with the highest T_c. The wide size distributions of the produced liposome dispersions were usually narrowed down by (low) pressure extrusion or ultrasonication [reviewed e.g., in (11,33,49)].

"At the bedside" generation of liposomes was proposed in 1978 in an effort to treat joint rheumatoid arthritis with intraarticularly injected liposomes

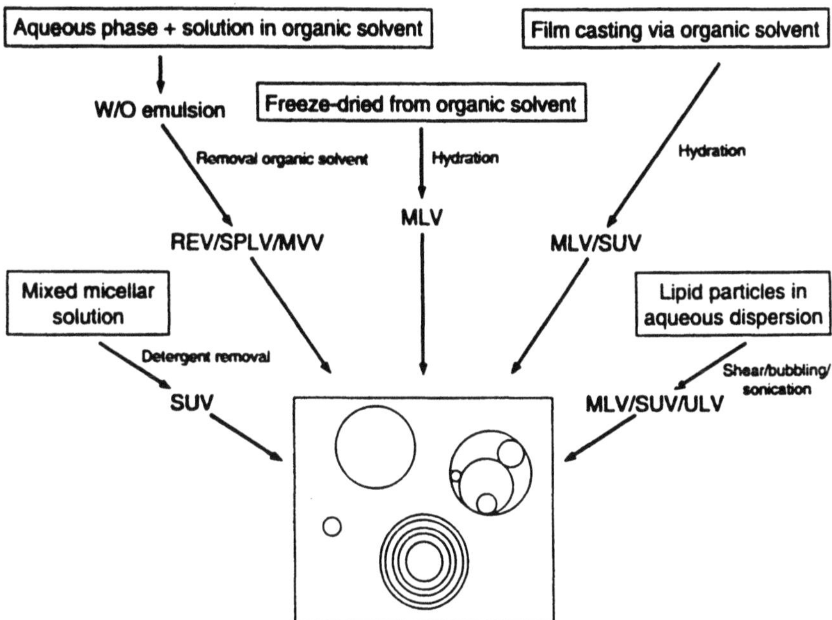

Figure 6 Schematic diagram of regularly used methods for liposome preparation. The commonly obtained type of vesicle is indicated. REV = reverse phase evaporation vesicles. ULV = unilamellar vesicles. SUV = small unilamellar vesicles. MLV (classical) = multilamellar vesicles. MVV = multivesicular vesicles. SPLV = stable plurilamellar vesicles.

Table 3 Methods of Vesicle Preparation and Type of Liposome Product Formed[a]

Hydration Stage

A. Mechanical methods[b]
  a. Vortexing or hand shaking of phospholipid dispersions (MLV)
  b. "Microfluidizer" technique (mainly SUV)
  c. Bubbling inert gas through aqueous phospholipid dispersions (MLV, LUV)
  d. High shear homogenization (mainly SUV)
B. Methods based on replacement of organic solvent(s) by aqueous media[b]
  a. Removal of organic solvent(s) before hydration (MLV, OLV, SUV)
  b. Reverse-phase evaporation (LUV, OLV, MLV)
  c. Use of water immiscible solvents: ether and petroleumether infusion (solvent vaporization) (MLV, OLV, LUV)
  d. Use of water miscible solvents such as ethanol injection (MLV, OLV, SUV)
C. Methods based on detergent removal by[b]
  a. Gel exclusion chromatography (SUV)
  b. "Slow" dialysis (LUV, OLV, MLV)
  c. Fast dilution (LUV, OLV)
  d. Miscellaneous related techniques (MLV, OLV, LUV, SUV)
D. Methods based on size transformation and fusion[b]
  a. Spontaneous fusion of SUV in the gel phase (LUV)
  b. Freeze-thawing (MLV)
  c. Freeze-drying (MLV)
  d. Dehydration of SUV followed by rehydration with or without sizing
  e. Ca^{2+} ion induced fusion (LUV, OLV, MLV)
  f. Detergent-induced growth (LUV, OLV)
E. Methods based on pH adjustment (SUV and possibly LUV)[b]

Sizing Stage

A. High pressure extrusion (see also section: "Mechanical methods")
B. Low pressure extrusion
C. Ultrasonic treatment

Removal of Nonencapsulated Material

A. Dialysis
B. Ultracentrifugation
C. Gel permeation chromatography
D. Ion-exchange resins

[a]For definitions and abbreviations see Table 1.
[b]For more details see Refs. 11, 15, 22, 33, 47, 49, 111 and many relevant references listed in these reviews.
Source: Refs. 15, 33, and 48.

containing derivatized corticosteroids (50). These liposomes were prepared by shaking freeze-dried lipids (from an organic phase) with a water phase at temperatures above the phase transition temperature of the chosen lipid bilayer (main component DPPC, T_c 41 °C). The elevated temperature needed for proper dispersion of these lipids was an obstacle for the successful introduction of this treatment modality in therapy. Hoogevest and Fankhauser (51) followed a similar approach. However, they avoided the heating step by selecting lipid bilayers with phase transition temperatures below room temperature [1-palmitoyl-2-oleyl-sn-glycero-3-phosphocholine, POPC, and 1,2 dioleyl-sn-glycero-3-phospho-L-serine (DOPS)]. They first lyophilized the bilayer constituents and the drug (muramyltripeptide-PE) from a tert butanol solution. The dry lyophilizate can be simply reconstituted with a buffered aqueous solution and after vortex shaking administered to the patient (in situ preparation). The dispersions contain large liposomes with an average size between 2 and 3.5 μm. This meets the criteria defined by the authors, who aim for increased lung localization after intravenous injection. The proliposome concept as developed by Payne and coworkers (52, 53) is also an example of in situ liposome preparation. They deposited lipids from an organic solvent on finely powdered sodium chloride or sorbitol. Upon hydration, liposome dispersions were formed. A certain degree of particle size control over the liposome dispersions can be achieved by proper selection of lipid film thickness, nature of the lipid, and type of solid carrier. However, with the above mentioned in situ liposome preparation methods, rather wide particle size distributions will be encountered. To narrow down the particle size distribution, Chen and Alli (54) proposed the use of fluidized bed technology for casting of the lipid film on the carrier.

High-shear homogenizers [Microfluidizer® (55,56), Stansted Cell Disruptor® (57), Gaulin Homogenizer 15M® (58), and Gaulin Micron Lab 40® (59)] have been used to reduce the average particle size of MLV and to narrow down the particle size distribution.

Attempts have been made to circumvent the organic film formation/lyophilization stage by direct introduction of powdered lipid particles into these high-shear homogenizers (11,59). Brandl et al. provided evidence that—after an equilibration period—lipid mixing occurs between hydrogenated soybean PC and cholesterol (1:1 molar ratio) in liposomes directly produced by high-shear homogenizer treatment of the two powders. However, no quantitative data were presented on the extent of this lipid mixing process. Apart from the obvious advantage that no organic solvent has to be used during preparation, these high-shear homogenizers can handle high concentrations of lipids, produce the product fast, and presumably are designed to work under GMP conditions. The question of the occurrence of surface erosion of these machines during operation and the consequences for the resulting liposome dispersions has not been addressed yet. In conclusion, high-shear homogenizers may turn out to be valuable in producing in one step (highly concentrated) liposomes with diameters smaller than 100 nm. The relatively low captured volume of the aqueous phase for these small vesicles does not exclude high encapsulation efficiencies in the case of bilayer associated compounds.

A different approach for generating liposomes in the absence of organic solvents or detergents (see below) and under low-shear conditions was developed by Talsma et al (60). The liposomes are produced at temperatures above the phase transition temperature by hydrating solid phospholipid particles in a stream of bubbling N_2 through the aqueous phase. The shear forces are much smaller than with the homogenizers introduced above, and the size range of the resulting liposomes is substantially larger than that of liposomes produced under high-shear conditions. This results in a relatively high aqueous captured volume. The authors hypothesize that the liposomes are generated upon collapse of the N_2 bubbles coated with (phospho)lipid (mono)layers at the water/air interface (60).

b. Methods Based on Replacement of Organic Solvents by Aqueous Media. Several options are listed in Table 3. In all options, except B.a, the lipid constituents are first dissolved in an organic solvent, which is subsequently brought into contact with an aqueous phase. The organic solvent is removed later. During removal of the organic phase, liposomes are formed. Their characteristics (size, organization of bilayers) depend on the protocol used. If the organic solvent with the dissolved lipids is not miscible with the aqueous phase (ether, chloroform, freons), then the intermediate stage is an emulsion (immiscible solvents). Other organic solvents containing the dissolved lipid(s) can be mixed homogeneously with the aqueous phase (e.g., ethanol) in the first stage. Then liposome formation occurs when the organic solvent concentration drops below a certain critical value (miscible solvents).

The content of residual organic solvent that is acceptable in the finished product depends on the solvent in question and the route of administration. Apart from evaporation, techniques similar to those used to remove nonencapsulated material can be selected: gel permeation, ultracentrifugation, or dialysis (cf. Sec. II.D.3). Organic solvent may contain impurities with a high affinity for bilayers; they may be enriched in the bilayer and cause safety or stability problems. Diethylether, for instance, can be contamined with peroxides that accumulate in bilayers. Freshly (from bisulfite) distilled ether should therefore be used. The techniques based on the use of water-immiscible and water-miscible solvents will be dealt with separately below.

Immiscible Solvents: The Aqueous Phase in Excess. Solvent injection or solvent infusion techniques were described by Deamer and Bangham (61).

Ether or ether/methanol solutions of lipids are slowly injected into a water phase. The organic solvents are continuously removed by evaporation, either at elevated temperatures and/or under reduced pressure, and LUV are formed. Care must be taken to avoid impurities in solvents, in particular peroxides in those solvents liable to oxidative damage. Chlorofluorocarbons have been used as organic solvents as well (62). The method of Martin allows the production of liposome dispersions with high lipid concentrations (>250 μmol/mL) and high encapsulation efficiencies (>50%). A special situation has been described by McGurk (63). A combination of water, ethanol and chlorofluorocarbon(s) has been proposed as a vehicle for the extemporaneous formation of a liposome dispersion in an aerosol spray.

Immiscible Solvents: The Organic Solvent Phase in Excess in the Initial Stage. The characteristics of liposome dispersions generated by using protocols where the lipid is dissolved in water-immiscible organic solvents and where the organic solvent is initially present in excess can vary widely: from cell-size, large unilamellar liposomes down to stable, plurilamellar liposomes (SPLV) and REV. The structure of the liposomes strongly depends on bilayer constituents, ratio organic phase/lipid/water and application of shear (e.g., ultrasonication). The preparation of cell-size unilamellar liposomes was described by Kim and Martin (64). Here double emulsions are formed. First, W/O emulsions are made with water drops emulsified inside the organic phase; this is followed by emulsifying the W/O emulsion in a water phase creating W/O/W emulsions and evaporation of the organic phase. After centrifugation the cell-size liposomes are found in the pellet; the yield is low. The authors claim that only with a combination of four ingredients, (1) neutral amphipatic lipids, (2) cholesterol, (3) amphipatic lipids with negative charge, and (4) neutral lipids without a hydrophilic head group (triolein, α-tocopherol, cholesteryloleate), are these cell-size liposomes formed. By manipulating the water droplet size in the W/O emulsion, either unilamellar (see above) or multivesicular, large (»1 μm) vesicles could be obtained (65).

Reverse phase evaporation vesicles (REV) are unilamellar or oligolamellar liposomes. They are formed after emulsifying a lipid-containing water-immiscible solvent (chloroform, ether) with a water phase by sonication and subsequent controlled removal of the organic phase by evaporation (66,67). Essential for reproducible formation of REV is the proper emulsion formation by sonication. A critical step in the preparation procedure of REV occurs when most of the organic solvent has been removed. A gel is formed that needs vigorous vortexing to be converted into a viscous fluid state with the liposomes. The collapse of the gel supposedly coincides with the conversion of the W/O emulsion into the liposomal form. Many variations have been proposed on the basic scheme of reverse phase evaporation technology, e.g., by Handa and coworkers (68). Some variations have been shown to produce liposomes that are structurally very different from REV.

Unilamellar REV are only generated when the ratio of the organic solvent: water:(phospho)lipid is chosen within certain limits. In the initial stage a typical formulation contains 3 mL of ether, 1 mL of aqueous buffer, and 60 μmol lipid. When the water content drops considerably and/or higher lipid concentrations are selected, so-called MLV-REV are formed. With MLV-REV a large aqueous core is surrounded by many (phospho)lipid bilayers (69,70). The encapsulation efficiency of these MLV-REV is relatively high compared to that of MLV. The authors claim that the number of bilayers can be controlled, within certain limits, by experimental conditions: proper selection of lipid/water ratio and droplet size in the emulsion.

The generation of SPLV (stable plurilamellar vesicles, cf. Fig. 1) and MLV-REV requires similar relative amounts of organic solvent, water, and lipid. The difference between MLV-REV and SPLV preparation procedures is that for the SPLV the removal of the organic solvent is performed under sonication (71).

Structurally, MLV-REV differ from SPLV. SPLV lack the large aqueous core. Moreover, the encapsulation efficiency of the interbilayer spaces is more important for SPLV than for MLV, as the bilayer repetition distance in SPLV is wider than in MLV (with neutral lipids).

Solvents Miscible with Water. Batzri and Korn (72) developed the ethanol injection method for preparation of liposomes. The method stands out because of its simplicity. A solution of the lipid(s) in ethanol is injected in the aqueous phase. As the ethanol is diluted, the precipitating lipids form liposomes. Small liposomes with diameters as small as 25 nm (rapid injection) can be formed. Slower injection results in larger vesicles (73). Major drawbacks are the low solubility of many lipids in ethanol limiting the liposome concentration, the low encapsulation efficiency for hydrophilic, nonbilayer interacting drugs, and the difficulty of quantitatively removing ethanol from the liposomal bilayers (e.g., by dialysis). Variations on the ethanol injection technique have been published to avoid these problems. Perrett and coworkers (74) published a protocol in which a proliposome mixture with lipid, ethanol, and water is converted into a liposomal dispersion by dilution. Remarkably high encapsulation efficiencies were reported.

c. *Methods Based on Detergent Removal.* (Phospho)lipids, lipophilic compounds, and amphipathic proteins can be solubilized by detergents forming mixed micelles. Upon removal of the detergent, vesicle formation can occur. Since the early 1970s this preparation technique for vesicles is used by biochemists to study proteins in reconstituted, well-defined membranes, e.g., (75,76). Detergent removal techniques are also well established for preparation of reconstituted virus envelopes, e.g., (77) or reconstituted tumor membrane material, e.g., (78). Schreier and coworkers describe a two-step strategy for insertion of proteins (e.g., recombinant HIV surface protein gp 160) into the outer layers of liposomes. First, liposomes are formed by the detergent dialysis method, and subsequently proteins are inserted by partial resolubilization of the membrane by detergent (deoxycholate) in the presence of the protein (79). Detergent depletion can be induced by dilution, gel filtration, dialysis, or the addition of polymeric adsorbents (e.g., Bio-beads). Allen discussed the pros and cons of the different methods described (80). In general, dispersions of liposomes generated with this technique have a relatively low encapsulation efficiency for hydrophilic nonbilayer-interacting low molecular weight compounds. On the other hand, detergent removal techniques are the first choice for incorporating integral membrane (glyco)proteins into liposomes. Concern is expressed about the safety of the residual detergent in liposomes prepared by this technique. As this aspect strongly depends on the type of detergent used, no general rule can be given. Experience with detergents in other clinically used parenteral formulations is available. For instance, for a number of years glycocholate/soybean PC mixed micelles have been used as solubilizers for lipophilic, poorly water soluble drugs to be administered parenterally, e.g., (81).

Although the theory behind vesicle formation is qualitatively well understood, no exact predictions about particle size and other properties of the formed vesicles can be made, as often nonequilibrium conditions prevail during vesicle formation, and shielding capacities are detergent type dependent. As a rule, fast detergent depletion results in smaller vesicles than slow detergent removal (82).

Extensive studies on factors influencing liposome characteristics, reproducibility of preparation, and scaling-up potential were performed by Weder and collaborators [reviewed in (83)]. They showed that the presence of cholesterol or charge inducing agents in the bilayer influenced liposome size. Under well-controlled experimental conditions, liposome dispersions with narrow particle size distributions (average size $<0.2\ \mu m$) can be obtained.

The detergent removal technique can also be used for the preparation of immunoliposomes. Instead of coupling antibodies or antibody fragments to preformed liposomes with anchor groups sticking out as described above, Huang and coworkers first modified antibodies by attaching palmitic acid molecules to them and solubilized them in deoxycholate (84). Palmitic acid was utilized as bilayer anchoring moiety for the antibody. A similar approach was chosen by Harsh et al. (85). Both groups report specific target cell binding in vitro. Huang and his group reported on successful targeting of immunoliposomes prepared by this technique to lung endothelium in vivo as well [e.g., (86)].

Philippot and coworkers (87) described an extemporaneous preparation method of unilamellar vesicles with diameters of around 240 nm. Octylglycoside/phospholipid mixed micelles are converted into liposomes by addition of Amberlite XAD-2 beads. Son and Alkan followed a similar approach (88,89). They proposed to store mixed micelles consisting of a lipophilic drug, (phospho)lipids, and glycocholate either as an aqueous dispersion or in dried form. Just before administration liposomes are formed in situ by dilution. Glycocholate is not removed from the formulation before injection. No toxicological data are available on this formulation. However, as mentioned above, toxicity is expected to be low, as for many years mixed micellar systems consisting of glycocholate and soybean PC have been used in therapy to solubilize poorly water soluble drugs (diazepam) for intravenous administration (81).

d. Methods Based on Size Transformation and Fusion. Gel State SUV Transformations. Size transformations in liposomal dispersions have been described many times. Several articles dealt with the phenomenon of the growth in size of sonicated SUV of pure DPPC upon storage at temperatures below the phase transition temperature [e.g., (33)]. This size growth can be prevented by adding charged lipids to the bilayer.

Sonication of phospholipids below their phase transition temperature T_c results in vesicles with defects in the bilayers. Heating the dispersion to T_c (or slightly above T_c) eliminates these structural defects (annealing) and causes fusion, resulting in large unilamellar liposomes with a wide size distribution (90).

The main disadvantages of this process for the preparation of drug containing liposome preparations are the limited number of bilayer compositions that react as described above and the poor reproducibility of the particle size distribution of the liposome dispersion that is formed.

Freezing/Thawing. Strauss discussed the behavior of liposomes undergoing a freezing/thawing cycle (91). During the cycle, exchange of material is possible from the liposomal aqueous core to the external phase and vice versa. Proper selection of the experimental conditions is extremely important. Leakage and structural changes strongly depend on freezing time, freezing temperature, freezing rate, bilayer composition, physical nature of encapsulated compounds, and presence of cryoprotectants [e.g., sugars (92)]. When MLV are exposeed to a series of freezing/thawing cycles, structural changes occur. Multivesicular structures are frequently found; stacked bilayer structures are rare (93). The frozen/thawed MLV are called FATMLV. An increase in the encapsulation efficiency results from improved swelling and disappearance of bilayer stacks. Another important observation is that with this technique an equilibrium distribution of solute is achieved (see Sec. II.F.1). This technique allows the loading of proteins and other compounds that are sensitive to organic solvents or detergents to be incorporated into liposomes, provided that freezing (with the accompanying solute concentration effects) is not destructive. The freezing-thawing cycle can be repeated if required. The trapped volume initially increases with the number of freezing-thawing cycles but reaches a plateau upon continuing cycling. The presence of cryoprotectants, discussed in more detail in Sec. II.G, may interfere with efficient loading of the vesicles in a freezing-thawing cycle. Oku and MacDonald modified the freezing/thawing procedure to prepare giant unilamellar or oligolamellar vesicles with diameters of over 10 μm. These giant vesicles are formed upon freezing/thawing SUV in the presence of a high concentration of electrolytes and subsequent dialysis against a low electrolyte concentration (94).

Dehydration/Rehydration Vesicles. Kirby and Gregoriadis described a preparation method based on hydration of freeze-dried liposomes (95). Upon hydration of the freeze-dried liposomes in a small volume of water, large liposomes were generated again, and high encapsulation efficiencies were encountered (dehydration/rehydration vesicles, DRV). Ohsawa et al. published in the same year a similar procedure for liposome formation and loading (96). DRV can be used as pharmaceutical formulations for in situ preparation of liposomes for parenteral delivery, if the nonencapsulated drug does not have to be removed, or if the encapsulation efficiency for the drug is close to 100%. This method differs from the proliposome approach in that with DRV liposomes are already present prior to the freeze-drying stage. The material to be encapsulated can be added either before freeze drying (95,96,97) or in the rehydration medium (96). As cryoprotectants tend to protect cells and liposome structures, their presence may interfere with the formation of the desired liposome structure during the freeze-drying/rehydration procedure. The liposomes formed by dehydration/rehydration in the absence of cryoprotectants are heterogeneous in size. Homogenizing this dispersion with a Microfluidizer® reduces the average size and narrows down the particle size distribution. During this process a decrease in the encapsulation efficiency is observed. Thus if a narrow particle size distribution is required, the encapsulation efficiency is likely to be poor. Details on this delicate balance between size and encapsulation efficiency were published by Gregoriadis and coworkers (98). Instead of freeze-drying, Shew and Deamer (99) prepared SUV by direct sonication of powdered POPC in water with a probe sonicator and dried these dispersions (plus added protein solution) at 37°C in a rotary evaporator; the films were subsequently hydrated. For the chosen lipid composition the encapsulation efficiency did not increase linearly with the lipid concentration but instead reached a plateau at a level much lower than would be expected when freeze-drying would have been used instead of rotary drying. Moreover, the presence of ions interfered with the rehydration process after rotary evaporation. This further limits the applicability of this approach for pharmaceutical purposes.

e. Methods Based on pH Adjustment. Upon transient exposure to alkaline pH values, phosphatidic acid (PA) dispersions spontaneously form small unilamellar vesicles (23,24). Apart from the small unilamellar vesicles in the 20–60 nm range, larger vesicles are formed as well. The extent of vesiculation depends on the experimental conditions such as pH and the ratio PC/PA. The higher this ratio is, the higher the fraction of large vesicles. The authors could exclude chemical degradation and lysophospholipid formation as a reason for the generation of these small vesicles. They ascribe the spontaneous formation to the high surface charge density. At higher ionic concentrations, aggregation and fusion may occur. This concept was further investigated by Li and Haines (100). Phospholipid structure (PA, PG), phospholipid mixture (PC/PG), cholesterol content, ionic strength, and timing of the pH adjustment procedure substantially influenced the particle size of the resulting dispersion. Hauser (24) proposed that the pH gradient over the bilayer (generated by the pH jump) acted as the energy source driving the spontaneous formation of the small vesicles with high curvature. Until now this preparation method has not been used frequently for pharmaceutical preparations, probably because of the rather low encapsulation efficiency, size heterogeneity, and the relatively high costs of negatively charged lipids such as PA and PG.

2. Sizing Stage

The size characteristics of liposomes have a major effect on their fate in vivo [e.g., (101)] and in vitro [e.g., (102)]. Therefore liposome production procedures must generate predictable and reproducible particle size distributions within an accepted size range. Availability of validated techniques for proper definition of the particle size distribution of a batch of liposomes is a prerequisite to determine whether the liposome batches meet the qualifications. These techniques will be discussed in the Sec. II.H. Here, different approaches to meet the above demands concerning liposome size distribution and its reproducibility will be discussed. The approaches can be divided in two categories, one without a special sizing step (1) and one including a special sizing step (2).

(1) In the liposome formation process circumstances are selected and controlled in such a way that particle size distributions with an acceptable width are produced. Examples of this approach are encountered with the major routes of liposome formation as discussed under II.D.1. For example, the liposome size distribution generated by high-shear homogenization depends on the operational pressure (59). Several articles (83,103) give examples of pronounced effects on liposome size distribution as a result of changing parameters in a preparation process based on detergent removal.

In the above-mentioned examples, the produced average particle diameter is usually below 0.2 μm and the dispersions are generally claimed to be acceptable for injection. With approaches where organic solvents are used to establish lipid hydration, vesicle diameters of liposomes in the produced dispersions as a rule exceed 0.2 μm. The bilayer composition should be carefully chosen. Talsma et al. (104) reported on the influence of the presence of charge-inducing agents in the liposome bilayer on the liposome size distribution (average diameter over 0.2 μm), captured volume, and number of bilayers upon lipid film hydration.

(2) In the second category a special sizing stage is introduced in the preparation process. For example, in dispersions where the "film method" is used, unacceptable heterogeneity in size is often observed after manual shaking of the film in the hydration stage. Different approaches are available to narrow down the particle size range by selecting certain fractions. First, for small dispersion volumes the liposome dispersion can be fractionated by (ultra)centrifugation as liposome density usually differs from the density of the medium (11). Secondly, gel permeation chromatography with e.g., Sepharose 6B or Sephacryl S-1000 columns has been used for subdividing wide particle size distributions on an analytical or semipreparative scale. The selection of the pore size of the chromatographic material provides an opportunity to manipulate the size class resolution within certain limits.

Instead of selecting certain fractions, the liposomes can be subjected to a treatment where the vesicles are reshaped to new structures. Already in the 1960s sonication was used to reduce liposome particle size (105). Because of the higher energy power densities that can be reached, probe sonicators are more effective than bath sonicators. Small unilamellar vesicles with diameters as small as 20 nm can be produced. However, probe sonicators have a number of potential drawbacks, which may limit their use in pharmaceutical preparations: (1) exclusion of oxygen during sonication is difficult. Peroxydation reactions may be induced in unsaturated acyl chains during the sonication process; (2) titanium probes tend to shed metal particles, contaminating the pharmaceutical product; (3) probe sonicators can generate aerosols; this excludes them from use with hazardous or toxic agents. With bath sonicators these drawbacks can be avoided. However, for reproducible results with bath sonicators, experimental conditions should be kept constant (water level in bath, position of liposome dispersion in bath, temperature). The development of high-power cup horn sonifiers has substantially enhanced the efficiency of bath sonication [e.g., (33,49)].

The French press was used in the late 1970s for reducing the particle size of liposomes by exposing them to high shear forces [e.g., (106)]. Later, the more user-friendly high-shear homogenizers (55,56,59) were introduced to narrow down the size range and average diameter of the liposomes as well. Dependent on the experimental circumstances (type of equipment, bilayer composition, temperature), average sizes from 0.2 down to 0.05 μm can be obtained. As a rule these high-shear homogenizers need over ten mL of liposome dispersion to operate properly; this minimum threshold quantity excludes their use in high-cost, small-scale lab operations.

Another option is "low pressure" extrusion of the liposome dispersion under pressures up to about 1 MPa through polycarbonate membranes (107,108). Polycarbonate filters with pore diameters ranging from 8 down to 0.05 μm are available. If necessary, the extrusion process can be repeated. Five consecutive extrusions of MLV through 0.2 μm pore membranes produced unilamellar vesicles (109). Two (stacked) filters with 0.1 μm pores and pressures up to 4 MPa ("high pressure extrusion") were used by Hope et al. (110) to convert highly concentrated MLV dispersions into unilamellar liposomes in the 60–100 nm range. For the extrusion of large volumes equipment with a built in stirrer may be advantageous (111). Alternatively, ceramic extrusion systems may be used; the advantage being the relative ease of unclogging the system through back flushing and the possibility of cleaning the filters (112).

As a rule, sonication, high-shear homogenization and extrusion (<0.2 μm pore membrane) produce particles in the size range below 0.2 μm. These dispersions can be considered for pharmaceutical applications where terminal sterilization (through 0.2 μm pores) is required.

3. Removal of Nonencapsulated Material

Many lipophilic drugs exhibit a high affinity to the bilayer and are completely liposome associated. However, for other compounds the encapsulation efficiency is less than 100%. The nonencapsulated fraction of the active compound can cause unacceptable side effects (e.g., doxorubicin) or physical instability (e.g., doxorubicin) (113). For removal of the nonencapsulated material the following techniques have been described: (a) dialysis and ultracentrifugation, (b) ultracentrifugation, (c) gel permeation chromatography, (d) ion exchange reactions.

a. Dialysis and Ultrafiltration. Conventional dialysis membranes can be used with molecular weight cutoff characteristics dependent on the molecular weight of the "free" compound to be removed from the liposome dispersion. Typically, membranes with a molecular weight cutoff between 10 and 100 kDa are used. An impressive arsenal of dialysis and ultrafiltration equipment is available from lab scale up to industrial production scale (hollow fiber, spiral wound). By selecting proper experimental conditions, separation is fast and, if desired, concentration of the liposome dispersion can be achieved simultaneously. During ultrafiltration the dispersion is stirred or circulated by a pump. This convection process must not induce leakage of encapsulated material.

b. Ultracentrifugation. Ultracentrifugation can be used for removal of dissolved nonliposome associated material, for separation of heterogeneous colloidal material, be it liposomes with different sizes and/or subfractions with different densities, and finally for concentrating dilute liposome dispersions [e.g., (114)]. Discontinuous or continuous density gradient ultracentrifugation techniques are powerful tools for separating different colloidal structures (11,115). Centrifugation techniques can be extremely helpful for purifying liposome dispersions on a lab scale. However, they are difficult to integrate into large-scale liposome production schemes.

c. Gel Permeation Chromatography. Sephadex® (e.g., G-50), Sepharose®, or Bio-Gel® columns are regularly used to separate liposome-associated material from nonencapsulated material [see (11)]. The liposomes are not retained by the gel bed and elute in the void volume; the free material elutes in later fractions. This technique is frequently used for assessment of the encapsulation efficiency. Special care must be taken that (1) the bead size and column packing allow large liposomes to pass through the column, (2) the bilayer retains the encapsulated material under the chosen conditions of temperature, ionic strength, etc., and (3) the column is preequilibrated with lipid and the column material does not interact with the liposomes or liposome constituents causing destabilization. The destabilization effect may result from unoccupied "active" sites on the gel and may be avoided by preequilibration. For preparative purposes, major drawbacks limit the use of gel permeation chromatography: dilution occurs during the elution process, and the technology is difficult to scale up, in particular for the production of parenteral products (sterile and pyrogen-free).

d. Ion Exchange Reactions. Storm and coworkers studied the potential of ion exchange resins to remove nonencapsulated material from dispersions with neutral and negatively charged liposomes (116). Several different resins have been proposed. Dowex 50W-X4®, a cation exchange resin, is used regularly (117,118). Dowex® removed free doxorubicin, vincristine, cisplatin, cytarabine, and methotrexate from liposome dispersions even under (pH) conditions where the drug was not present in the cationic form. This led to the conclusion that the interaction with Dowex® may not be simply electrostatic in nature but may also include a hydrophobic component (116). In general, ion exchange resins can be utilized for efficient and fast removal of nonliposome-associated material without dilution of the dispersion, if they have a high affinity for the nonliposome-associated compound.

E. Liposome-Solute Interaction

1. Bilayer-Solute Interaction and Bilayer Permeability

The interaction between the bilayer and the agent to be entrapped depends both on bilayer characteristics such as rigidity and charge and on the physicochemical characteristics of the agent itself. The relationship between the chemical structure of the agent and the degree of partitioning in the bilayer or "adsorption" to the bilayer has been studied by several groups, and interactions based on electrostatic, van der Waals, and/or hydrophobic forces could be identified (119,120,121). Barenholz and Crommelin (15) discern three patterns of agent-liposome interactions. It is therefore also possible to identify three groups of agents.

Group I concerns hydrophilic, water-soluble agents with very low (close to zero) oil/water or octanol/water partition coefficients. These compounds do not interact with the bilayer, and their encapsulation efficiency is directly related to the trapped water volume. Gel state bilayers or bilayers with major fractions of cholesterol offer a substantial barrier; leakage kinetics are not dependent on dilution of the external water phase as with group II agents.

Group II includes compounds with a low oil/water partition coefficient. However, their octanol/water partition coefficient can vary considerably. Here, the liposome bilayer/aqueous medium partition coefficients are not directly related to the oil/water and octanol/water partition coefficients, because of the limited space in the bilayer and the rather fixed orientation of the bilayer components compared to the oil or octanol phases. Partition coefficients of the members of this group depend on pH and ionic strength. Amphipathic weak bases such as the cytostatic doxorubicin belong to this group. It interacts with the bilayer through electrostatic, van der Waals, and hydrophobic forces. An increase in negative charge density of the bilayer induces stronger electrostatic interactions and a higher encapsulation efficiency (122). Barenholz and Amselem report a rapid leakage of doxorubicin from the liposomes upon extensive dilution (cf. intravenous bolus injection) (123). They conclude that a liposome-associated compound can only be kept liposome associated when either a high kinetic barrier exists or the associated compound has an extremely high liposome/aqueous phase partition coefficient.

Group III includes compounds with high oil/water and octanol/water partition coefficients. Some of them mix very well with bilayer constituents. However, for other members of this group the affinity for bilayers is limited because of the earlier-mentioned geometrical constraints. For instance, for cholesterolesters and triacylglycerols phase separation occurs at relatively low incorporation levels in phospholipid bilayers (15).

Formulation of intravenous injectables for lipophilic drugs is often a problem, as the number of pharmaceutically acceptable solubilizing systems is limited (e.g., mixed micelles and emulsions). Successful attempts have been made to produce liposome-based injectable colloidal dispersions for lipophilic drugs (group III) (124,125).

Peptide and protein interactions with bilayers have been extensively studied, e.g., (126,127,128). Many of the therapeutic proteins produced via rDNA or hybridoma technology (e.g., cytokines) are soluble globular proteins. To optimize the liposome encapsulation efficiency of these proteins by rationale, a proper insight into the mechanism of interaction between these molecules and bilayers is essential. Recently, a study was completed on the mechanism of interaction of four soluble globular (model) proteins with liposomal bilayers demonstrating the importance of electrostatic interactions for bilayer adsorption (129). The insights gained could be applied directly when optimizing the encapsulation efficiency of interleukin-2 (IL-2) containing liposomes (130). In a subsequent study the occurrence of irreversible conformational changes in the protein structure (myoglobin) was described after adsorption to the bilayer (131).

2. Triggered Release of the Encapsulated Agent

The concept of targeted drug delivery required that upon attachment to the target site, or delivery into the target cell, the drug has to be released from its carrier (the liposome) to exert its action. If the targeted liposomes are taken up by the target cells through endocytosis, they are localized first in endosomes and subsequently in lysosomes, where the bilayer components are degraded. Both in endosomes and lysosomes acid conditions prevail. For peptides and proteins it may be essential to escape from the liposome and the endosome and to enter the cytosol before the liposome reaches the lysosomal structure with its highly efficient degradation machinery. In order to accomplish this, destabilization of liposomes under acidic conditions and a tendency to fuse with the endosomal membrane are required properties. For example, Huang and coworkers have utilized this "destabilization/fusion by acid pH" approach in their efforts to manipulate the immune response and promote the expression of antigens in combination with MHC class I molecules to trigger a cytotoxic T-cell response (class I restricted antigen presentation) (cf. Sec. III) [e.g., (132,133)]. Several approaches to trigger the release of drugs from liposomes will now briefly be discussed: bilayer composition controlled release, pH-dependent release, destabilization by removal of bilayer components, complement-induced leakage, and temperature-induced destabilization of liposomes.

a. Bilayer Composition Controlled Release. Proper selection of the bilayer structure will ideally allow release of the drug over the desired length of time, preferably after reaching the target site. Gel state bilayers or bilayers with high cholesterol contents release drugs slowly; fluid crystalline state bilayers without cholesterol are relatively permeable and liable to rapid biodegradation in vivo.

b. pH-Dependent Release. An overview of literature dealing with liposomes that destabilize and fuse at low pH is given by Szoka and coworkers (134). The first pH-sensitive liposomes consisted of *N*-palmitoyl-L-homocysteine (PHC) as the acid-sensitive component and phosphatidylcholine. At neutral pH, PHC is in the charged form and mixes with the other bilayer components. However, at weakly acidic pH, PHC has a neutral thiolactone structure; phase separation occurs and the liposomal bilayer structure is destabilized (135).

Later, liposomes composed of oleic acid (OA) and dioleylphosphatidylethanolamine (DOPE) were developed. These bilayers destabilize at $pH < 6.5$ because of protonation of the OA and the subsequent instability of the protonated OA/PE bilayers [e.g., (136)]. Other pH-sensitive liposomal bilayer structures consist of combinations of *N*-succinyldioleylphosphatidylethanolamine and DOPE (137). These liposomes released the water-soluble marker molecule upon a drop in external pH but did not fuse under those conditions. Combining cholesterolhemisuccinate (CHEMS) and PE results in pH-sensitive liposomes ($pH < 5.5$) that also tend to fuse with other bilayers in the presence of Ca^{2+} (138). The interaction of a protein with the bilayer can affect the bilayer structure. The pH-dependent variation of the secondary structure of peptides (e.g., GALA) has been used by Szoka and coworkers (139) to destabilize liposomal membranes upon acidification of the aqueous medium.

c. Destabilization by Removal of Bilayer Components. Hydrated PE forms liposomes only in combination with other bilayer components, e.g., OA/PE liposomes are stable. Destabilization occurs upon extraction of OA from the bilayer when water-soluble molecules with a high affinity for OA, such as delapidated albumin, are added to the aqueous medium. Finally, cholesterol in liposomal bilayers can easily exchange with serum components such as HDL. Therefore bilayers that are stabilized by cholesterol can loose their integrity in the presence of serum or ascites, and the liposomal contents can be released. Scherphof and coworkers reported that the exchange kinetics depend not only on the components of the bilayer but also on the size and number of bilayers of the liposomes. Lipids in SUV are more susceptible to exchange than lipids in MLV with the same chemical constituents. The highly curved surface of the SUV and the resulting irregularities in the surface are responsible for this phenomenon (140).

d. Complement-Induced Drug Leakage. The concept of complement-induced leakage of marker molecules is used in diagnostic tests. Relevant literature can be found in early articles by Kinsky's group (141) and others (142,143).

One approach, the so-called Liposome Immune Lysis Assays (LILA) is based on the following scheme. The liposomes expose an antibody on their surface. This antibody reacts with a soluble antigen in the bathing medium. Upon the formation of this antigen-antibody-liposome complex, the complement cascade is activated, the liposome is lysed, and the marker molecule or marker enzyme is released from the liposomes.

In vivo one can envision a different situation. Immunoliposomes interact specifically with cells or tissues carrying the corresponding antigens, e.g., in the peritoneal cavity or the blood compartment. These targeted immunoliposomes carry, apart from their homing device, an antigen X. After attachment of the immunoliposomes (plus antigen X) to their target site, an antibody (usually IgM) against antigen X is added to the medium bathing the target site; then complement-mediated lysis of the liposomes occurs, and the liposome encapsulated compound is released.

e. Temperature-Induced Destabilization of Liposomes. Imperfections occur in bilayer structures in the temperature range where the liquid-crystalline-to-gel phase transition is found; then both liquid-crystalline and gel state structures coexist. Bilayer permeability reaches a maximum under those conditions; at temperatures both below and above this phase transition lower permeabilities are found. Exchange of lipids is facilitated at the phase transition temperature as well (144). By selecting lipids with phase transition temperatures in the range just above body temperature (DPPC and DPPG: 41 °C), the site-specific release of liposome encapsulated drugs can be triggered by local heating of the tissue.

F. Techniques for Improved Loading Efficiencies

1. Loading Liposomes with Therapeutically Active Compounds: Passive Loading Strategies

Passive loading of liposomes with the required compounds means that the compound is entrapped in the liposome during liposome formation, or in a stage of the preparation process when the liposome structure is severely weakened.

For an evaluation of the success of the loading process of liposomes with an active compound, different definitions can be used. The captured volume (L/mol lipid) gives the volume of water encapsulated per mol lipid. The encapsulation efficiency (%) is the fraction of solute that is encapsulated in the liposome. The encapsulation efficiency for a compound can be much higher than predicted by calculations based on the captured volume. It can reach 100% for compounds with a high affinity for bilayers. For water soluble, nonbilayer interacting agents one would expect that the encapsulation efficiency could be calculated on the basis of the product of captured volume · (mol lipid · L^{-1}) (× 100%). However, Gruner and coworkers (71) reported that MLV prepared with the film method showed lower encapsulation efficiencies than expected, because exclusion of solutes can occur during liposome formation. Apart from osmotic pressure differences between the external phase and the water phase in between the swollen lamellae, this results in lower encapsulation efficiencies than calculated from captured volume data. A number of freezing/thawing cycles induces a transmembrane equilibrium of solute concentrations and an increase in encapsulation efficiency (93,145).

Many other factors influence captured volume and encapsulation efficiency. For example, for unilamellar liposomes the captured volume decreases with particle size. Multilamellarity (equal size and concentration) also reduces the captured volume compared to unilamellar liposomes. Moreover, the captured volume of liposome dispersions depends on the lipid concentration; it tends to decrease with increasing lipid concentration. Therefore it is impossible to predict encapsulation efficiencies of highly concentrated liposome dispersions by extrapolation from captured volume data obtained for diluted dispersions.

An interesting phenomenon was observed by Chapman et al. (146,147). Ca^{2+} and sucrose in the liposome dispersion medium were concentrated to a considerable extent (up to 12- and 7-fold, respectively) inside the liposomes (70 nm size range) upon a freezing/thawing cycle. This concentration effect strongly depended on the experimental conditions. It was attributed to the presence of high solute (Ca^{2+} or sucrose) concentrations in the early thawing stage. The presence of compounds that prevent osmotic lysis later on during the thawing process might preserve this concentration difference between the inside and outside milieu of the liposome. Although it is clear that the phenomenon exists, a full understanding of the underlying principle is still not available.

2. Active Loading Strategies: pH Gradient Driven

Active loading or remote loading encompasses techniques that allow us to load liposomes with the desired compound after the formation of the liposomes with full conservation of the integrity of the liposomal bilayer. The driving force for accumulation in the liposomes is either a pH gradient or a membrane potential over the bilayer.

A prerequisite for successful active loading is the possibility of diffusional transfer of the uncharged species of a basic or acidic drug through the bilayer, while the ionized form is not able to do so. By applying and maintaining a pH gradient over the bilayer, the membrane permeable, uncharged species of the drug will equilibrate on both sides of the bilayer. Following the Henderson-Hasselbalch equation, pH conditions inside and outside the liposomes can be chosen in such a way that the ionic species accumulates inside the liposomes. In principle, equilibrium is reached (for drug $DH^+ \leftrightarrow D + H^+$) when $DH^+_{in}/DH^+_{out} = H^+_{in}/H^+_{out}$.

Nichols and Deamer (148) and later Cullis and coworkers (149,150,151, 152) generated the pH gradient by preparing the liposomes in a buffer at low pH. The buffer species should have a negligible tendency to penetrate through the membrane as then the pH gradient would not be maintained. When loading of the liposomes is requested, the external pH is raised. The accumulation rate is enhanced by working at elevated temperatures with the bilayers, if possible, in the fluid state. The pH gradient also induces a membrane potential (negative inside). This membrane potential may play a role in the uptake process as well. Uptake percentages range between 100% (e.g., doxorubicin, daunorubicin, mitoxantrone, propanolol, dopamine) and <1% (e.g., pilocarpine, codeine). Table 4 [by Vingerhoeds, based on data provided by Madden et al. (153)] gives a list of drugs and their internal accumulation under the influence of a pH gradient in uncharged PC LUV.

One might speculate why drugs such as doxorubicin and mitoxantrone (group 1) are completely taken up, as this is more than predicted on the basis of an analysis using the Henderson-Hasselbalch equation. It has been suggested that precipitation or complex formation occurs due to the high concentrations inside the vesicles. As the Henderson-Hasselbalch equation only applies to the concentrations of soluble species, these extremely high encapsulation efficiencies supposedly result because of internal precipitation e.g., of doxorubicin. In this case, the above equation has to be modified to *soluble* $DXR^+_{in}/DXR^+_{out} = H^+_{in}/H^+_{out}$. In addition, partitioning of the compound into the bilayer also may play a (minor) role in governing uptake in the presence of a pH gradient. No clear relationship between partition coefficient and accumulation surpassing the predictions of the Henderson-Hasselbalch equation can be established for neutral PC liposomes.

In group 2, the accumulation is in accordance with the expectations based on calculations with the Henderson-Hasselbalch equation. In group 3, drugs are listed that may effectively partition into the bilayer and by doing so increase the ion permeability. This in turn results in a reduction or even disappearance of the pH gradient. In category 4, the uptake of drug is very low. The reason for this low uptake has not yet been fully elucidated. It has

Table 4 Extent and Stability (Uptake after 2 Hours) of Accumulation of Various Drugs by Vesicles Exhibiting a pH Gradient Adjusted by M. Vingerhoeds

Drug	Class	Uptake 15 min (nmol/μmol PL)	Uptake 15 min %	Uptake 2 h (nmol/μmol PL)	Uptake 2 h %	Partition coefficient[1]	Apparent solubility[2] (mM)	pK[3]
Group 1								
Mitoxantrone	Antineoplastic	200	100	198	99	–	<0.01	–
Epirubicin	Antineoplastic	201	101	200	100	–	0.26	–
Daunorubicin	Antineoplastic	200	100	204	102	3.5	9.10	8.2
Doxorubicin	Antineoplastic	202	101	203	102	1.1	0.24	8.2
Dibucaine	Local anesthetics	194	97	176	88	4.4	>700	8.5
Propanolol	Adrenergic antagonists	198	99	187	94	1.3	326	–
Dopamine	Biogenic amines	190[a]	95	177	89	–	1400	–
Imipramine	Antidepressant	182	91	188	94	4.6	4.43	9.5
Group 2								
Lidocaine	Local anesthetics	87	44	87	44	–	240	7.9
Chlorpromazine	Local anesthetics	98	49	96	48	1.5	–	9.2
Timolol	Adrenergic antagonists	95	48	97	49	−0.1	135	–
Serotonin	Biogenic amines	80[b]	400	78	39	–	–	–

Chloroquine	Antimalarial	104[c]	52	88	44	—	585	8.1
Quinacrine	Antiprotozoan	73[a]	37	71	36	—	90	8.0/10.2
Group 3								
Vincristine	Antineoplastic	178	89	130	65	2.8	>35	—
Vinblastine	Antineoplastic	175[c]	88	127	64	—	19.1	—
Quinidine	Antiarrythmic agents	203	102	74	37	—	5.83	4.2/8.3
Quinine	Antimalarial	148[c]	74	88	44	1.7	1.05	4.2/8.8
Diphenhydramine	Antihistamine	176[c]	88	87	44	3.4	—	9.0
Group 4								
Pilocarpine	Cholinergic agents	<1	<1	<1	<1	—	—	1.6/7.1
Physostigmine	Cholinergic agents	<2	<1	<1	<1	0.2	—	2.0/8.1
Codeine	Analgesic	<1	<1	<1	<1	1.2	—	7.9

[a]Maximum uptake taken at 30 min.
[b]Maximum uptake taken at 90 min.
[c]Maximum uptake taken at 5 min.

[1]Log. octanol/water partition coefficient.
[2]Apparent solubility in 300 mM citrate pH 5.0.
[3]Ref. 153 and references therein.

Large unilamellar vesicles (egg phosphatidylcholine, 1 mM lipid) were incubated with the drug (0.2 mM) in 300 mM NaCl, 20 mM HEPES (pH 7.5) at 25°C. The vesicles have a trapped volume of 1.5 μL/μmol phospholipid. For more details see Ref. 153.

Source: Ref. 153.

been suggested that the unionized drug is not able to pass through the membrane. Alternatively, the drug might cause a major increase in membrane permeability. The last possibility has been proven incorrect for certain drugs (153).

An elegant alternative way to generate a pH gradient over liposomal bilayers to achieve remote loading was described by Barenholz and coworkers (154). The principle of this technique, the ammonium sulphate technique, is described in Fig. 7. Basically, the internal pH drops upon passage of the unionized NH_3 molecule through the bilayer from the internal to the external aqueous medium. This transport is the result of a concentration difference over the bilayer and is diffusion driven.

Active or remote loading procedures can only be used successfully for certain classes of compounds. Entrapment depends on the lipid bilayer composition, physical characteristics of the encapsulated compounds (pK, tendency to interact with bilayer), and experimental conditions (e.g., temperature). Active loading offers advantages over conventional loading techniques, e.g., (1) enhanced encapsulation efficiencies, making removal of nonentrapped material often unnecessary; (2) no exposure of the encapsulated compound to organic solvents, detergents, or high-shear forces; (3) encapsula-

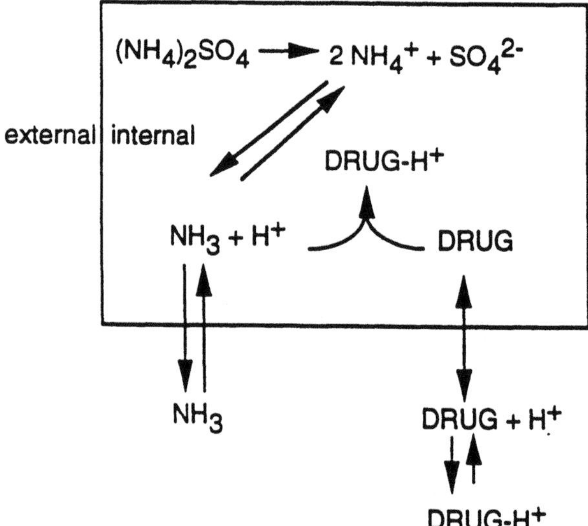

Figure 7 The mechanism by which an ammonium gradient induces a pH gradient and active loading of drugs (weak bases) in liposomes.

tion can be achieved "at the bedside," avoiding degradation of labile compounds during (long-term) storage in the form of aqueous dispersions.

The above mentioned (first) remote loading methodology where the pH gradient is generated by changing the external pH (148,149,150,151,152) in reality implies that for basic compounds such as doxorubicin, aqueous liposomal dispersions should be stored at low pH. This may cause chemical stability problems (155,156,157), because of increased hydrolytic reaction rates compared to aqueous dispersions with pH values around 7. Liposomes containing ammonium sulphate (the "ammonium sulphate technique") can be stored at neutral pH before starting the loading process.

3. Active Loading Strategies: Loading of Metal Cations into Liposomes

This loading technique is based on carrier-mediated transport of metallic cations across the membrane. These ions are "lipophilized" by complexing them with a lipophilic carrier such as quinolate (^{67}Ga-oxine, ^{111}In-oxine). In the internal phase of the liposomes a water-soluble chelator with high affinity for the metal ion (nitrilotriacetic acid, ethylene diamine tetraacetic acid, or deferoxamine) of the cation is entrapped. The chelators form a stable complex with the metal cations, thus trapping them inside the liposome (158, 159,160). Encapsulation efficiencies of over 90% have been described. This efficiency is affected by the concentration of the lipophilic carrier, the pH of the medium, and the presence of chelators in the loading incubation mixture (through leakage or incomplete removal).

G. Stability of Liposomes

As industrially produced drug or antigen containing liposomes will reach the patient or person to be vaccinated only after a prolonged time, the liposome dispersion should not change its characteristics or lose the associated drug or antigen during storage or transport. In general, a shelf life of at least one year is a minimum prerequisite in the pharmaceutical industry.

Pharmaceutical liposomes face a number of chemical and physical destabilization processes. In the following sections, first possible chemical aging reactions will be addressed. Subsequently, physical aging processes relevant to pharmaceutical liposomes will be dealt with.

1. Chemical Stability

As phospholipids usually form the backbone of the bilayer their chemical stability is important. Two types of chemical degradation reactions can affect the performance of phospholipid bilayers: (1) hydrolysis of the ester bonds; (2) peroxidation of unsaturated acyl chains (if present). Analytical procedures to monitor the state of chemical degradation are discussed by Grit and coworkers (157) and by Kemps and Crommelin (161,162).

a. Hydrolysis of the Ester Bonds. Phosphatidylcholine possesses four ester bonds. The two acyl ester bonds are most liable to hydrolysis. The glycerophosphate and the phosphocholine ester bonds are more stable. Therefore the major route of hydrolytic degradation is the one depicted in Fig. 8. Interestingly, the 1-acyl-lysophosphatidylcholine isomer (1-acyl LPC) is

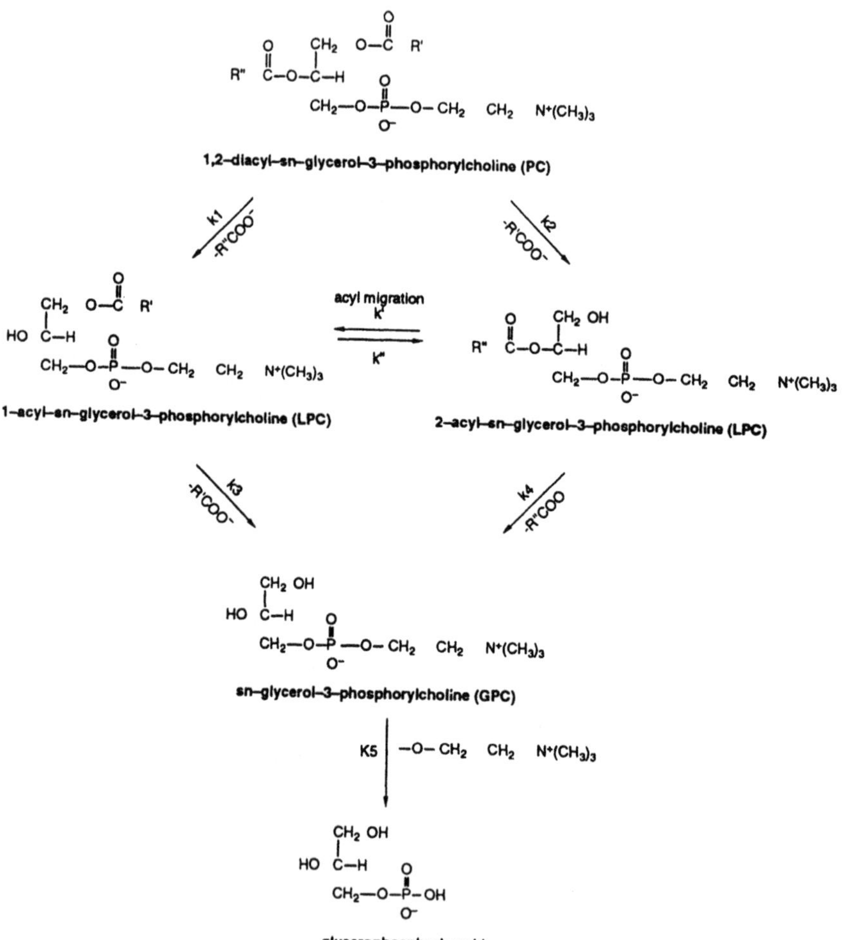

Figure 8 Hydrolysis reactions of phosphatidylcholine in aqueous liposome dispersions. R′ and R″ are acyl chains. (From Ref. 156.)

the major lyso isomer found. However, detailed monitoring of hydrolysis reactions suggests that both 1-acyl LPC and 2-acyl LPC are formed at comparable rates, and that accumulation of mainly 1-acyl LPC occurs because of rapid conversion of 2-acyl LPC into 1-acyl LPC (acyl migration reaction, Fig. 8). Following the pioneering work by Frøkjaer (163), phospholipid hydrolysis kinetics were extensively investigated by Grit and coworkers (155, 156,157,164). Temperature, pH, bilayer rigidity, and buffer species influence the hydrolysis rate. Optimum pH conditions were found around pH 6.5 (Fig. 9).

For the fluid crystalline state bilayers consisting of phosphatidylcholine, the temperature dependence of the hydrolysis rate constants can be adequately described by the Arrhenius equation. A linear relationship was found when plotting the log hydrolysis rate constant versus $1/T$. For these liposomes, the possibility should be examined to predict room or refrigerator temperature stability on the basis of accelerated stability data. However, for bilayers with a phase transition in the experimental temperature range (e.g., hydrogenated soybean PC, phase transition temperature 52°C), a discontinuity in the Arrhenius plot was observed (Fig. 10). The nature of this discontinuity depended on the pH. Here accelerated stability studies based on temperatures above the phase transition temperature are of little use in predicting shelf life stability under real-life conditions.

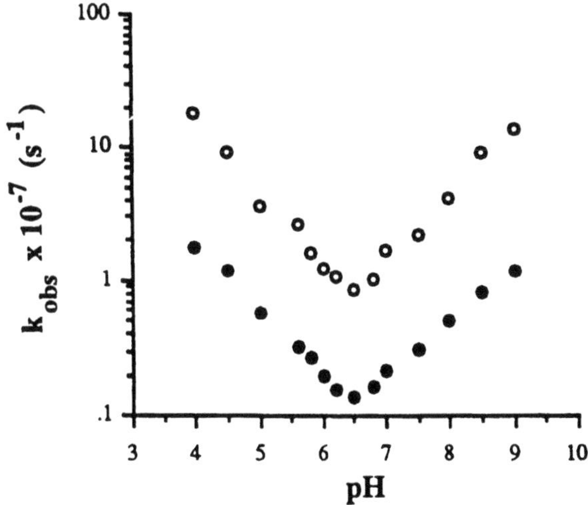

Figure 9 The effect of pH on the hydrolysis of saturated soybean phosphatidylcholine. Buffer concentration = 0.05 M; each point represents the mean of at least two separate determinations. ● = 40°C; ○ = 70°C. (From Ref. 156.)

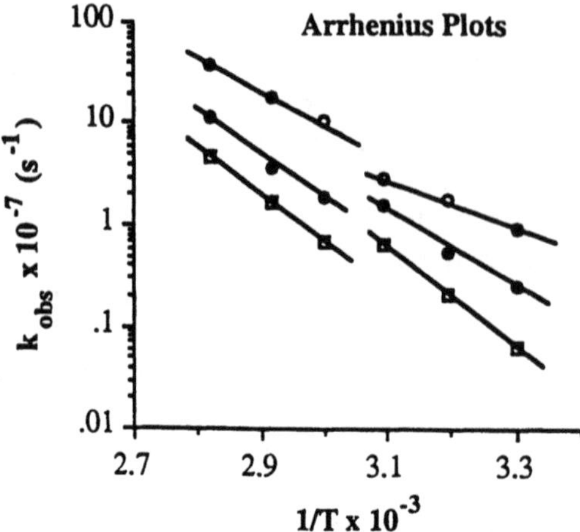

Figure 10 The effect of temperature on the hydrolysis of saturated soybean phosphatidylcholine. The lines were calculated by linear regression analysis. Each point represents the mean of at least two separate determinations. ○ = pH 4.0; ● = pH 5.0; □ = pH 7.0. (From Ref. 156.)

Liposomal surface pH values can differ significantly from bulk pH values in the presence of charged lipid bilayer components. For negatively charged bilayers the bulk pH will be higher than the surface pH. This difference between surface pH and bulk pH increases when the ionic strength in the aqueous phase is decreased or the surface charge density is raised (165). As the surface pH controls hydrolysis kinetics of phospholipids, the bulk pH with maximum phospholipid stability will shift to higher pH values.

The effect of phospholipid hydrolysis on the permeability of liposome bilayers was investigated by Grit and Crommelin (166). When phospholipids hydrolyse, lysophosphatidylcholine (lyso-PC) and fatty acids (FA) are formed. Bilayer permeability for the nonbilayer interacting, hydrophilic probe, calcein, decreased with increasing lyso-PC/FA contents until about 10% of the phospholipid had decomposed. This behavior is clearly different from the behavior seen when lysophospholipids alone (no FA present) are added to phospholipid bilayers. Then bilayer permeability for calcein increases steeply with the concentration of lyso-PC.

In conclusion, when aqueous liposome dispersions are designed for use as drug delivery systems, their pH, buffer, ionic strength, bilayer charge, and bilayer fluidity should be carefully selected. Otherwise unacceptably short shelf lives will be the result.

b. Lipid Peroxidation of Phospholipids. The polyunsaturated acyl chains of phospholipids are sensitive to oxidation via free radical reactions. Cyclic peroxides, hydroperoxides, malondialdehyde, and alkanes are among the decomposition products produced via complex degradation pathways. Oxidation can be induced by sonication, radiation, photooxidation, or autooxidation (162,167). Low oxygen pressure, the absence of heavy metals, the addition of antioxidants [α-tocopherol, butylated hydroxytoluene (BHT), vitamin C], complexing agents (e.g., EDTA), and quenchers (β-carotene) of the photooxidation reactions have been reported to improve resistance against lipid peroxidation. Collagen also was mentioned as a stabilizer of bilayers (168). However, one has to be cautious, as under certain conditions tocopherols and vitamin C have been reported to act as prooxidants. The sensitivity of polyunsaturated lipids to ionizing radiation has excluded γ-radiation as a method for sterilization of liposome dispersions. Recently, it was reported that not only unsaturated but also saturated acyl chains were cleaved from phospholipids upon γ-radiation (169).

Lipid peroxidation causes an increase in the permeability of the bilayer [e.g., (170)]. The degradation process produces a number of products with a highly different chemical nature. At present no quantitative data are available on the relationship between lipid peroxidation and bilayer permeability. The impact of lipid peroxidation on liposome behavior in vivo and in vitro, and on safety upon parenteral administration, has not been studied in detail yet (30).

2. Physical Stability

Physical processes that affect shelf life include loss of liposome-associated drug and changes in size: aggregation and fusion. Aggregation is the formation of larger units of liposomal material; these units are still composed of individual liposomes. In principle, this process is reversible, e.g., by applying mild shear forces, or by changing the temperature or by binding metal ions that initially induced aggregation. Fusion indicates that new colloidal structures were formed. As fusion is an irreversible process, the original liposomes can never be retrieved.

Drug molecules can leak from the liposomes. The leakage rate strongly depends on the bilayer composition and the physicochemical nature of the drug. Bilayers in the gel state or those containing substantial (molar) fractions of cholesterol tend to lose the associated drug only slowly; liquid state bilayers are more prone to drug loss and are less stable during storage. Bilayer permeability is not necessarily a constant parameter. Changes in bilayer permeability can occur as a result of chemical degradation processes, such as the formation of lyso-PC and FA (as described above).

3. Approaches to Improve Shelf Life

Loss of drug and changes in particle size will change the fate of the drug upon administration. Therefore the stability should be carefully monitored and, if necessary, measures have to be taken to make the liposome formulation meet the demands of the pharmaceutical industry: a shelf life on the order of years. Approaches that can be chosen to ensure the chemical and physical stability of liposomes are (1) the use of an aqueous dispersion: proper selection of medium and of bilayer components; (2) the use of a dry formulation: (freeze) drying of liposomes; and (3) the use of the proliposome approach.

a. Selection of Bilayer Components: Possibility 1. With aqueous liposome dispersions the selection of a proper pH range (pH 6–7) and low temperatures will reduce hydrolysis and lipid peroxidation mediated damage. Phosphatidylcholine with saturated lipid chains (e.g., phosphatidylcholine with palmitic and/or stearic acid chains, DPPC, and DSPC) tend to be rather stable against peroxidation. If unsaturated lipids are used, addition of antioxidants and removal of oxygen can be considered to minimize peroxidation reactions (see above). Loss of retention of water soluble, nonbilayer interacting active agents from liposomes on storage can be avoided by choosing either gel state bilayers (DPPC and DSPC) or liposomes with unsaturated lipids that contain cholesterol; addition of substantial fractions of cholesterol rigidifies a bilayer composed of phospholipids with unsaturated acyl chains [cf. (171)].

To reduce the probability for liposome aggregation or fusion, a charge inducing agent is often included into the bilayer. Phosphatidylglycerol (PG), phosphatidic acid (PA), and charged cholesterol esters, such as cholesterol hemisuccinate (CHEMS), are negatively charged lipids at neutral pH that have proven to be good candidates for inducing a negative charge on the liposome bilayers and thereby providing extra stability against aggregation and fusion. A phenomenon that can interfere with liposome stability is the occurrence of high bilayer permeabilities in the temperature range where the bilayer changes from the gel state into the liquid-crystalline state [see above or (144)]. Therefore lipid bilayer phase transitions should not occur during liposome storage, if the encapsulated drug tends to leak. Clearly, the use of aqueous liposome dispersions limits the choice of lipids and allows only rather rigid membranes to be used. These rigid bilayers may be undesirable for use in vivo, because of their (too) slow release characteristics when a fast release pattern is desired. This means that alternative strategies (e.g., freeze-drying) must be considered to achieve an acceptable stability.

b. (Freeze)-Drying of Liposomes: Possibility 2. (Freeze)-drying of liposomes intended to be used for pharmaceutical purposes has been extensively investigated over the past years. Reviews can be found in the literature (172,

173,174). Limited work has been done on drying of liposomes at temperatures above 0°C with a spray drying technique (175).

A number of studies are aimed at gaining insight into the behavior of liposomes under freezing conditions. As the storage of cellular material under frozen conditions poses similar problems as the storage of liposomes, insights from cryobiology can be used to optimize liposome stability and vice versa. This also explains why a substantial number of studies focus on the freezing of liposomes and cells instead of on freeze-drying, as freezing cellular material for long-term conservation is an acceptable procedure in cryobiology. Although freezing of drug-laden liposomes is not a preferable form of stabilizing these pharmaceutical formulations, a proper understanding of the processes occurring during the freezing process has been shown to be helpful in the optimizing of freeze-drying conditions of liposomes by rationale instead of by trial and error.

Variables that have been shown to affect the success of the freeze-drying of liposomes are listed in Table 5.

Cryoprotectants. Cryoprotectants can be chemically quite different in nature. Saccharides, proteins and amino acids, and polyalcohols have been shown to exert cryoprotective effects. However, so far most attention has been paid to disaccharides, because of their (expected) acceptability in parenteral formulations. Saccharose, lactose and trehalose have been shown to be able to protect liposomes from aggregation and fusion during freezing/thawing (F/T) and freeze-drying/rehydration (F/R) cycles (92,176). Sugars also tend to improve the retention of encapsulated nonbilayer interacting, hydrophilic compounds such as the probe carboxyfluorescein (CF). The mechanism of cryoprotection is still not fully understood. A number of effects are proposed to play a role in cryoprotection by sugars. First, the cryoprotective sugars form amorphous glass structures upon freezing; the liposomes would be embedded in this glass structure. This would prevent mechanical damage by growing ice crystals. Secondly, the interaction of the sugars with

Table 5 Variables Affecting Freeze-Drying of Liposome Dispersions

Presence of cryoprotectants
Nature of the bilayer and of the liposome-associated compound (interaction/no interaction with bilayer)
Liposome size
Technological parameters
Freezing rate
Freezing temperature
Freezing time

the polar head groups of the phospholipids would be important. This may change liquid-crystalline state/gel state transitions in membranes. Moreover, through this interaction a pseudohydration layer may be formed, and/or loss of hydration water that stabilizes the bilayer structure may be prevented.

Nature of the Bilayer and of the Liposome-Associated Compound (Interaction/No Interaction with Bilayer). There is no standard rule for the selection of bilayer compositions that are stable upon freezing/thawing or freeze-drying/rehydration. Cholesterol addition to bilayers does not necessarily improve liposome integrity (176,177).

Phase transitions during the cooling phase of the freeze-drying process should be avoided. One should realize that cryoprotectants and low water activities can change the behavior of phospholipid bilayers dramatically. Therefore during the freezing and rehydration step phase transition (T_c) temperatures can differ significantly from T_c's that one observes when the liposome is fully hydrated and when no cryoprotectants are present [cf. (173)].

If the liposome-associated compound is lipophilic and dissolves in the bilayer, no decrease in retention or change in particle size is observed after a freeze drying/rehydration cycle with a proper cryoprotectant (cf. above). For compounds such as doxorubicin that can interact electrostatically with negatively charged bilayer components, the retention after rehydration depends on the chosen conditions (176), but high degrees of retention are possible. The most challenging situation is met with hydrophilic, nonbilayer interacting compounds with a low molecular weight. Some successful results were reported by the groups of Hauser and of Crowe (175,178). Other groups report full retention of the probe CF upon freezing/thawing, if the proper conditions are chosen, but considerable decreases in retention upon freeze-drying (179).

Systematic work to elucidate stabilizing and destabilizing mechanisms and to identify exact conditions for maximum retention of CF-like compounds is required in order to rationalize this field.

Liposome Size. The importance of particle size on the retention of the probe CF (as a model for a hydrophilic, nonbilayer interacting, low molecular weight molecule) in gel state liposomes (hydrogenated PC and dicetylphosphate) after an FT-cycle has been discussed by Talsma et al. (180,181, 182). With differential scanning calorimetry they could monitor the crystallization behavior of the internal water volume of liposomes. The lowest crystallization temperatures reached were similar to the homogeneous nucleation temperature for water as reported in the literature (around −42°C). The smaller the size of the liposomes, the lower was the crystallization temperature and the larger was the fraction of the encapsulated volume that solidified around the homogeneous nucleation temperature. Interestingly, both Fransen et al. (92) and Talsma et al. (180) demonstrated that even at low (freezing) storage temperatures (e.g., −25°C) leakage from the liposomes is not completely stopped.

In the presence of cryoprotectants, such as saccharose, retention data after an F/T cycle were found to be size dependent. For dispersions with a mean particle size of 0.12 μm, full CF retention was found; for dispersions with a mean diameter of 0.2 μm, 50% CF retention occurred (181). However, retention of CF after freeze drying showed low retention values for both the 0.12 and the 0.2 μm diameter liposome dispersions.

Technological Parameters: Freezing Time, Freezing Rate, Freezing Temperature. These variables have been shown to be of critical importance for a successful outcome of the F/T or F/R process. Internal and external ice formation depends on the freezing rate, as does the degree of solidification. Özer et al. (173) presented an overview of existing literature. A diffuse picture emerges from these data.

The influence of a number of technological variables on CF retention and liposome size after F/T cycles has been investigated by Crommelin and coworkers (92,182). In an attempt to monitor the variables systematically only one bilayer type (hydrogenated soybean PC/DCP) and probe molecule (CF) were used. When the dispersion was not completely solidified (−25 and −30°C), leakage continued to occur during storage at the selected freezing temperature. With the proper cryoprotectants (saccharose and trehalose), freezing at −50 and −75°C and boiling liquid nitrogen resulted in a CF retention of over 90% for 0.14 μm liposomes. Increasing the temperature to −25°C caused CF loss again. This information is useful to rationalize freeze-drying protocols.

c. The Proliposome Approach: Possibility 3. The proliposome approach has been discussed in Sec. II.D.1. With the proliposome concept the liposomes can be prepared "at the bedside." It is technologically a feasible production procedure. However, this approach can only be used for compounds that are fully liposome associated upon hydration and if a rather wide particle size distribution is acceptable in the clinical situation.

H. Characterization of Liposomes

Both physical and chemical characteristics of liposomes influence their behavior in vivo and in vitro [e.g., (101)]. Several more examples demonstrating the importance of proper selection of liposome structures to obtain optimum and reproducible therapeutic effects have been published (122,183). Therefore it is essential to characterize liposomes properly. In many studies poorly defined liposomes have been used, either because the importance of working with well characterized liposomes was not appreciated or because of lack of equipment and expertise.

Liposome characterization should be performed immediately after preparation. One should also ensure that no major changes occur on storage, so that a well characterized product is injected and the liposome dispersion warrants optimal reproducibility of clinical effects.

Relatively little attention has been paid to the development of validated protocols for the chemical characterization of liposomes, compared to the vast amount of existing literature on physical characterization. New (11) provides detailed information on methods to estimate hydrolysis and peroxidation of lipids. The pros and cons of different analytical methods used to identify and quantitate phospholipids, (free) fatty acids, and other lipids have been reviewed by Grit et al. (157) and Kemps and Crommelin (161, 162). Different types of chromatography can be used to separate bilayer components (TLC, GLC, HPTLC, HPLC). One major problem is detection and quantitation, as UV molar absorptivity of lipids is low and dependent on the degree of saturation of the acyl chains. In particular, for those phospholipids with only saturated acyl chains, alternative detection systems not based on UV absorption are described, i.e., systems based on differential refractometry (RI detector), light scattering, and flame ionization (FID). Derivatization techniques may provide improved means of separation and/or detection. Examples of this approach are methylation of free FA or attachment of a label that can be detected with high sensitivity by standard detectors.

The physical properties of liposomes have a direct impact on the behavior of the liposome with its contents in vivo. The following factors are usually considered to play a role in the in vivo disposition, and information on these parameters should therefore be available both for the freshly prepared product and the "aged" product at the moment of administration (cf. above) to ensure reproducible results (Table 6).

Inappropriate conclusions are readily drawn if only average readings for size, charge, etc., are considered. This applies in particular to data collecting systems, where the individual response per particle is weighed with a complex weighting factor. The complexity of proper averaging and data handling is clearly demonstrated when analyzing dynamic light scattering data of a heterodisperse liposome dispersion. Another example concerns the assessment

Table 6 Physicochemical Parameters Influencing Liposome Disposition

Size
Number of lamellae, internal morphology
Charge
Bilayer fluidity

of bilayer rigidity. Probes to monitor fluorescence polarization provide an insight into bilayer rigidity on the basis of the "environment" of that particular probe (see below). This information is only relevant for bilayers consisting of perfectly mixed lipids. Lipid demixing or coexistence of two different colloidal systems should be excluded, as it is difficult to evaluate fluorescence polarization data properly under those conditions.

Techniques in use to characterize size, number of lamellae, charge, and bilayer fluidity are listed in Table 7.

Background information on these techniques (experimental procedure, interpretation, possible artifacts) can be obtained from reviews and books (4,11,33,109). In addition to the above-mentioned techniques, experimental information can be collected that does not give detailed information about one of the above parameters (Table 6) alone; it is related to a set of parameters. For instance, a change in the "captured volume"—defined as the volume of water encapsulated per mole of phospholipid—of water-soluble, nonbilayer interacting probe molecules, such as CF, in a liposome dispersion can be the result of changing sizes or numbers of lamellae of the liposomes. In the following paragraphs some aspects of the above-mentioned characterization techniques will be dealt with.

The advantage of electron microscopy for size determination is that an impression of mean size, size distribution, and number and form of the lamellae can be obtained. Disadvantages are the costly equipment and the laborious evaluation of the data. These disadvantages make electron microscopy less favorable as a routine technique to follow size stability or assess batch-to-batch reproducibility compared to e.g. dynamic light scattering (see below).

Table 7 Techniques for Physical Characterization of Liposomes

Parameter	Technique
Size	Electron microscopy (e.g., freeze fracture, bare grid method)
	(Dynamic) light scattering
	Ultracentrifugation
	Molecular sieve chromatography
	Coulter counter
Number of lamellae	NMR spectroscopy
	Small angle x-ray scattering
	Electron microscopy
Bilayer fluidity	Fluorescence polarization
	ESR spectroscopy
Charge	Microelectrophoresis

Negative staining is simple to perform, and relatively simple equipment can be used. Artifacts should be excluded by careful validation of the process (11). Freeze-fracture is a more complex technique, and the liposomes are presumably better fixed in their original shape than with negative staining. Recently, new cryofixation methods for liposomes were described (184).

To obtain reliable information on the particle size distributions, the freeze-fracture data should be corrected for both nonmidplane cleavage of the liposomes and the size-dependent chance for a liposome to be found in the fracture plane (185).

Scanning tunneling microscopy pictures of hydrated and dehydrated phospholipid bilayers have been published recently. The exact potency of this exciting technique for liposome characterization has to be assessed in the near future (186).

Dynamic light scattering has become a regularly used technique for liposome size measurement. Dynamic light scattering is usually used for "sized" liposomes with diameters between a few nm and a few μm. For routine measurements of sized liposome dispersions, commercially available equipment can be used. The results of the experiments should be interpreted with caution. Only for narrow particle size distributions are precise and accurate results obtained. Because of the overemphasis of the contribution of large particles in a distribution, the size analysis of heterogeneous dispersions can easily be affected by a few aggregates in the dispersion. However, as mentioned before, this technique is very useful when dealing with sized liposome dispersions and when, for instance, liposome stability is investigated or for "in process control" during liposome production (11,187).

For heterogeneous liposome dispersions in the supra-μm range coulter counter techniques can be used (51). Other techniques that may have advantages for analyzing heterogeneous dispersions in the nm range are based on size exclusion chromatography (33,188). The analytical ultracentrifuge has rarely been used for liposome size analysis since the late 1970s (189).

The estimation of the number of bilayers can be done with a number of techniques: electron microscopy, ^{31}P-NMR, small angle x-ray scattering (SAXS). The "bare grid" electron microscopy method (184) allows an easy analysis of the number of lamellae of individual liposomes; validated negative staining techniques are also claimed to provide reliable information on the number of lamellae. With the ^{31}P-NMR method the fraction of the phospholipid molecules at the exposed outer leaflet of the bilayer is determined by broadening the NMR signal of the outer leaflet molecules beyond detection by interaction with (nonbilayer permeable) metal ions, such as Mn^{2+} (103,109). The fraction of PE exposed to the outer surface of the liposome can also be estimated by reacting the exposed PE with trinitrobenzene sulphonic acid (TNBS) that has been added to the outer water phase; a colored product is formed. The limitations of this technique were discussed by Lichtenberg and Barenholz (33) and New (11). SAXS data also allow us to estimate the average number of bilayers in a liposome dispersion (109). Talsma and coworkers (190) combined ^{31}P-NMR and SAXS analyses of the same liposome dispersion and suggested that this liposome dispersion contained multivesicular structures. Charge inducing agents such as PG, PA (phosphatidic acid), or PS are regularly added to bilayers, for example, to improve their physical stability against aggregation and/or fusion. Moreover, charge has been shown to be a parameter that plays a role in determining the disposition of liposomes after administration (101). Sometimes a charge is introduced nondeliberately. This happens when PE (no net charge at neutral pH) molecules are derivatized to act as coupling and anchor units for homing devices or antigens (41). The charge density on liposomes can be estimated from mobility measurements of liposomes in an electrical field (microelectrophoresis). The mobility data can be converted into ζ-potentials by using the Helmholtz-Smoluchowski equation or the Henri equation. From the ζ-potentials the charge density at the hydrodynamic plane of shear can be calculated with the Gouy-Chapman equation (191,192). Methods that determine the mobility of individual liposomes in an electrical field visually (e.g., cylindrical cell microelectrophoresis apparatus, Mark II, Rank Brothers) are now replaced by methods where liposome mobility in an electrical field in a capillary is followed by dynamic light scattering (e.g., 165).

The bilayer "fluidity" depends on the bilayer composition, the temperature, and the aqueous environment. In particular, in the fluid-crystalline state molecular motion occurs in the bilayer. This motion can have different characteristics (e.g., flexing of the acyl chains, wobbling of the molecule itself, or lateral motion of the molecule). Information on molecular motion in a bilayer can be collected by fluorescence polarization techniques. Probes such as diphenylhexatriene (DPH), TMA (trimethylamino-DPH), or transparinaric acid (TPA) are used (193,194,195). Manipulation of the physicochemical nature of the probe offers the opportunity to gain information on particular parts of the bilayer structure. For instance, TMA-DPH is located at the lipid-water interface and provides therefore information on the motion of molecules in that particular region. The proper performance and interpretation of fluorescence polarization experiments requires experienced investigators, as artifacts and overinterpretation of results readily occur (195).

I. Sterile and Pyrogen-Free State

Liposomes to be administered via the parenteral route, on damaged skin, or in the eyes must be sterile. The following routes to achieve sterile products have been considered:

Heat sterilization by autoclaving
Sterilization by γ-irradiation
Filtration sterilization
Aseptic production procedures

Heat sterilization by autoclaving can have a detrimental effect on the product. Loss of liposome-associated agent (retention loss) and chemical degradation of liposome components and/or associated agent can occur. Relatively little work has been published considering both aspects in detail (196, 197,198). Sterilization through autoclaving (121°C, 15 min) is possible if certain conditions are met as in the case of α-tocopherol-containing liposomes: (1) the associated agent should be lipophilic and be associated with the bilayer; (2) this agent should be stable during autoclaving; and (3) proper pH conditions should be chosen.

If autoclaving is not possible, sterilization by γ-irradiation as indicated in the USP XXVII (absorbed dose 2.5 Mrad = 25 kGy) might be considered. However, this treatment causes unacceptable chemical breakdown of bilayer components in aqueous liposome dispersions in the absence of radical scavengers; addition of radical scavengers prevents the damage to some extent. However, it is not clear yet whether this decrease in damage for the liposomes is not paralleled by a reduction in sterilizing power efficiency (167, 169,199,200). For (freeze) dried products no data on damage by γ-irradiation are available at the present time.

Sterilization by filtration through 0.2 μm pores before aseptic filling into the final vial is an accepted technique in the pharmaceutical industry for sterilization of heat labile products such as many biotechnological products. A detailed description of a filter-sterilization procedure is described by Amselem and coworkers (111). As the liposomes should be able to pass the 0.2 μm pores of the filter, liposomes with diameters over about 0.2 μm cannot be sterilized by filtration.

Large liposomes that cannot be sterilized by heat, therefore, should be manufactured aseptically. Aseptic production of parenterals is a standard operation in the pharmaceutical industry. However, the more complex the manufacturing process—and liposome production is relatively complex compared to aqueous drug solutions—the more difficult and expensive it is to validate the system (e.g., through challenge tests) and to meet the required specifications of the finished product (201).

Assessment of the absence of pyrogens in liposome dispersions is difficult. Gram-negative endotoxins (lipopolysaccharides) have a molecular weight between 10 and 20 kDa. They are amphipathic and form larger aggregates; they can also be taken up into bilayers. It is not clear whether the LAL (limulus amebocyte lysate) test can be used for pyrogen detection. If chosen for quality control purposes, it should be carefully validated (202). To avoid problems with pyrogens in the final product, the use of depyrogenated [cf. (111)] lipids and equipment is required. Depyrogenation of fluids (including organic solvents holding the lipids) is possible by ultrafiltration through filters with cutoffs of 10 kDa. The manufacturing procedure should be tested for its depyrogenation ability by endotoxin challenge tests.

J. Scaling Up of Liposome Preparation Procedures

Liposomes are widely used in cosmetics. A liposome-based formulation containing the drug econazole, to be used on the skin for the treatment of fungal skin infections, is marketed in several countries. This means that large-scale production methodology of nonsterile products has been developed successfully. Much of the knowhow is proprietary information and has not been released to the public domain. However, some publications have appeared over the years discussing different options for large-scale liposome production. The econazole liposome production process (based on the ethanol injection technique, see above) has been discussed by Kriftner (203). A large-scale production scheme for dermatological liposome dispersions with the detergent removal technique has been discussed as well (204).

Because of the higher quality standards that in general apply to parenteral products in comparison to dermatological products (the former have to be sterile and pyrogen free), the technology involved in liposome production for parenteral use is more complex. Martin published a chapter on pharmaceutical manufacturing of liposomes and addressed questions related to sterility and pyrogenicity (201). Some information is available on manufacturing procedures of specific types of liposomes designed to be administered intravenously. Schwenderer described the preparation of liposomes containing lipophilic cytostatics with the detergent removal technique (205). Doxorubicin liposome manufacturing is discussed by Barenholz and coworkers (111, 206). Van Hoogevest and Fankhauser (51) reported on MTP-PE liposome manufacturing. Information on parts of the production process, such as filter and seal selection, is found in a wide variety of specialized articles [e.g., (207)].

II. PART II. THERAPEUTIC APPLICATIONS

In the second part of this chapter, therapeutic applications of liposomes will be reviewed, with an emphasis on

1. New parenteral delivery systems including sterically stabilized long-circulating liposomes, "functionalized" liposomes for immunotherapy and vaccination, and cationic liposome-DNA complexes
2. Topical delivery of liposomes including ocular, dermal, pulmonary, and oral delivery systems

Within each of the selected areas the focus will be on recent developments, providing a historic perspective only where appropriate.

A. Parenteral Route

1. Introduction

The great majority of early and recent published work on liposomes as drug delivery systems employed the parenteral route of application. In addition to intravenous (i.v.) administration, liposomes have also been administered by the intraperitoneal (i.p.), intramuscular (i.m.), and subcutaneous (s.c.) route. While i.v. injected liposomes are generally rapidly cleared from the circulation by cells of the mononuclear phagocyte system (MPS; reticuloendothelial system) in liver, lung, and spleen (except sterically stabilized liposomes; see Sec. A.2 below), liposomes embedded in the peritoneal cavity, muscle, or subcutaneous tissue may serve as a slow-release depot but also accumulate in draining lymph nodes and enter the circulation via the lymphatic route. While the literature on these alternative routes is less abundant, useful information on liposome disposition and fate of liposome-associated drugs following i.p., i.m., or s.c. administration can be found in a number of original articles (208–212) and reviews (47,213,214).

Independent of the route of administration, the in vivo fate of liposomes and the pharmacokinetics of encapsulated drug are mainly determined by three factors: (1) liposome size; (2) lipid composition, which determines surface hydrophilicity/hydrophobicity and net charge; and (3) lipid bilayer fluidity (101). It is generally accepted that liposomes cannot escape the vascular system except in organs with fenestrated vascular membranes, e.g., liver, spleen, and, to a minor part, bone marrow, or under pathologic conditions, e.g., in inflamed or tumorous tissue, when the vascular membrane has become "leaky." Recently, there has been discussion that particulates may also be taken up by endothelial cells lining the blood vessels.

The following topics were selected for review and discussion because, in our opinion, they comprise recent milestones in the progress of liposome technology and will continue to challenge liposome researchers over the next decades. These include

1. Sterically stabilized liposomes that escape removal by the MPS and thus may serve as systemic targetable carriers or "circulating reservoirs"
2. Liposomes as carriers for peptide and protein drugs
3. "Functionalized" liposomes (liposomes with surface-grafted or surface-coupled antibody, antibody fragment, lectin, glycoprotein, etc.) as powerful and safe adjuvants in immunotherapy and vaccination
4. Cationic liposome DNA-complexes for gene delivery

2. Sterically Stabilized Liposomes

It was known, mostly through the work of Davis and Illum and coworkers (215) with colloidal systems, that in addition to size, the surface characteristics, specifically charge and affinity, play a crucial role in the fate of systemically injected colloids. Hence, the development of sterically stabilized long-circulating liposomes (37–39,216–218) represented a milestone in the search for liposomes that can bypass the MPS. This stabilization can be achieved by incorporation of natural hydrophilic components such as gangliosides (G_{M1}) or phosphatidylinositol, essentially mimicking the outer surface of red blood cells, or of synthetic hydrophilic polymers, specifically polyethylene glycols (PEG) [for review see (219)]. (*This new type of liposome is sometimes referred to as a Stealth® liposome (220,221); alternative coinages include "cryoptosomes" and "ninjasomes"; which one of these will survive the test of time remains to be seen.*) The physicochemical basis underlying the observed prolonged circulation in plasma has been provided by Lasic and colleagues (222,223). A recent issue of the *Journal of Liposome Research* (Vol. 2, No. 3, 1992) is entirely dedicated to a forum on "Covalently Attached Polymers and Glycans to Alter the Biodistribution of Liposomes."

Prolonged circulation results in increased probability of liposomes extravasating in areas where the permeability of the endothelial barrier is increased, specifically in tumor tissue. Penetration of cancer tissue in mice by PEG-modified liposomes has been demonstrated with rhodamine B isothiocyanate-dextran fluorescence-labeled liposomes (224) and liposome-incorporated colloidal gold label (225). Reduction of cancer mass and increased survival were reported for PEG-liposome-incorporated doxorubicin in the mouse mammary carcinoma model (226) and for epirubicin in mouse colon 26 tumor in vivo (227).

In another example of targeted delivery due to prolonged circulation of hydrated phosphatidylinositol-containing liposomes, Bakker-Woudenberg et al. (228) found tenfold higher concentrations of liposomes in lung lobes infected with *Klebsiella pneumoniae* compared to the uninfected lobe, while conventional liposomes did not generate a selective delivery to the infected area.

A recent study (221) compared survival of L1210/C2 leukemia-bearing mice following three treatment regimens: (1) ara-C-carrying SM:PC:CH:GM_1 or HSPC:CH-PEG-DSPE liposomes; (2) conventional PC:CH liposomes (all 0.4 μm REV); or (3) a slow infusion of ara-C. A rather complex picture as to the benefits of long-circulating liposomes emerged. While the authors maintain that the "pharmacokinetics" of G_{M1}-modified and PEG-modified liposomes are slower than those of PC:CH liposomes, and that the advantages of long-circulating liposomes are found in their "dose-independent pharmacokinetics" and their decreased MPS uptake, the data indeed show otherwise. Interestingly, terminal in vivo elimination rates appear to be identical for all types of liposomes, with the expected difference that PC:CH liposomes encounter a rapid initial elimination due to MPS uptake. Furthermore, MPS uptake of PEG-containing liposomes appeared to be saturable and unexpectedly high, 69% and 61% for a 0.5 and a 4 μmol dose respectively (liver and spleen combined; lung data not provided).

Thirdly, the pharmacologic effect (survival) was clearly a function of the unencapsulated dose (which was not determined with the liposome preparations): liposomes with a shorter half-life provided better survival; slow infusion as well as conventional liposomes, although at higher doses, and a combination of long-circulating liposomes and unencapsulated drug, provided comparable improvement of survival.

Employment of long-circulating liposomes as a "circulating reservoir" for prolonged release of peptides has been proposed. To date the only experimental demonstration of this concept has generated positive yet puzzling results. Woodle et al. (229) examined long-circulating liposomes to deliver vasopressin in the Brattleboro rat. A formulation consisting of PG:SM:PC:

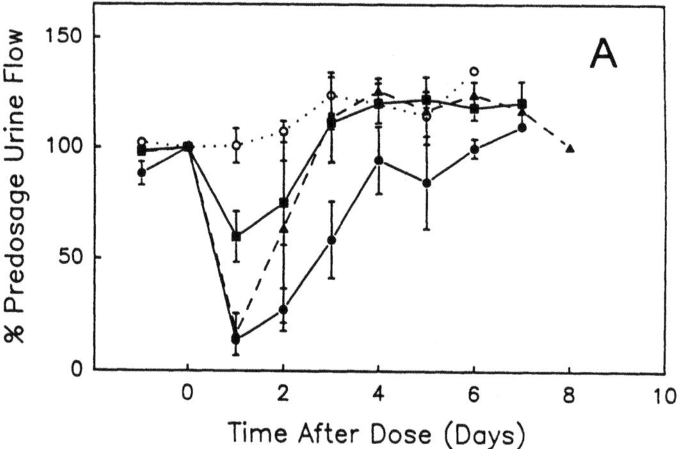

Figure 11 Diuresis as a percentage of the predosage rate of Brattleboro rats after surgery and i.v. dose administration of liposomal vasopressin (VP) with liposomes composed of PG:SM:PC:Chol (A), PEG-DSPE:SM:PC:Chol (B), and PEG-DSPE:SM:PC with increasing cholesterol content (8 μg VP dose) (C). Symbols for A and B: open circles = saline control; filled squares = 2 μg VP; filled triangles = 8 μg VP; filled circles = 24 μg VP (mean ± SD; n = 3). Symbols for C: open circles = saline control; filled circles = 0% Chol; filled triangles = 16% Chol; filled squares = 33% Chol (mean ± range; *n* = 2). (From Ref. 229, with permission.)

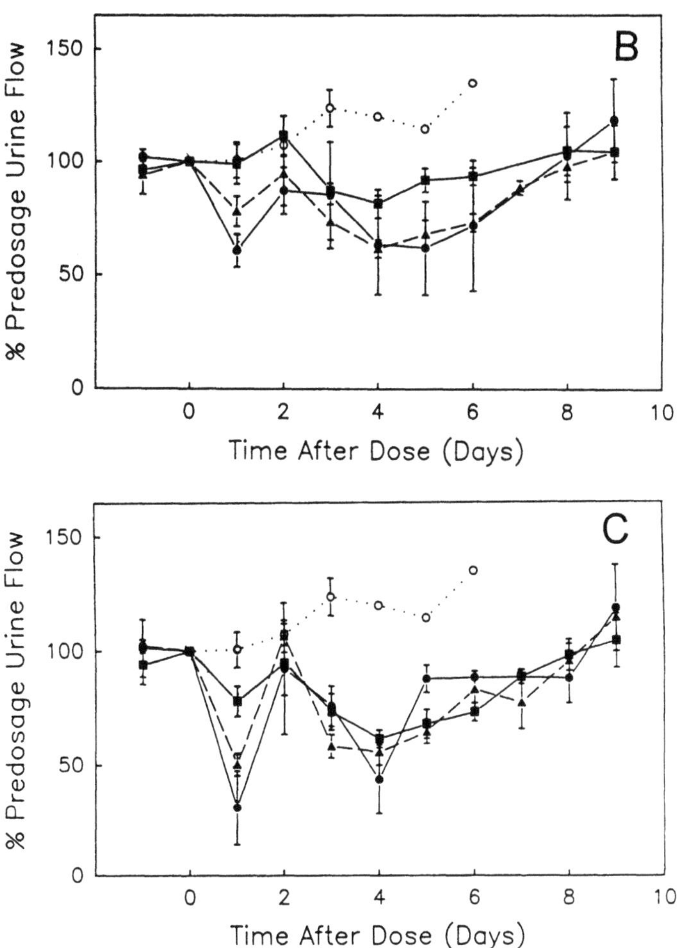

Figure 11 (Continued)

Chol (0.2:1:1:1) indeed generated a prolonged antidiuretic effect that was maintained for about 4 days at the highest dose employed and was attributed to prolonged circulation. However, with a formulation consisting of PEG-DSPE:SM:PC:Chol in the same molar ratio, an unexpected pattern was found: a combination of immediate, cholesterol-dependent, and delayed, cholesterol-independent effect emerged that could no longer be attributed to prolonged circulation. In fact, it was obvious that the cholesterol-containing liposomes did not release enough, if any, vasopressin during circulation, and the immediate response therefore was caused entirely by unencapsulated drug (Fig. 11).

In an exemplary analysis, the authors rule out several explanations including a biologic response to the PEG liposome formulation, receptor downregulation, or a sudden increase in the release rate, to conclude that a fraction of the liposomes must have distributed intact to "another anatomical compartment where a delayed release of vasopressin occurs." The authors speculate as to the nature of the compartment as a most likely non-RES compartment, although they seemingly are not entirely certain that RES deposition was not a factor in the observed delayed response.

This offers an interesting new view of these long-circulating liposomes; not so much prolonged circulation as perhaps accumulation in an as yet undefined, likely nonphagocytic deep body compartment promises some useful future applications.

A combination of long-circulating liposomes and immunotargeting has been employed by Maruyama et al. (230) with some unexpected synergistic results. PC:CH:GM_1 (10:5:1) liposomes were labeled with a monoclonal IgG antibody with a specific affinity to pulmonary endothelial cells (see Sec. 5 below). A larger fraction of these liposomes, compared to a control formulation consisting of PC:CH:PS in the same molar ratio, was bound to the lung endothelium, and the residence time in the lung was prolonged, leading the authors to speculate that conventional PS-containing liposomes were perhaps removed from the target site by pulmonary macrophages, while GM_1-containing liposomes may be capable of dissociation and reattachment to the target tissue during prolonged circulation.

Questions as to the toxicologic effects of such liposomes (once their in vivo fate is better understood) as well as their objective advantages over conventional liposomes directly injected into subcutaneous, intraperitoneal, or intramuscular reservoirs for prolonged activity remain to be answered. Also, the targeting potential of long-circulating liposomes when combined with a "homing device" (antibody, surface glycoprotein, etc.) remains to be explored further.

3. Protein and Peptide Delivery

Due to the labile nature and rapid systemic elimination of peptides and proteins, carrier systems that protect them from metabolic degradation and prolong their plasma half-lives are needed. However, the requirements of peptides and proteins compared to "small" drugs are radically different and immensely challenging:

1. The active molecule, due to lipophilic domains, may interact with lipophilic domains of the carrier and cause discontinuities (phase separation) in the matrix, which may severely reduce physical stability and shelf-life.
2. Electrostatic and hydrophobic interactions of active molecule and carrier may compromise the conformational integrity, hence the biological activity of the protein.

3. The active molecule is generally highly potent and often exhibits severe systemic toxicity at high plasma concentrations, resulting in a very small therapeutic index.
4. Prolonged receptor occupancy may trigger receptor down-regulation; some agents, e.g., human growth hormone, are principally active only when applied in a pulsed rather than a continuous fashion.

These peculiarities of peptides and proteins are extremely difficult to overcome with any delivery system, including liposomes. A critical review pointing out both opportunities and pitfalls of liposomal peptide and protein formulations has been published recently by Storm et al. (231).

It is our frank opinion that much of the early work with liposomes and macrophage-activating factor, various types of interferons, and insulin [for review see (231)] suffers from poor characterization of liposome-protein interaction and lacks quantitative evaluation of lipid and/or protein recovery following the various preparative steps. In addition, use of biologic in vitro assays (e.g., antiviral plaque assay) and in vivo response (e.g., glucose depression) often lacks the sensitivity and reproducibility desirable at the early stages of development to allow reasonable and objective conclusions as to the potential benefit of the liposomal dosage form.

It would be desirable to generate more fundamental information on the molecular arrangement of proteins and peptides and phospholipid bilayers that may provide a more rational basis for dosage form design in the future.

Examples for this type of useful basic investigation of liposome-protein interaction exist. The nature and stoichiometry of the physical interaction of acidic phospholipids with calcitonin has been studied (232–234). The molecular arrangement of recombinant α-interferon in phospholipid monolayers has also been elucidated (235). Bergers et al. (131) characterized the interaction (kinetics of binding and protein unfolding) of myoglobin and negatively charged liposomes (PC:PG 1:1) as a function of pH. Bergers et al. (130) studied the interaction of interleukin-2 (IL-2) with liposomes under different conditions of pH, ionic strength, and lipid composition, and found nearly quantitative (81%) encapsulation efficiency with negatively charged liposomes (PC:PG 9:1) and IL-2 dissolved in acetate/glycerol buffer at pH 5, although release of IL-2 from liposomes was rapid when the preparation was diluted with buffer.

4. "Functionalized" Liposomes

"Functionalized" liposomes are defined by us as liposomes whose surface has been modified so that they exhibit a specific ability to interact actively with the biologic environment, i.e., liposomes containing or exhibiting tissue-selective enzymes, (monoclonal) antibodies and antibody fragments, surface glycoproteins, carbohydrates, or haptens, all of which are being explored either to achieve tissue/cell-selective targeting for drug delivery or nonspecific immunostimulation or to elicit a specific immune response. (*This definition of "functionalization" excludes long-circulating liposomes, which are, of course, also "functionalized" in their own way but remain intentionally passive in vivo.*)

Such functionalized liposomes may induce

1. Nonspecific macrophage immunostimulation
2. Cellular and humoral antibody response
3. Cytotoxic T-lymphocyte response
4. Mucosal immunity (IgA response)

Liposomes as adjuvants for vaccination have first been proposed by Allison and Gregoriadis in 1974 (236). Early work has been reviewed by van Rooijen and van Nieuwmegen (237,238). A synopsis of activities in the area of immunomodulation and vaccine development up to 1986 can be found in a chapter entitled "Liposomes in Immunomodulation/Vaccines" in the monograph *Liposomes as Drug Carriers* (9). A more recent collection of concepts and experimental work was published as the 41st Forum in Immunology entitled "Liposomes and Macrophage Functions" in the journal *Research in Immunology* (Vol. 143, 1992). Reviews have been published by Gregoriadis (239), Alving (240), Storm et al. (231), and Buiting et al. (241).

a. Nonspecific Macrophage Immunostimulation. As an alternative to chemotherapy, immunomodulation, i.e., activation of monocytes to a tumoricidal state using liposome-incorporated immunostimulants, has been suggested, with the rationale that liposomes are avidly taken up by macrophages. Hence they should efficiently deliver the immunostimulating agents to their cellular targets: a unique application of the concept of "passive" targeting.

Potent immunostimulants such as interleukin-2 (130), muramyl dipeptide (MDP) (242,243), and lipophilic derivatives thereof including muramyl dipeptide L-alanyl cholesterol (MDP-CHOL) (244), muramyl dipeptide glycerol dipalmitate (MDP-GDP) (245,246), disaccharide tripeptide glycerol dipalmitate (DTP-GDP; Immther) (247), and muramyl tripeptide phosphatidylethanolamine (MTP-PE) (130,248–250) have all been used successfully to elicit strong macrophage stimulation.

b. Cellular and Humoral Antibody Response. It is generally accepted that macrophage uptake of antigen-carrying liposomes and digestion of the antigen(s) to peptides that bind to class II major histocompatibility (MHC) molecules is the mechanism responsible for enhanced T-cell-dependent humoral immune response [for an in-depth discussion, see (251)]. Following lysosomal degradation, the processed antigens are presented in conjunction with class II MHC molecules, which trigger $CD4^{+}$ T cell and B cell response.

Recent experimental evidence suggests that liposomes may indeed play a double role as safe carriers and potent adjuvants for (subunit) vaccines against viral, bacterial, and parasitic infections, and various forms of cancer.

An arsenal of chemical methods is available to couple functional moieties to the surface of liposomes, or to modify them with hydrophobic "anchors" for bilayer insertion (41,45,46,252–258).

Viral surface glycoproteins carry their natural lipophilic "anchor" in form of protein subunits that can be incorporated into liposome membranes by detergent dialysis techniques. This has been employed to insert surface glycoproteins of Sendai (259,260), influenza (259), rubella (261), Friend leukemia (262), herpes simplex (263,264), human immunodeficiency (79) (Fig. 12), and reovirus (265) into liposomes.

Surface antigens from parasites, e.g., *Leishmania* surface antigen gp63 (266) or tumor cell surface antigens (78) can similarly be reconstituted in liposome membranes by detergent dialysis.

For enhanced adjuvant effect, antigens may also be incorporated within the liposomal inner compartment(s). This strategy has been employed for synthetic malaria sporozoite antigen (267), tetanus toxoid (268,269), influenza virus glycoproteins (270,271), herpes simplex (264), and polio virus peptides (272). Enhanced immune response to a cloned synthetic malaria sporozoite antigen (R32tet$_{32}$) was found in rabbits and monkeys when the protein was encapsulated in liposomes. Further enhancement was achieved by incorporating lipid A as immunostimulant in the lipid membrane, and/or alum adsorption of the liposomal antigen preparation (267). It has also been reported that the liposome composition modulates the immune response, although the results reported are partially contradictory. Liposomes consisting of DOPC, egg PC, DMPC, and DPPC combined with equimolar amounts of cholesterol were found to elicit a strong immune response to incorporated tetanus toxoid in Balb/c mice, while DSPC failed to show the same effect (269). Later, a stronger response of a similar preparation consisting of DSPC with entrapped polio virus peptide, when compared to a DMPC formulation, was reported (272).

Alternatively, antigens may be nonspecifically adsorbed to the liposome surface, which has been employed for hepatitis A virus (273), rabies glycoprotein (274–276), and human immunodeficiency virus (HIV) gp160 (277). A comparison of antigen-specific IL-2 production in Balb/c mice following injection of either complete inactivated rabies virus or purified rabies glycoproteins in the form of immunosomes showed equal activity, which was not found for aggregated glycoproteins alone (275,276). Immunosomes with adsorbed HIV gp160 exhibited superior efficacy with respect to humoral and cell-mediated immune response in mice compared to the immune response generated with gp160 injected alone (277). The authors point out

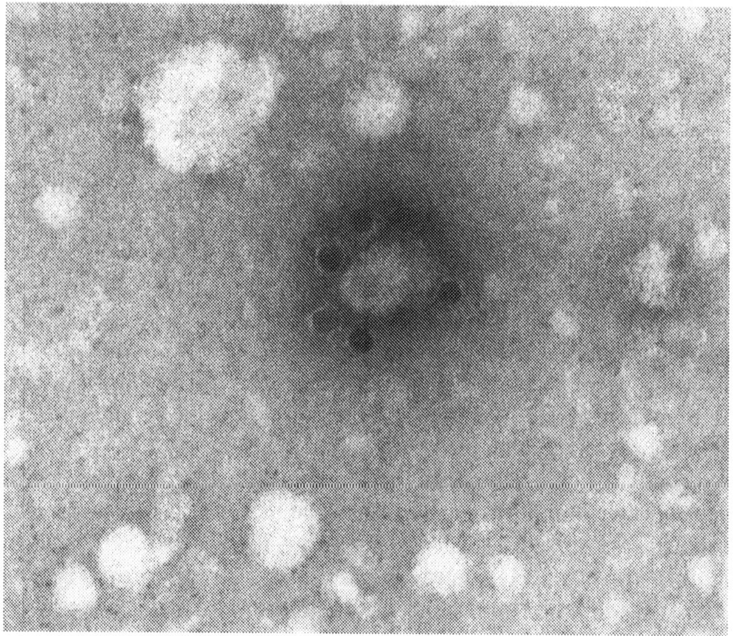

Figure 12 Liposome exhibiting the human immunodeficiency virus (HIV) surface glycoprotein gp160 inserted in its outer surface (artificial viral envelope) following sandwich immunolabeling with anti-gp160/41 antibody and goat anti-mouse IgG coupled to 15 nm colloidal gold. Artificial viral envelopes were adsorbed to Formvar coated nickel grids and incubated for 1 h on a 1:250 dilution of the monoclonal anti-HIV gp160/41. Grids were incubated on a 1:20 dilution of goat anti-mouse IgG coupled to 15 nm colloidal gold for 1 h, negatively stained and observed with a Jeol 100 CX electron microscope operated at 60 kV. (Sample preparation by R. Chander, Drug Delivery Laboratory, University of Florida Progress Center, Alachua, Fla.; electron microscopic preparation by G. Erdos, Interdisciplinary Center for Biotechnology Research, University of Florida, Gainesville, Fla.; with permission.)

the apparent importance of presenting the immunogen in a membrane-associated, three-dimensional conformation in order to elicit a strong immune response.

As with the development of antiviral subunit vaccines, reconstituted vesicles carrying tumor antigens are also under investigation in cancer immunotherapy to elicit specific antitumor immune responses. In several studies, reconstituted membranes of cancer cells containing tumor-associated antigens were shown to provide some degree of protection against tumors [for a review see (278)]. Critical design factors to obtain optimal antitumor effects were studied by Bergers et al. (78,115,279). The enhancement effect of immunostimulants such as MTP-PE and IL-2 was also demonstrated: a 100-fold increase in rejection ability of the syngeneic lymphosarcoma SL2 was demonstrated in DBA/2 mice when MTP-PE and IL-2 were coadministered in the vaccination protocol (279).

c. Cytotoxic T-Lymphocyte Response. In addition to eliciting a humoral immune response, liposomes can be engineered so that the antigenic peptides are complexed with the class I MHC glycoproteins, resulting in the induction of $CD8^+$ cytotoxic T lymphocytes (CTL). A CTL response is triggered when instead of conventional (pH-insensitive) liposomes pH-dependent fusogenic liposomes are employed as antigen carriers. Following uptake into the phagolysosomal vacuole, pH-dependent liposomes fuse with the lysosomal membrane as a consequence of the pH drop, and peptides are released into the cytosol. This has been demonstrated in vitro (132,133) and in vivo (280,281) (Fig. 13) with ovalbumin-containing pH-sensitive liposomes in mouse EL4 thymoma cells that were sensitized for ovalbumin-specific CTL killing [for an in-depth discussion, see (282)].

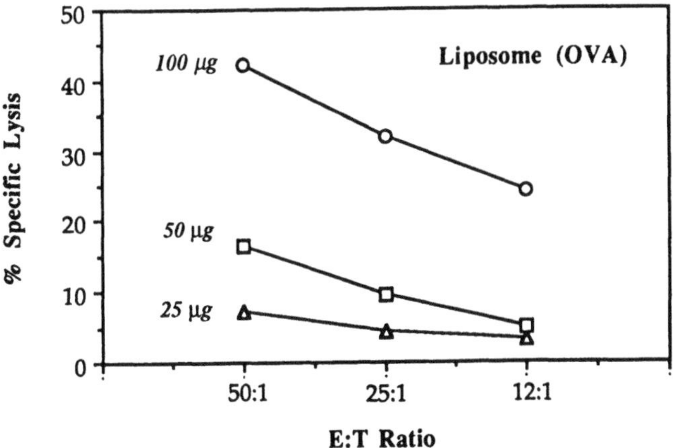

Figure 13 Dose-dependent cytotoxic T-lymphocyte (CTL) induction by OVA entrapped in liposomes composed of dioleylphosphatidylethanolamine (DOPE) and dioleylsuccinylglycerol (DOSG) (1:1 molar ratio). C57BL/6 mice were immunized with liposome-encapsulated OVA at the indicated dose 7 days before the effector cells were prepared. Specific lysis of EG7-OVA cells was plotted against effector: target cell ratio as the average of triplicate samples. Spontaneous release was less than 12%. (From Ref. 281, with permission.)

d. Mucosal Immunity (IgA Response). Oral administration of antigen-carrying liposomes may be a unique means to stimulate IgA secretion, specifically in the oral mucosa where it would aid in the generation of a salivary IgA response to sero-type specific carbohydrates of *Streptococcus mutans* and thus reduce colonization of the oral cavity by virulent bacteria and protect against caries.

It is known that mucosal secretion of IgA is induced in the so-called gut-associated lymphoid tissue (GALT) or Peyer's patches in the small intestine. Specifically, the so-called M cells in the Peyer's patches phagocytize antigens from the gut lumen, providing a target for antigen-carrying liposomes in the small intestine (283).

Since a number of viruses are actively removed by M cells, Rubas et al. (265) have proposed a targeted oral liposome exhibiting the reovirus M cell attachment protein and have demonstrated in vitro that uptake of such a construct by rat Peyer's patches was 10- to 20-fold increased compared to plain liposomes. Michalek and collaborators (283–285) have designed liposomes carrying soluble antigens of *S. mutans* for oral delivery and have demonstrated a significant increase in salivary IgA levels, reduction in *S. mutans* colonization and, concomitantly, improved caries protection. This effect could be augmented when MDP was present as coadjuvant. In a rat model, an oral liposome-based antiidiotype vaccine against *S. mutans* was shown to induce protective immunity to *S. mutans* at the oral mucosa (285).

5. Cationic Liposome DNA Complexes

According to a recent *Science* report, as of October 1992, there were 18 clinical gene therapy trials in progress (286). While practically all of the early (*"early" is in this case equivalent to "two years old"*) work was performed with modified retroviral (287) and adenoviral (288) carrier systems, synthetic carriers including cationic polyamine-DNA (289) and cationic liposome-DNA complexes are gaining rapid acceptance.

Liposomes have been explored as delivery systems for RNA and DNA as early as 1977 (290). The encapsulation of plasmid DNA in liposomes (291, 292), introduction of poliovirus RNA and SV40 DNA into cells via liposomes (293,294), and expression of liposome-encapsulated β-lactamase gene in mycoplasma (295,296), as well as expression of the thymidine kinase gene in mouse cells (297) following delivery in liposomes, were reported between 1979 and 1982. Nicolau et al. (298) presented preliminary evidence of in vivo gene transfer via liposomes and expression of the insulin gene for the first time in 1982.

Gene delivery by means of liposome-DNA complexes has been reviewed by Mannino and Gould-Fogerite (299), Nicolau and Cudd (300), Hug and Sleight (301), and Felgner and Rhodes (302). An issue of the *Journal of Lipo-*

some Research (Vol. 3, No. 1, 1993) is entirely dedicated to the topic of cationic liposomes.

With most early methods, the frequency of transfection was low and comparable to established methods, e.g., calcium phosphate precipitation (which may have been the active principle in many reported cases, due to the precipitation of negatively charged liposomes in the presence of calcium). Delivery was principally via endocytosis or phagocytosis (303), which greatly reduced the fraction of DNA delivered intact to the cell cytosol. A prerequisite for fusion was the presence of both calcium as a "bridge-forming" divalent ion and a "fusogen," e.g., glycerol or polyethylene glycol, to induce close contact of liposome and cell membranes. Itani et al. (304) reported highly efficient delivery of the thymidine kinase gene encapsulated in phosphatidylserine liposomes that were fused with the receptor cells via Ca^{2+}-induced fusion. Gould-Fogerite et al. (305) designed proteoliposomes for gene-transfer that contained Sendai virus or influenza glycoproteins as fusogenic moieties ("chimerasomes") and reported transfection efficiencies of plasmids six orders of magnitude higher than those achieved with the conventional calcium phosphate precipitation method. In vivo, they reported stable gene transfer to 50% of mice injected subcutaneously with "chimerasomes" containing a polyoma virus-based plasmid.

In 1987, Felgner et al. (306) reported a new technique, "lipofection," using liposomes that consisted of the positively charged synthetic lipid *N*-[1-(2,3-dioleyloxy)propyl]-*N,N,N*-trimethylammonium chloride (DOTMA) in a 1:1 weight mixture with dioleylphosphatidylcholine (DOPE). Several other versions of positively charged liposomes and lipid mixtures respectively have been introduced since, including a methyl sulfate salt of DOTMA (DOTAP) (307), and positively charged cholesterol derivatives such as cholesteryl-(4′-trimethylammonio)-butanoate (ChoTB) (308), cholesteryl hemisuccinate choline ester (ChoSC) (309), and 3β-[*N*-(*N*′,*N*′-dimethylamminoethane)-carbamoyl]-cholesterol (DC-chol) (309), which are also embedded in a matrix of DOPE, similar to DOTMA and DOTAP. Furthermore, positively charged lipid- or cholesterol-amino acid complexes including lysinyl phosphatidylethanolamine and cholesteryl-β-alanine have been successfully employed (310). Alternatively, Trubetskoy et al. (311) combined *N*-terminal-modified poly(L-lysine) antibody conjugates with DNA and cationic liposomes. Another alternative protocol using a mixture of the cyclic cationic amphipathic peptide gramicidine S and DOPE has been introduced by Legendre and Szoka (312).

This class of lipid-based transfection agents has been employed very successfully for a variety of cell lines and is currently the method of choice for in vitro cell transfection, although the exact mechanism of action is unknown. While it is clear that the necessary close proximity of liposomal and cell membranes is brought about by the opposite electrostatic force of the two interacting membranes, the process may be best described by a "chaotic" electrostatic interaction of positively charged liposomal lipids with negatively charged cell membrane phospholipids. This results in an indiscriminate exchange of membrane components, which apparently transfers significant quantities of liposome-associated DNA into the cell cytosol as demonstrated experimentally by Stamatatos et al. (307) and Düzgünes et al. (313).

Luciferase mRNA was transfected into NIH 3T3 cells using the DOTMA-DOPE liposome (Lipofectin®) complex (314). Brigham et al. (315) were able to express the chloramphenicol acetyl transferase (CAT) gene inserted into a Rous sarcoma virus plasmid (pRSVCAT) in bovine pulmonary artery endothelial cells using "lipofection" (Fig. 14).

Debs et al. (310) studied the time course and duration of expression of CAT in rat lung alveolar type II cells, alveolar macrophages, and several human lung carcinoma cell lines following transfection with DOTMA-, lysinyl phosphatidylethanolamine- and cholesterol-β-alanine-DNA complexes. CAT activity was found maximal in type II cells 5 to 11 days after transfection

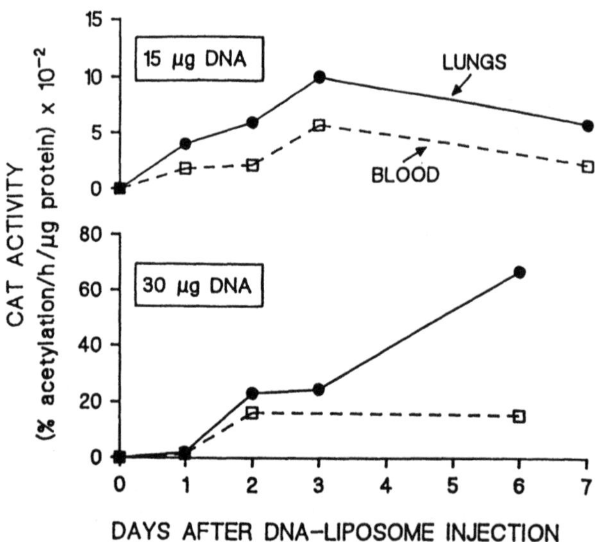

Figure 14 In vivo time course of expression of chloramphenicol transferase (CAT) activity in blood and lung at two different doses of plasmid-liposome complex. Following tail vein injection of the pSV2CAT/Lipofectin (1:5 w/w) complex, mice were killed on the days indicated, organs were removed, extracted, and CAT activity quantified with an automatic radioactivity imaging system (Bioscan). (From Ref. 315, with permission.)

but persisted up to 35 days. Plasmid was found largely in extrachromosomal form and had not inserted into host-cell DNA. A 10- to 20-fold enhancement of specific transfection of mouse lung endothelial cells in culture was reported by Trubetskoy et al. (311). Comparing Lipofection® and their gramicidin S-DOPE-DNA complex, Legendre and Szoka (312) reported a 2- to 20-fold enhancement of transfection of the luciferase gene into nonphagocytic cell lines including CV-1, HeLa, and HepG2, using their own construct.

With respect to in vivo use, positively charged liposomes have met with skepticism as they interact spontaneously and indiscriminately with plasma components, red blood cells, and cell membranes of the vasculature upon intravenous administration. Plasma turbidity, hemolysis, and clot formation have been observed in vitro (218). While a systematic in vivo toxicity study has not been reported, Brigham et al. (316) have successfully used Lipofectin® in vivo to deliver a reporter gene to the lungs of mice without observing any toxic side effects and have recently confirmed the acute safety of pulmonary delivery of such DNA-liposome complexes in the rabbit (317). Similarly, Stribling et al. (318) aerosolized mice with CAT expression plasmid and found high levels of CAT activity in the lung, both the airway epithelium and the alveolar lining, without any detectable damage. Other in vivo applications of this technology include transfection of arterial endothelial cells in the heart and systemic vessels (319,320). A recent report found that intravenous injection of DNA-cationic liposome complexes of a biodegradable DC-chol derivative (309) was nontoxic and resulted in expression of the transgene in the lungs (321).

Examples for successful targeting of gene-carrying liposomes in vivo by selecting appropriate routes of administration include delivery via intratracheal instillation to the lungs of rats (322) and sheep (323) and inhalation via aerosol in rabbits (323) and mice (318). In these cases, selective delivery of DNA-liposome complexes to the lung epithelium and a corresponding expression of marker genes has been shown. Recently, Hyde et al. (324) have for the first time demonstrated correction of the ion transport defect in cystic fibrosis transgenic mice by delivering a CFTR expression plasmid complexed with Lipofectin® to the lungs of these mice via instillation.

Catheterization has been employed to target DNA-carrying liposomes to selected segments of the vasculature as shown by Lim et al. (319) and Nabel et al. (320).

Earlier attempts to impart selectivity on gene-carrying liposomes by surface modification included incorporation of glycosyl moieties (e.g., lactosylceramide) to target hepatocytes and endothelial cells in the liver (325) and coupling of transferrin to the liposome surface to target the bone marrow (326). A more recent approach employing immunoliposomes (see Sec. A.4 above), tagged with monoclonal antibodies against the lung endothelium, has resulted in their selective accumulation in the lung (86,230).

The major obstacle of liposome surface modification with antibodies is resulting endocytosis, which prevents efficient delivery of intact DNA. However, Wang and Huang (327) have designed a metastable lipid mixture consisting of DOPE, cholesterol, and oleic acid that, upon exposure to low pH, fuses with the lysosomal membrane and releases its content into the cytosol. Using this approach, they demonstrated accumulation of antibody-coated pH-sensitive immunoliposomes in an ascites tumor which expressed the corresponding surface antigen and concomitantly less uptake in liver and spleen. Similarly, liposomes have been engineered such that their bilayer is destabilized in a pH-dependent fashion via a pH-sensitive amphipathic peptide (GALA) (328) or pH-sensitive liposomes consisting of DOPE and cholesterylhemisuccinate (134). However, Legendre and Szoka (329) compared the efficiency of transfection by pH-sensitive liposomes versus cationic liposome-DNA complexes and found that the latter were significantly more efficient in vitro, mostly due to a rapid and up to eightfold higher uptake of DNA (Table 8).

Table 8 Transfection of Different Mammalian Cell Lines With Cationic Liposomes or pH-Sensitive Liposomes

	Liposomes		
Cell line[a]	Cationic[c]	pH-sensitive[d]	Factor[b]
CV-1	21.5 ± 13.2	1.2 ± 1.2	17
CV-1 + 10% serum	4.0 ± 2.0	0.8 ± 0.1	5
HepG2	827 ± 333	7.5 ± 7.0	110
HepG2 + 10% serum	925 ± 547	6.3 ± 4.0	147
HeLa	196 ± 18	6.0 ± 2.9	33
p388D1	5.6 ± 1.0	1.9 ± 0.7	3
KD83	0.13 ± 0.03	0	–

[a]Cells are treated with 4 μg of plasmid DNA per 60 mm culture dish. DNA is complexed with 10 μg Lipofectin reagent or encapsulated in pH-sensitive liposomes. Luciferase activity at 48 h is expressed as 10^5 light units per mg of cell protein. Light units background was subtracted from each value. Results are the mean ± SD of three to five experiments.

[b]The factor is computed as the ratio of luciferase activity per mg of cell protein induced by cationic liposomes to luciferase activity per mg of cell protein induced by pH-sensitive liposomes.

[c]Cationic liposomes consisted of dioleyloxytrimethylammonium (DOTMA) and dioleylphosphatidylethanolamine (1:1 w/w) (Lipofectin reagent).

[d]pH-sensitive liposomes were composed of dioleylphosphatidylcholine (DOPC) and cholesterol hemisuccinate morpholino salt (CHEMS) (2:1 molar ratio).

Source: Ref. (329), with permission.

The use of liposomes to deliver other genetic material such as antisense constructs is only emerging at this point (330,331). Inhibition of viral replication (332,333) and introduction of liposome-encapsulated oligodeoxynucleotides into tumor cells (334,335) in vitro have been demonstrated.

Open questions remain concerning the safety of cationic liposome-DNA complexes, although safety data have been generated on a limited scale. Also, in vivo kinetics and the fate of these complexes are essentially unknown. Systems that prove safe and efficacious will then have to be scaled up so that sufficient supplies can be generated for clinical trials to commence.

B. Topical Route

1. Introduction

In addition to the parenteral route of administration, topical application of liposomes including the eye, lung, and skin as target sites, and oral application, have been proposed and investigated with increasing intensity over the last decade. The performance as well as the fate of parenterally administered liposomes is relatively uniform and predictable. However, topical target sites are more unique with respect to the (patho)physiologic conditions encountered, the nature of uptake, and the metabolic fate of drug and carrier material. In addition, suitable analytical techniques must be established to monitor the kinetics and metabolic fate of liposomes in specific tissues, for instance in the cornea, the lumen of the lung, or the various layers of skin.

In this part, the status of ocular, (trans)dermal, pulmonary, and oral drug delivery via liposomes will be examined.

2. Ocular Application

The use of liposomes as an ocular drug delivery system has been proposed to achieve

1. Sustained release of drugs applied to the cornea and corneal lesions, or administered by subconjunctival injection
2. Prolonged retention or targeting of drug to selected intraocular cell populations in lens and vitreous

Lee et al. (336) and, more recently, Niesman (337) have reviewed the literature on this subject.

The eye is protected by three highly efficient mechanisms: (a) an epithelial layer that is a formidable barrier to penetration; (b) tear flow; and (c) the blinking reflex. All three mechanisms are responsible for poor drug penetration into the deeper layers of the cornea and the aqueous humor, and for the rapid wash-out of drugs from the corneal surface.

Enhanced efficacy of liposome-encapsulated idoxuridine in herpes simplex infected corneal lesions in rabbits was first reported in 1981 (338). Early work until 1985 included the study of corneal accumulation and transcorneal flux of liposome-associated penicillin G (339), indoxole (339), triamcinolone acetonide (340), dihydrostreptomycin sulfate (341), epinephrine (342), pilocarpine (343), and gentamicin (by subconjunctival injection) (344), as well as the model compound inulin (342,345). These compounds were incorporated in a great variety of liposomes, both with respect to size and structure (MLV, REV, SUV), as well as lipid composition (positively/negatively charged, neutral).

Perhaps not unexpectedly, the major conclusion drawn from these studies by Lee in 1985 (336) was that ocular delivery of drugs could be either promoted or impeded by the use of liposomal carriers, depending on the physicochemical properties of the drug and the lipid mixture employed.

A variety of strategies has been explored to facilitate adhesion of liposomes on the cornea, thus improving corneal accumulation and penetration.

Ganglioside-containing liposomes and wheat germ agglutinin, a lectin that has a high binding affinity for both cornea and gangliosides, were tested for corneal adhesion (346). Corneal binding as well as accumulation and transcorneal "flux" of carbachol (*"flux" is used misleadingly here; what is measured is the fraction transported or, in a wider sense, "bioavailability"*) was enhanced 2.5- to 3-fold over 90 min exposure time. However, one must be aware that the differences in absolute terms ranged from 0.018% to 0.46% of the administered dose!

Norley et al. (347) employed immunoliposomes (Sec. A.4 above) targeted against herpes simplex virus glycoprotein D to treat corneal herpetic keratitis. The validity of the concept was demonstrated in vitro (347,348), with acyclovir showing a 1000-fold enhancement of inhibition of viral replication at the 0.001 μg/mL level in cultured cornea cells. However, the strategy failed to eliminate acute herpetic keratitis in vivo in the mouse model, which was supposedly due to weak binding, hence due to rapid mechanical removal from the cornea (348).

Fitzgerald et al. (349,350) employed gamma-scintigraphy to study corneal retention time of various types of liposomes (349) and nanoparticles (350). Stearylamine prolonged the corneal half-life time of liposomes by a pronounced yet clinically insignificant increase from 1.06 min to 3.77 min (349). However, stearylamine was also found to be a strong irritant, acting similar to a "penetration enhancer" by altering the permeability of the corneal epithelium (351). Increased tear flow due to irritation offset prolonged retention gained through electrostatic surface binding.

More recently, benzyldimethylstearyl ammonium chloride, a homolog of benzalkonium chloride, which is widely used in ophthalmic formulations as antimicrobial agent, as well as dimethyldioctadecyl ammonium bromide have been used to impart a positive surface charge in liposomes (352). These acted as bioadhesives on the cornea surface (353,354) and were tolerated well in the rabbit eye irritancy test (353). No drug studies have been reported with these compounds to date.

Davies et al. (355) proposed the use of mucoadhesive polymers, Carbopol 934P and Carbopol 1342, to retain liposomes at the cornea. While precorneal retention times were indeed significantly enhanced under appropriate conditions (i.e., at pH 5.0), liposomes, even in the presence of the mucoadhesive, had migrated toward the conjunctival sac with very little activity remaining at the corneal surface. Furthermore, liposome-encapsulated tropicamide, with or without mucoadhesive present, failed to improve the pupil dilatory effect found with tropicamide solution in the rabbit eye.

3. Dermal/Transdermal Application

Liposomes have also been considered for the dermal or transdermal delivery of drugs [for reviews see (14,356–359)]. Dermal liposome products have since 1987 been exploited by the cosmetics industry, with more than 100 liposome and nonionic surfactant vesicle (niosome) products on the cosmetics market, and are aggressively promoted as superior moisturizing, "anti-aging," and "anti-wrinkling" skin preparations. The first therapeutic topical liposome preparation, the antifungal econazole "liposome gel" (Pevaryl® Lipogel; Cilag AG, Schaffhausen, Switzerland) has been introduced only recently (see also Part I, Sec. II.J).

The rationale for the use of liposomes as topical drug carriers is fourfold:

1. As solubilizing matrix for poorly soluble drugs, e.g., corticosteroids, to apply higher concentrations of drug at the thermodynamic activity maximum.
2. As local depot for the sustained release of dermally active compounds including antibiotics, corticosteroids, or retinoic acid.
3. As unique "penetration enhancer" via improved hydration of the stratum corneum, facilitating dermal delivery of drugs.
4. Hypothetically, liposomes may serve as rate-limiting membrane barriers for the modulation of systemic absorption, i.e., they may serve as controlled transdermal delivery systems.

Mezei and Gulasekharam (360,361) suggested that liposomes loaded with triamcinolone acetonide could be formulated as topical dosage forms, both as a $CaCl_2$ lotion (360) and a hydrogel (361). However, the authors' claim of liposomes apparently acting as "transdermal carriers" was considered improbable by most investigators.

Henceforth, Ganesan et al. (362) and Ho et al. (363) demonstrated unequivocally in an in vitro hairless mouse skin system (finite dose diffusion cell) that neither liposomes nor phospholipid molecules can traverse intact skin. Skin transfer coefficients of hydrocortisone and progesterone were practically identical in the presence and absence of liposomes. Other investigators (364,365) confirmed these data in vitro in the hairless mouse skin model with progesterone liposomes that had been immobilized within an agarose matrix. The rate-limiting step was determined to be phase transfer from the liposome bilayer (unhindered by the presence of agarose gel) into the aqueous bulk medium (364). Transdermal delivery of progesterone was shown to be greatly dependent on the presence of unsaturated fatty acids (oleic acid) in the phospholipid mixture, serving as "fluidizing" agent in the skin, i.e., in the widest sense as penetration enhancer, and faciliating transdermal flux of lipophilic compounds such as hydrocortisone (365).

Further insight into the distribution of liposomes in the skin was recently provided by Lasch et al. (366), using liposomes labeled with FITC-dextran and a fluorescent lipid marker, *N*-(7-nitro-2,1,3-benzoxadiazol-4-yl)-dipalmitoylphosphatidylethanolamine (NBD-DPPE). While liposomes dispersed rapidly within the stratum corneum, no further penetration of either label into epidermis, dermis, or deeper layers of the skin was found.

Following exposure of skin to niosomes (nonionic polyoxyethylene ether surfactant vesicles), Hofland et al. (367) observed the appearance of structural changes in the stratum corneum, resembling multilamellar vesicular structures (Fig. 15).

The authors speculated that either intact niosomes migrated into the stratum corneum, or molecularly dispersed high local concentrations of nonionic surfactants could form curved lamellar structures within the lipid interstitial spaces of the stratum corneum.

In another effort to identify the role of lipids in the skin, Abraham et al. (368) prepared liposomes from stratum corneum lipid, consisting of ceramides, cholesterol, cholesteryl sulfate, and free fatty acids, and showed the calcium-dependent conversion to lamellar sheets, apparently being dependent on the presence of fatty acids. Golden et al. (369) have further characterized the physicochemical characteristics of these lipids including their phase transitions and water barrier properties, i.e., a phase transition around 60–80°C and a concomitant dramatic increase in water flux. Egbaria and Weiner (357) used "skin" lipid mixtures (bovine brain ceramide, cholesterol, cholesteryl sulfate, and palmitic acid) and compared them to conventional egg lecithin, cholesterol, and phosphatidylserine containing liposomes, made by either the dehydration/rehydration or the reverse-phase evaporation technique. While ceramide-containing liposomes made by dehydration/rehydration penetrated deeper and more efficiently into skin than conventional liposomes

Figure 15 Freeze-fracture electron micrograph of nonionic surfactant vesicle (NSV)-treated human skin. A cluster of vesicular structures (V) is found in the region of intercellular lipid lamellae (ILL) between corneocytes (C) at a depth of ≈ 10 μm. The arrow indicates the direction of Pt evaporation. (From Ref. 367, with permission.)

and liposomes made by the reverse-phase evaporation method, no mechanistic explanations for this phenomenon were offered.

The focus of dermal application of liposome drug products has been in steroid therapy, including liposomal formulations of triamcinolone acetonide (360,361), triamcinolone acetonide-21-palmitate (370), hydrocortisone (361, 371), betamethasone dipropionate (372), cortisol (373), progesterone (362–365), and dihydrotestosterone (374). Transdermal delivery of liposomal forms of the α_1-blocker bunazosin HCl (375) and the nonsteroidal antiinflammatory agent flufenamic acid (376) have also been reported. Improved stratum corneum penetration of liposomal formulations of two lipophilic compounds, tocopherol nicotinate and 2-(*t*-butyl)-4-cyclohexylphenyl nicotinate (L440), was recently reported by Michel et al. (377).

Cevc and Blume (378) have formulated so-called "transfersomes," consisting of a mixture of phosphatidylcholine, sodium cholate (20–50 mol%), and ethanol (3–7%) (379), which are supposedly absorbed into the skin by a hydration driving force, generated by the large hydration gradient across the skin.

While in many of the above studies liposomes have been shown to be superior to either commercial or ad hoc topical formulations prepared with other bases (cremes, propylene glycol, etc.), one must caution that, as a rule, investigators have neglected to compare preparations with equal thermodynamic activity; rather they have compared equiconcentrated formulations. Clearly, a difference in relative thermodynamic activity will apparently favor the formulation with the higher thermodynamic activity independent of the nature of composition of the carrier.

The application of liposomes for the topical delivery of proteins has emerged only recently. Superoxide dismutase (SOD) activity on skin has been shown to be retained better following UV radiation when SOD was applied incorporated within liposomes (380). Brown et al. (381) showed that liposome incorporation prolonged the exposure of incisions to epidermal growth factor and to transforming growth factor-beta (TGF-β), resulting in increased tensile strength, while application of the factors in solution failed to improve treatment over controls. As a means to improve treatment of cutaneous Herpes simplex virus (HSV) infections, Egbaria et al. (382) evaluated the deposition of interferon-alpha (IFN-α) formulated with "skin lipids" and showed that liposome-associated IFN-α was delivered to deep skin layers. Similarly, Ho et al. (264) succeeded in treating herpes simplex genitalis infection in guinea pigs with liposomes presenting the recombinant glycoprotein D antigen of the herpes simplex virus.

4. Pulmonary Application

Pulmonary delivery of liposomes has been explored as a target-selective alternative to systemic administration of antiasthmatic and antiallergic compounds, and for antibiotics used against pulmonary infections. Mihalko et al. (383) reviewed the small body of preliminary experimental data on liposome aerosols available by 1987. More recently, Kellaway and Farr (384) and Schreier et al. (385,386) have critically reviewed aspects of pulmonary liposome delivery.

Liposomes are useful tools for pulmonary delivery of drugs due to their solubilization capacity for poorly water-soluble substances, rendering them more practical for aerosolization. Their biodegradability allows for (intermediate) prolonged pulmonary residence times without danger of allergic or other deleterious side effects. The targeting capacity to infected or immunologically impaired alveolar macrophages is a unique feature of liposomes.

The pharmaceutical aspects, specifically retention of encapsulated material and physical integrity (size changes), of liposome aerosols have been investigated in some detail, although the observed behavior is not always well understood mechanistically. The effects of liposome-related parameters such as size and lipid composition, as well as operating parameters, including airflow pressure, temperature, osmotic pressure, and pH, on the retention of encapsulated material have been investigated by Niven et al. (387–389) and Taylor et al. (390). A frequently observed phenomenon is the "processing" of large liposomes during aerosolization, resulting in a significant reduction of the overall size and size distribution (390–392). However, Farr et al. (392) determined in human volunteers that not the size of the liposome but the size of the aerosol droplet determines the pulmonary distribution of the liposome aerosol.

The toxicity of liposome aerosols has been investigated systematically. Gonzalez-Rothi et al. (393) found no inhibition of phagocytic activity or viability upon prolonged exposure of alveolar macrophages to liposomes. Myers et al. (394) exposed mice to a 4-week chronic treatment with liposome aerosols without any significant changes in alveolar macrophage function or pulmonary histology. Lung function measurements in sheep (395) (Fig. 16) and human volunteers (396) during and following liposome aerosol inhalation indicated that liposome aerosols may indeed be innocuous and not cause any acute or delayed adverse pulmonary effects.

While drug-carrying liposome aerosols are still an experimental pharmaceutical dosage form, phospholipid powders have been instilled successfully in the treatment of respiratory distress syndrome (RDS) in newborns for many years [for review see (384,397)]. Recently, Forsgren et al. (398,399) have also suggested the use of phospholipid compositions for Adult Respiratory Distress Syndrome (ARDS).

A variety of drugs has been investigated for pulmonary delivery via liposomes including antiasthmatic and antiallergic compounds, anticancer, antifungal, and antibiotic agents, and glutathione and superoxide dismutase.

McCullough and Juliano (400,401) published seminal work on the pulmonary delivery and localized pharmacologic effects of liposome-incorporated [^{3}H]-ara-C in the rat. Pulmonary instillation of liposomal ara-C resulted in greatly prolonged retention with no systemic peak radioactivity. Most elegantly, the authors demonstrated effective inhibition of [^{14}C]-thymidine incorporation in the lung, whereas 10–100 times higher pulmonary liposomal drug doses were necessary to induce tissue "damage" in the peripheral organs intestine and bone marrow. Although efficacy, i.e., tumor regression, was not demonstrated, the comprehensive approach of these early studies is remarkable, specifically the demonstration of the pharmacokinetic modulation of drug input and its effect on the pharmacodynamic outcome (i.e., tissue toxicity).

Padmanabhan et al. (402) demonstrated high enzyme activity and prolonged tissue protection following pulmonary instillation of liposome-incorporated superoxide dismutase (SOD) and catalase, although the mechanism responsible for the observed protection remained obscure. Similarly,

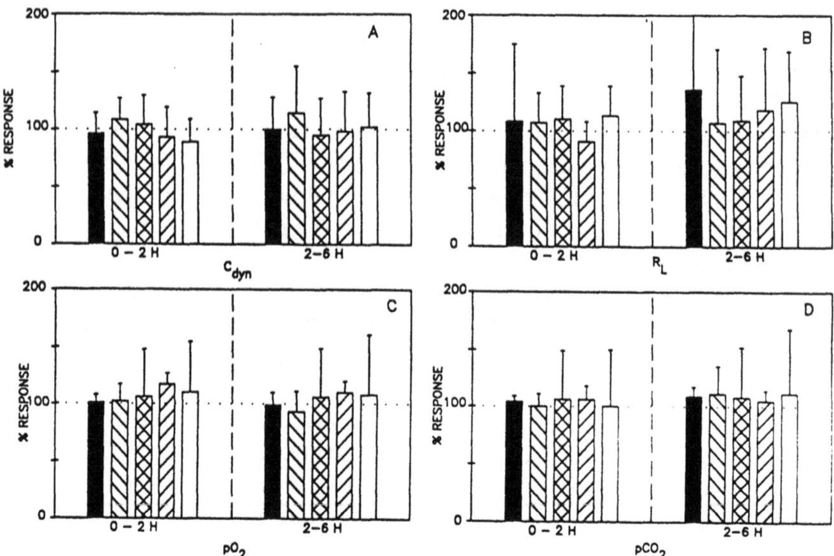

Figure 16 Lung function (C_{dyn}; R_L) and blood gas (p_aO_2; p_aCO_2) analysis after 30 min liposome aerosol and saline (control) aerosol administration in the awake sheep. Liposomes consisting of soy phosphatidylcholine (SPC) or hydrogenated SPC (HSPC) were administered as "low dose" (LD = 15 mg total lipid/mL saline) or "high dose" (HD = 150 mg total lipid/mL saline). The area under the response versus time curve (AUC) from 0–2 h and from 2–6 h is plotted relative to the baseline response (100%; dotted line). The 0–2 h response was considered the acute response, and the 2–6 h value the late response. Saline (control) (■); HD-SPC (▧); HD-HSPC (⊠); LD-SPC (▨); LD-HSPC (□). Error bars indicate mean ± SD (n = 4 for LD-SPC, HD-SPC, LD-HSPC; n = 5 for HD-HSPC; n = 8 for PBS). (From Ref. 395, with permission).

Shek et al. (403) and Jurima-Romet et al. (404) could measure plasma concentrations of instilled glutathione encapsulated within liposomes over 48 hours in the lungs of rabbits. The application of pentamidine-liposome aerosols in mice and rats was reported by Debs et al. (405), although pentamidine without carrier performed similarly well, most likely due to its low aqueous solubility or to potential ionic interactions with macromolecules in the pulmonary cell lining.

The delivery of the poorly water-soluble antiviral agent enviroxime was greatly improved by utilizing the liposome membrane as a solubilization matrix (391,406). High and persistent pulmonary drug concentrations were demonstrated in mice infected with rhinovirus. Human volunteers exposed to a 5 mg/mL liposome-enviroxime aerosol for 1 hour were estimated to have received 7–10 mg total drug, corresponding to a concentration of ≈100 μg/mL in the respiratory secretion fluid (391), although the therapeutic superiority of the liposomal dosage form remains to be demonstrated.

Aerosolized liposomal amphotericin B was found to be active against *Cryptococcus neoformans* in the lung as well as systemically in infected mice (407). Single 2-hour inhalations were found more efficacious in reducing the pulmonary *Cryptococcus* infection than i.v. treatment over 3 days. Interestingly, the authors also found an effect on the brain burden following a pulmonary dose 21 days after infection when the organisms had spread to the brain. The nature of this response remains to be elucidated.

Protection from bronchoprovocation following pulmonary administration of liposome-incorporated metaproterenol sulfate has been demonstrated in dogs (383), rabbits (408), and guinea pigs (409). While drug solution and liposome formulation were equivalent in their protective effect, tachycardia was significantly lower with the liposomal dosage form. Pulmonary clearance could be manipulated over a range of 5 to 20 hours by selecting different liposome formulations (409).

Taylor et al. (410) administered liposome-incorporated sodium cromoglycate to five healthy human volunteers at a dose of 20 mg with a Hudson air-jet nebulizer. Very low peak plasma concentrations and a biphasic lung clearance, with a very slow terminal half-life of 56.9 hours, were reported.

Promising future applications of liposome aerosols may include localized immunomodulation, antiviral and antimycobacterial therapy, and gene therapy of cystic fibrosis.

An area with great potential for improving current pulmonary anticancer therapy is immunomodulation using muramyl dipeptide (MDP) or its lipophilic derivative muramyl tripeptide-phosphatidylethanolamine (MTP-PE) (see also Sec. III.A.4 above). Sone et al. (411) demonstrated the potentiation of the tumoricidal activity of alveolar macrophages upon exposure to liposomal MDP.

Another promising application is the pulmonary application of liposome-incorporated antimycobacterial drugs directed against *Mycobacterium avium-intracellulare* (MAI)-infected alveolar macrophages. Wichert et al. (412) demonstrated the superior killing efficacy of liposomal amikacin against MAI in alveolar macrophages in vitro. Schreier et al. (395) have recently determined in the sheep model that the mean residence time of instilled liposomal amikacin in the lung could be extended from 4–7 hours to 17 hours when cholesterol-containing liposomes were used. Experimental evidence has been provided in vivo in pigs (399) and mice (394) that liposome aerosols reach the alveolar space and are taken up by macrophages.

A most exciting future perspective is the potential role of pulmonary application of DNA-carrying liposomes in the emerging area of gene therapy (see also Sec. III.A.5 above).

5. Oral Application

The suitability of liposomes for oral application has been controversial. Research in this area has not yielded consistent results, specifically with respect to the absorption of macromolecules, insulin being a prime target of investigation. Work until 1985 has been critically reviewed by Woodley (413). A more recent review (in German) by Schmidt and Michaelis (414) lists an unexpectedly large number of drugs (with over 100 references) that have been investigated for oral delivery via liposomes, although with very limited success.

The gastrointestinal environment may well be the most hostile environment liposomes could possibly encounter, exposing them to extreme pH conditions, high concentrations of detergent (bile salts), and high concentrations of enzymes in the gut lumen and intestinal cell walls. Yet phospholipid compositions have been found that can withstand such insults with remarkable stability. For instance, a formulation that has been shown repeatedly to be stable when exposed to pancreas lipase solution or physiologic bile salt solution is distearoylphosphatidylcholine/cholesterol in a 7:2 molar ratio (415,416).

Two aspects would seem to be critical prerequisites for a rational exploration of liposomes for oral use: (1) their stability in the unique gastrointestinal environment, and (2) the mechanism of interaction with the gut wall. While the former has been investigated rather systematically, virtually nothing is known about the latter to date.

In retrospect, one is tempted to conclude that perhaps liposomes are simply not a viable alternative carrier system for oral delivery of macromolecules. Yet liposomes may play an important future role in oral drug delivery as adjuvants for oral vaccination (see also Sec. III.A.4 above).

Some investigators (417,418) have recently proposed a unique oral application of liposomes, i.e., as a medicated dental lotion to treat oral ulcers. The mucosal uptake of a liposomal triamcinolone acetonide dispersion (418) when compared to other ad hoc or commercial formulations was found to be superior. However, as in a majority of all topical comparative studies published (see comments under Sec. III.B.3 above), performance was based on equal drug concentration rather than on equal thermodynamic activity, which unfortunately obscures the objective difference between the respective dosage forms tested. Nevertheless, therapy of mucosal ulcers may perhaps be a useful topical application for liposomal formulations.

C. Clinical Trials

1. Introduction

In this final part of the chapter, human experimentation with liposomes will be reviewed, with emphasis on recent clinical trials of liposome drug products, including parenteral delivery of antifungal, anticancer, and anti-arthritic agents, pulmonary delivery of bronchodilators and antiallergic agents, and topical delivery of steroids.

To date two large-scale clinical trials, of doxorubicin and amphotericin B, have been conducted, culminating in the market introduction of a liposomal form of amphotericin B (AmBisome®) in Ireland in early 1991. The topical antifungal Pevaryl Lipogel® has also been approved for marketing.

The increasing number of clinical trials as well as the launch of commercial liposome-based products have finally moved liposomes out of the realm of experimental obscurity (*some still call them "exotic"*) and onto the realistic stage of extensive clinical toxicity and efficacy trials. Concomitantly, pharmaceutical product development, specifically the development of scale-up technology and quality control protocols, has received greater and much needed attention, although little information can be found in the liposome literature.

2. Imaging

Perusing the literature on parenteral administration of liposomes to man [for review see (28)], one notices with great surprise that as early as 1974 (419) and 1976 (420) investigators have administered ^{131}I-labelled albumin (419,420) and ^{111}In-labelled bleomycin (420) liposomes to cancer patients to determine tissue (specifically cancer tissue) distribution. Between 1978 and 1985, a series of human studies using ^{99m}Tc (421–424), ^{111}In (425), a combination of ^{125}I-polyvinylpyrrolidone and ^{99m}Tc (426), or antibody-labelled (427,428) liposomes demonstrated an improvement in lymph node imaging, although tumor imaging was found to be a more elusive goal. More recently, successful imaging in cancer patients (429) and AIDS patients with Kaposi's sarcoma and lymphoma (430) with ^{111}In-labelled small liposomes (431) has been reported.

3. Enzyme Storage Diseases

Also in 1976, the first attempts at therapeutic applications of liposomes in man for the treatment of enzyme storage diseases commenced. A patient with type II glycogenesis (Pompe's disease) was treated for 7 days with liposomes containing amyloglycosidase (432). Surprisingly, a patient with type I Gaucher's disease received glucocerebroside β-glucosidase-containing liposomes over a period of 5 years (433–435). The patient was reported to have remained in stable condition, and no pronounced toxic effects were observed over the entire treatment period.

4. Antifungal Therapy: Amphotericin B

The mechanism of action of liposomal amphotericin B and the resulting reduction of systemic toxicity have been extensively investigated in the mouse model (436–439). Attaining comparable antifungal efficacy, liposomal amphotericin B was found to be up to 10 times less toxic (as a function of lipid composition and liposome size) than the conventional deoxycholate-solubilized formulation (Fungizone®) (439,440).

Amphotericin B has been in clinical investigation since 1985 (441,442). Early spectacular improvement of patients who had been resistant to conventional amphotericin B treatment, as well as dramatic alleviation of severe side effects, was indicative of the great beneficial role liposomes would play with this drug. Systematic large-scale clinical Phase I and Phase II studies in patients with disseminated fungal infections were conducted from 1988 on at a number of cancer centers in Europe (443–459) and at the University of Texas M. D. Anderson Cancer Center (460–462). In a large multicenter study of 126 patients with systemic fungal infections, the response rate was 83% for *Candida* infection but only 41% for *Aspergillus* infection (452). The investigators suspect that the systemic dose was insufficient to eradicate *Aspergillus* in the lung where rather low tissue concentrations were measured. Overall, the toxicity of amphotericin B was found to be greatly reduced with the liposomal dosage form, even when relatively high doses of 3 mg/kg per day were administered for extended periods of time (452,453).

Most recently, liposomal amphotericin B was successfully employed in infants with very low birth weight (458) and in immunosuppressed (452) and AIDS patients suffering from disseminated cryptococcosis (451,456, 459), as well as in bone marrow (448) and heart transplant (449) patients. Three cryptococcosis patients receiving 3 mg/kg liposomal amphotericin B for 42 days had a complete mycologic and clinical response (451), although in another study a patient with advanced pulmonary infection did not respond to a similar treatment regimen (459). Doses of 3–5 mg/kg were well tolerated by 10 organ transplant patients (bone marrow, kidney, liver) of which 8 responded to treatment. This is the more remarkable as such doses were prohibitive before in patients that were simultaneously treated with cyclosporin or other nephrotoxic drugs (457). The clinical pharmacokinetics of liposomal and other lipid formulations of amphotericin B have been investigated in detail by Janknecht et al. (29). An entire issue of the *Journal of Antimicrobial Chemotherapy* (Vol. 28, suppl. B, 1991) is dedicated to clinical studies with AmBisome®.

5. Cancer Therapy: Doxorubicin

A large number of investigators demonstrated in preclinical studies that liposomes modulated the biodistribution of doxorubicin resulting in a significant reduction of dose-limiting cardiotoxicity (118,463–469), although the mechanism of action of liposomal doxorubicin has never been entirely clarified. Gel state liposomes may release their contents following uptake by liver macrophages, which serve as a reservoir for prolonged release of doxorubicin into the circulation, whereas fluid-phase liposomes most likely release a major portion of the encapsulated drug during circulation in the blood (118,470–472). It is noteworthy that a remarkable effort went into the pharmaceutical aspects of formulation development. Accordingly, a rather extensive data base of the physicochemical characteristics and stability of doxorubicin liposomes is available (469,473–476).

Clinical studies with liposomal formulations of doxorubicin commenced in 1985 in Japan (477). Preliminary data on two patients in the U.S. were reported in 1986 by Gabizon et al. (478). Results of extensive Phase I and Phase II studies appeared in the literature from 1989 on (472,479–482). It was found in these clinical studies that not only rate-limiting cardiotoxicity but also peripheral side effects were reduced. In a dose escalation study from 20 mg/m^2 to 120 mg/m^2, delivered in a standard 3-weekly schedule to 32 patients with primary and metastatic liver cancer, nausea and vomiting were mild and infrequent, while hair loss was only severe in patients who received a dose of >50 mg/m^2. Severe side effects such as leukopenia and stomatitis were only found at extremely high doses of 120 mg/m^2 (479), which were considered to be dose-limiting. Hence, in addition to therapeutic advantages, the quality of life of cancer patients is positively affected by doxorubicin liposome therapy.

6. Cancer Therapy: Muramyl Tripeptide

As an alternative to chemotherapy, immunomodulation, i.e., activation of monocytes to a tumoricidal state has been suggested using liposome-incorporated muramyl dipeptide (MDP) (242,243), or its lipophilic derivatives (244–250) (see also Sec. III.A.4 above).

Clinical Phase I studies with MDP-PE have been conducted since 1989 (247,483–486), followed by a Phase II study in 1992 (487). Upon administration of 2 mg/m^2 MTP-PE in liposomes twice weekly for 12 weeks, followed by once weekly for 12 weeks, to 16 osteosarcoma patients who were disease-free by surgery, rapid induction of tumor necrosis factor (TNF-α) and interleukin 6 (IL-6) was found only after the initial dose (487). In dose escalation studies of up to 4 mg/m^2 no serious adverse effects were found. Patients responded mostly with fever, rigor, and nausea. However, no dose response in macrophage activation (484) or objective tumor response (488) has been observed in any of the published studies.

7. Cancer Therapy: Other Compounds

Attempts at treating various types of cancer with the experimental insoluble cancer agent NSC 251635 failed (489,490). Remarkable about these studies is that large volumes of up to 1 L of liposomes containing stearylamine at lipid concentrations of 20 mg/mL were administered intravenously without encountering major toxicity, although a host of side effects including sedation, lumbar pain, fever, chills, rash, and respiratory distress were reported. One must caution, though, that Coune et al. (489) irradiated their preparations prior to infusion with 1.5 Mrad from a ^{60}Cobalt source. Since there was no chemical analysis performed post-irradiation, it is unclear what degradation the treatment caused and therefore whether or to what extent irradiation contributed to the observed side effects.

Clinical Phase I and Phase II studies with liposome-complexed mitoxantrone are currently in progress at the University Hospital in Zurich (491). The toxicity of liposome-complexed mitoxantrone was evaluated in 22 women with metastatic breast cancer. A dose of 18 mg/m^2 (6xq3wk) was the recommended regimen for the ongoing Phase II study of seven patients. Two patients showed a >50% decrease of pulmonary and liver metastases lasting 3 months after the fourth treatment cycle. An antitumor effect of liposome-complexed mitoxantrone was observed in one of four cases of liver metastases and in three of five cases of soft-tissue cancer.

8. Vaccines

The concept of employing liposomes in their simultaneous role as carriers and adjuvants for viral, bacterial, or parasitic antigens has been validated (see also Sec. III.A.4 above), and clinical trials are now in progress. Thirty healthy volunteers have been inoculated with a liposomal malaria vaccine (492) that presented a recombinant peptide ($R32NS1_{81}$) containing epitopes of the circumsporozoite protein of *Plasmodium falciparum*. The liposomes contained lipid A and were adsorbed to alum, which had previously been found to be the most immunogenic combination (267). The vaccine was found both safe and more efficacious than peptide adsorbed to alum alone.

Just et al. (273) tested a new hepatitis A liposome vaccine in human volunteers and demonstrated rapid seroconversion, high antibody titers, and low reactogenicity, administering two doses of the vaccine within the same day.

9. Intraarticular Arthritis Therapy

In 1979, De Silva et al. (493) reported improvement of synovitis with liposomal hydrocortisone-21-palmitate injected intraarticularly into the knee joint of a patient. No current work in this area has been reported.

10. Pulmonary Therapy

Commencing in 1964 (494), pulmonary application of phospholipids (not necessarily liposomes) is most likely the longest and most extensively used mode of phospholipid therapy, i.e., phospholipid replacement in newborns suffering from respiratory distress syndrome (RDS) [reviewed in (397)]. It is also from studies related to RDS that we know a great deal about the biologic fate and kinetics of removal of exogenous phospholipids from the lung [for review see (383,386)]. However, very little has been known until recently regarding tolerance of liposome aerosols and the pharmacokinetics and dynamics of drugs administered via liposome aerosols.

Farr et al. (392) studied pulmonary deposition of ^{99m}technetium-labelled liposomes by gamma-scintigraphy in four healthy volunteers. They found the pattern of deposition and the kinetics of removal of both MLV and SUV to be identical, indicating that deposition and mucociliary clearance from the lung was not a function of the liposome but instead of the aerosol droplet size.

Gilbert et al. (391) reported "no adverse effects" in five human volunteers after inhaling enviroxime-liposome aerosols for 1 hour.

Taylor et al. (410) administered the antiallergic compound sodium cromoglycate as a liposome aerosol, or as a saline aerosol to five healthy human volunteers at a dose of 20 mg. The resulting pharmacokinetics were markedly different, with the plasma peak concentration of cromoglycate in the liposomal dosage form being about seven times lower than with the drug solution (4.7 ng/mL versus 34.9 ng/mL) (Fig. 17). Furthermore, liposomes remained the controlling factor for absorption as indicated by an initial $t_{1/2}$ of elimination (α phase) similar to control (1.7 hours), but a very slow terminal half-life of 56.9 hours.

Following a chronic inhalation study in mice (394) and acute lung function studies in sheep (395), Thomas et al. (396) exposed five human volunteers to 15 or 150 mg SPC/mL liposome aerosols for 1 hour. None of the subjects experienced objective acute or delayed changes in lung function or oxygen saturation. No nonspecific side effects such as coughing or throat irritation were noted.

Martin (495) reported results from a Phase I study and preliminary results from a Phase II study with a liposomal aerosol dosage form of metaproterenol (Metasome®) in healthy volunteers, asymptomatic asthmatics, and mild asthmatics. All groups tolerated up to 30 mg of the liposome-metaproterenol formulation as compared to 10–15 mg of the commercial dosage form, a solution of metaproterenol (Alupent®). Forced expiratory volume in 1 sec (FEV_1), heart rate, and blood pressure were monitored. In a follow-up three-way cross-over study of 13 patients with Metasome®, Alupent®, and saline, Metasome® at a 20 mg dose performed comparable to Alupent® at the 10 mg dose level, although the initial peak in heart rate, observed with the commercial formulation, was absent with the liposome formulation.

11. Dermal Therapy

The first commercial liposome product containing a pharmacologically active agent was the econazole product Pevaryl® Lipogel (see also Sec. III.B.3

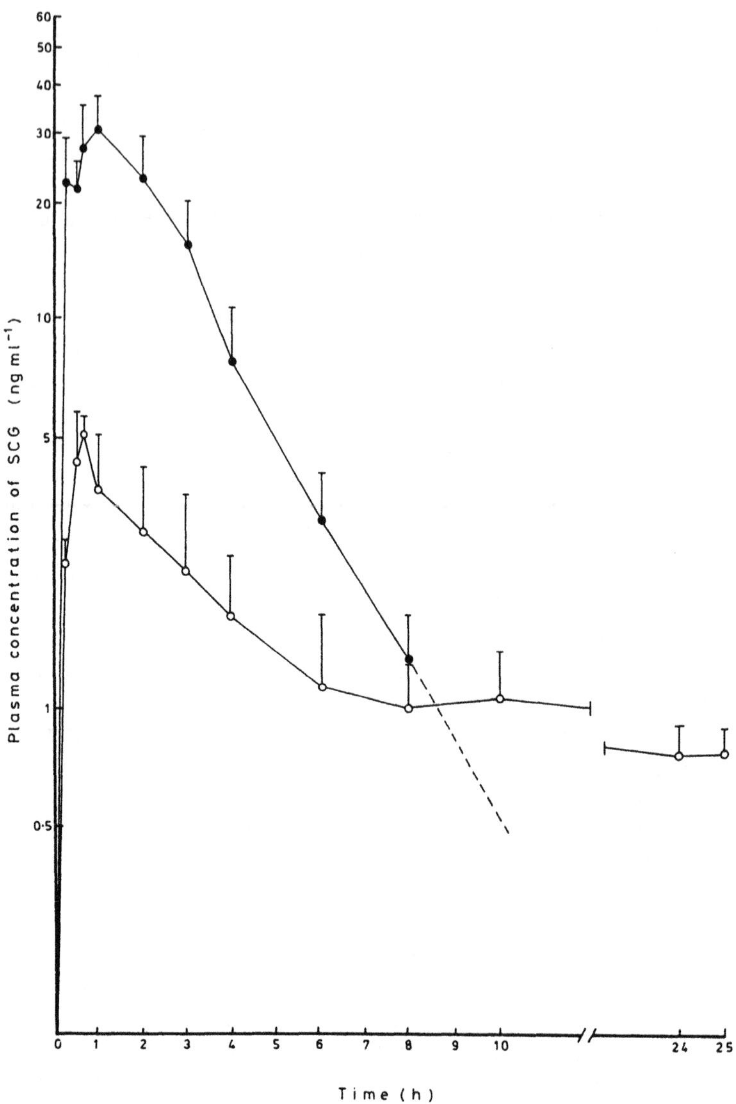

Figure 17 Plasma levels following nebulization of 20 mg sodium cromoglycate (SCG) to volunteers. SCG was delivered by an air-jet nebulizer at 172 kPa and inhaled through a mouthpiece as an aqueous solution in 0.9% saline (closed circles) or a DPPC/Chol (1:1 molar ratio) liposome formulation (open circles). Each point is a mean ± SE (free SCG n = 5; liposomal SCG n = 4). (From Ref. 410, with permission.)

above). However, the introduction of several other dermatologic compounds, specifically corticosteroids, may soon follow (358). Korting et al. (372) employed a liposomal preparation of 0.039% betamethasone dipropionate (BDP) in comparison to a 0.064% commercial propylene glycol gel for the treatment of two patient groups with atopic eczema and psoriasis vulgaris. While BDP liposomes reduced erythema and scaling in the eczema patients better than BDP in the propylene glycol gel, the latter gel showed superior efficacy in the psoriasis patients. The authors concluded that topical liposomal formulations of steroids may improve the antiinflammatory action but not the antiproliferative effects of these agents.

III. CONCLUSIONS AND OUTLOOK

Liposomes have developed into a viable pharmaceutical dosage form. Progress has taken place in quantum leaps, rather than in a continuum, over the last two decades. The milestones of this development have been summarized and were critically discussed here.

Progress in the field was not least propelled by a significant improvement in the availability and quality of lipids, concomitant with a dramatic drop in costs of the raw materials.

With advanced analytical methods and equipment, including differential scanning calorimetry, dynamic laser light scattering techniques for particle size analysis and zeta potential determination, and with a variety of electron microscopic methods becoming more generally available, liposomes have generally been better characterized and their physical state and stability have been monitored more carefully over the last decade than has been the case before.

While satisfactory loading and prolonged retention of small water-soluble drugs will continue to be a problem a priori, various techniques have been developed or are under development to overcome these problems, including chemical modification of drugs to render them more lipophilic ("anchoring" in the lipid bilayer) and remote loading that could be performed at the bedside if necessary. However, most of all, improved lyophilization protocols hold promise to retain the physical structure and the drug load for extended periods of time, in most cases up to the much desired (by the pharmaceutical industry) minimum of 1–2 years shelf life.

Manufacturing equipment including high-pressure homogenization and solvent injection/extrusion equipment has been developed and is now routinely used, even in research and development laboratories, to generate homogeneous and reproducible batches of liposomes of any desired size range and quantity.

On the therapeutic level, we have seen several major advancements. At the same time, we have not been able to overcome some of the principal conceptual bottlenecks, specifically the seemingly insurmountable task of targeting drugs systemically to selected tissues or cells. Combined efforts in the areas of immunoliposome and long-circulating liposome development may provide the much-needed impetus to advance in the area of drug targeting.

Clearly, vital progress has been made in the development of long-circulating liposomes that are not immediately recognized and removed by cells of the mononuclear phagocyte system. While the physicochemical basis of the surface modification of long-circulating liposomes has been explored in an exemplary in-depth fashion, we apparently do not quite understand the biological/biochemical processes determining the in vivo fate of these liposomes in any detail yet. Despite this, long-circulating liposomes have opened a new realm of therapeutic opportunities, and we will see a multitude of novel applications emerge in the near future.

A second area with wide-ranging implications for the future is the development of liposomes as safe and potent vehicles for antigen presentation to produce vaccines against viral, bacterial, and parasitic infections and perhaps even cancer. Liposomes will even become more valuable tools for immunization as we learn to engineer membranes with built-in fusogenic properties. Being able to elicit specific cytotoxic T-lymphocyte responses in addition to humoral immunity appears to be another milestone in the area of liposomal immunotherapy soon to be realized.

A third, very recent but rapidly developing area is the employment of cationic liposomes in gene therapy, to deliver plasmid DNA or antisense constructs. Obviously, the ability to target selective tissues or cell populations is an essential prerequisite in the area of gene therapy and is being tackled in different ways, including the use of immunoliposomes, combinations of long-circulating and immunoliposomes, and tissue-selective route of administration, e.g., inhalation to target the lung epithelium.

With respect to alternative routes of administration, we have seen major advancements in the development of topical liposome products for the delivery of dermally active agents, specifically antifungal agents and corticosteroids to the skin.

Investigators will continue to explore the validity of liposomes for the delivery of peptide and proteins, although progress in this particular area has been meager. The parenteral route of uptake appears currently to remain the route of choice for the delivery of these agents, although claims of successful delivery via the oral and, most recently, the dermal route have been made. The experimental basis for these claims needs to be evaluated carefully before final conclusions can be made.

A final optimistic observation is that apparently the field becomes more and more inter- and multidisciplinary, with basic physicochemical, biochemical, and (molecular) biological scientists forming strong interdisciplinary teams with pharmaceutical and clinical investigators. This development will hopefully safeguard against the overoptimistic and unrealistic ideas and promises of the past and lead into another highly productive and innovative phase of liposome research.

REFERENCES

1. Papahadjopoulos, D., ed. (1978). Liposomes and their uses in biology and medicine. *Ann. N.Y. Acad. Sci. 308.*
2. Gregoriadis, G., and Allison, A. C., eds. (1980). *Liposomes in Biological Systems.* New York: John Wiley.
3. Baldwin, T., and Howard, S. R., eds. (1980). *Liposomes and Immunobiology.* Amsterdam: Elsevier North Holland.
4. Knight, C. G., ed. (1981). *Liposomes: From Physical Structure to Therapeutic Applications.* Amsterdam: Elsevier/North Holland.
5. Gregoriadis, G., ed. (1984). *Liposome Technology.* 3 vols. 1st ed. Boca Raton, Fla.: CRC Press.
6. Gregoriadis, G., ed. (1993). *Liposome Technology.* 3 vols. 2d ed. Boca Raton, Fla.: CRC Press.
7. Yogi, K., ed. (1986). *Medical Application of Liposomes.* Tokyo: Japan Scientific Society Press.
8. Cullis, P. R., and Hope, M. J., eds. (1986). *Liposomes—Special Issue of Chem. Phys. Lipids. 40.* Shannon, Ireland: Elsevier.
9. Gregoriadis, G., ed. (1988). *Liposomes as Drug Carriers. Recent Trends and Progress.* Chichester: John Wiley.
10. Lopez Berestein, G. and Fidler, I. J., eds. (1989). *Liposomes in the Therapy of Infectious Diseases and Cancer.* New York: Alan R. Liss.
11. New, R. R. C., ed. (1990). *Liposomes: A Practical Approach.* Oxford: IRL Press at Oxford University Press.
12. Gregoriadis, G. (1990). *Liposomes in Drug Delivery—21 Years On.* Abstract book. London.
13. Rubas, W., and Schreier, H. (1991). *Liposomen: Fortschritte in Herstellungs-Technologie und Therapie. Pharmazie in unsere Zeit 20*: 255–270.
14. Braun-Falco, O., Korting, H. C., and Maibach, H. I. (1992) eds. *Liposome Dermatics.* Berlin: Springer-Verlag, 101–109.
15. Barenholz, Y., and Crommelin, D. J. A. (1994). Liposomes as pharmaceutical dosage forms. *Encyclopedia of Pharmaceutical Technology* (J. Swarbrick, ed.). New York: Marcel Dekker, pp. 1–39.
16. Hope, M. J., Bally, M. B., Mayer, L. D., Janoff, A. S., and Cullis, P. R. (1986). *Chem. Phys. Lipids 40*: 89–107.
17. Israelachvili, J., Marcelja, S., and Horn, R. G. (1980). Physical principles of membrane organization. *Q. Rev. Biophys. 13*: 121–200.

18. Israelachvili, J. N. (1985). *Innermolecular and Surface Forces*. New York: Academic Press.
19. Small, D. M. (1986). *Handbook of Lipid Research*. New York: Plenum Press, pp. 93–95.
20. Seddon, J. H. (1990). Structure of the inverted hexagonal (H_{II}) phase and non-lamellar phase transitions of lipids. *Biochim. Biophys. Acta 1031*: 1–69.
21. Safran, S. A., Pincus, P. A., Andelman, D., and MacKintosh, F. C. (1991). Stability and phase behaviour of mixed surfactant micelles. *Phys. Rev. 43*: 1071–1078.
22. Lasic, D. D. (1988). The mechanism of vesicle formation. *Biochem. J. 256*: 1–11.
23. Hauser, H., and Gaines, N. (1982). Spontaneous vesiculation of phospholipids: A simple and quick method of forming unilamellar vesicles. *Proc. Natl. Acad. Sci. (USA) 79*: 1683–1687.
24. Hauser, H. (1989). Mechanism of spontaneous vesiculation. *Proc. Natl. Acad. Sci. 86*: 5351–5355.
25. Lasic, D. D. (1982). *Biochim. Biophys. Acta 692*: 501–502.
26. Fromherz, P. (1983). *Chem. Phys. Lett. 94*: 259–266.
27. Fromherz, P., and Rüppel, D. (1985). *FEBS Lett. 179*: 155–159.
28. Zonneveld, G. M., and Crommelin, D. J. A. (1988). Liposomes: Parenteral administration to man. *Liposomes as Drug Carriers: Recent Trends and Progress* (G. Gregoriadis, ed.). Chichester: John Wiley, pp. 795–817.
29. Janknecht, R., De Marie, S., Bakker-Woudenberg, I. A. J. M., and Crommelin, D. J. A. (1992). Liposomal and lipid formulations of amphotericin B. *Clinical Pharmacokin. 23*: 279–291.
30. Storm, G., Oussoren, C., Peeters, P. A. M., and Barenholz, Y. (1993). Tolerability of liposomes in vivo. *Liposome Technology*. Vol. 3. 2d ed. Boca Raton, Fla.: CRC Press, Inc., pp. 345–383.
31. Spanjer, H. H., and Scherphof, G. (1983). Targeting of lactosylceramide-containing liposomes to hepatocytes in vivo. *Biochim. Biophys. Acta 734*: 40–47.
32. Nicolau, C., LeGrand, A., and Grosse, E. (1987). Liposomes as carriers for in vivo gene transfer and expression. *Methods in Enzymology*, Vol. 149 (R. Green and K. J. Widder, eds.). New York: Academic Press, pp. 157–176.
33. Lichtenberg, D., and Barenholz, Y. (1988). Liposomes: Preparation, characterization and preservation. *Methods of Biological Analysis 33* (D. Glick, ed.). New York: John Wiley, pp. 337–461.
34. Lang, J. K., Vigo-Pelfrey, C., and Martin F. (1990). Liposomes composed of partially hydrogenated egg phosphatidylcholine: Fatty acid composition, thermal phase behaviour and oxidative stability. *Chem. Phys. Lipids 53*: 91–101.
35. Eibl, H. Phospholipid synthesis. *Liposomes: From Physical Structure to Therapeutic Applications* (C. G. Knight, ed.). Amsterdam: Elsevier/North Holland, pp. 19–50.
36. Eibl, H., and Woolley, P. (1986). Synthesis of enantiomerically pure glyceryl esters and ethers. I. Methods employing the precursor 1,2-isopropylidene-sn-glycerol. *Chem. Phys. Lipids 41*: 53–63.

37. Allen, T. M., Hansen, C., Martin, F., Redemann, C., and Yau-Young, A. (1990). Liposomes containing synthetic lipid derivatives of poly(ethylene glycol) show prolonged circulation half-lives in vivo. *Biochim. Biophys. Acta 1066*: 29–36.
38. Klibanov, A. L., Maruyama, K., Torchilin, V. P., and Huang, L. (1990). Amphipathic polyethyleneglycols effectively prolong the circulation time of liposomes. *FEBS Lett 268*: 235–237.
39. Blume, G., and Cevc, G. (1990). Liposomes for the sustained drug release in vivo. *Biochim. Biophys. Acta 1029*: 91–97.
40. Abuchowski, A., Van Es, Th., Palczuk, N. C., and Davis, F. F. (1977). Alteration of immunological properties of bovine serum albumin by covalent attachment of polyethylene glycol. *J. Biol. Chem. 252*: 3578–3581.
41. Toonen, P. A. H. M., and Crommelin, D. J. A. (1983). Immunoglobulins as targeting agents for liposome encapsulated drugs. *Pharm. Weekbl. Sci. Ed. 5*: 269–280.
42. Poznansky, M. J., and Juliano, R. L. (1984). Biological approaches to the controlled delivery of drugs: A critical review. *Pharmacol. Rev. 36*: 277–336.
43. Illum, L., and Jones, P. D. E. (1985). Attachment of monoclonal antibodies to microspheres. *Methods in Enzymology*, Vol. 112 (K. J. Widder and R. Green, eds.). New York: Academic Press, pp. 67–84.
44. Crommelin, D. J. A., and Storm, G. (1990). Drug targeting. *Comprehensive Medicinal Chemistry*. Vol. 5. *Biopharmaceutics* (P. G. Sammes and J. D. Taylor, eds.). Oxford: Pergamon Press, pp. 661–701.
45. Martin, F., and Papahadjopoulos, D. (1982). Irreversible coupling of immunoglobulin fragments to preformed vesicles. *J. Biol. Chem. 257*: 286–288.
46. Martin, F. J., Heath, T. D., and New, R. R. C. (1990). Conjugation of liposomes. Liposomes: A practical approach. *Liposomes* (R. New, ed.). Oxford: Oxford University Press, pp. 163–182.
47. Nässander, U. K., Storm, G., Peeters, P. A. M., and Crommelin, D. J. A. (1990). Liposomes. *Biodegradable Polymers as Drug Delivery Systems* (M. Chasin and R. Langer, eds.). New York: Marcel Dekker, pp. 261–338.
48. Talsma, H., and Crommelin, D. J. A. (1992). Liposomes as drug delivery systems, Part I: Preparation. *BioPharm* 36–39.
49. Szoka, F., and Papahadjopoulos, D. (1981). Liposomes: Preparation and characterization. *Liposomes: From Physical Structure to Therapeutic Applications* (C. G. Knight, ed.). Amsterdam: Elsevier/North Holland, pp. 51–82.
50. Dingle, J. T., Gordon, J. L., Hazleman, B. L., Knight, C. G., Page Thomas, D. P., Phillips, N. C., Shaw, I. H., Fildes, F. J. T., Oliver, J. E., Jones, G., Turner, E. H., and Lowe, J. S. (1978). Novel treatment for joint inflammation. *Nature 271*: 372–373.
51. Van Hoogevest, P., and Fankhauser, P. (1989). An industrial liposomal dosage form for muramyl-tripeptide-phosphatidylethanolamine (MTP-PE). *Liposomes in the Therapy of Infectious Diseases and Cancer* (G. Lopez-Berestein and I. J. Fidler, eds.). New York: Alan R. Liss, pp. 453–466.
52. Payne, N. I., Timmins, P., Ambrose, C. V., Ward, M., and Ridgway, F. (1986). Proliposomes: A novel solution to an old problem. *J. Pharm. Sci. 75*: 325–329.

53. Payne, N. I., Browning, I., and Hynes, C. A. (1986). Characterization of proliposomes. *J. Pharm. Sci. 75*: 330–333.
54. Chen, C. M., and Alli, D. (1987). Use of fluidized bed in proliposome manufacturing. *J. Pharm. Sci. 76*: 419.
55. Mayhew, E., Conroy, S., King, J., Lazo, R., Nikolopoulos, G., Siciliano, A., and Vail, W. J. (1987). High-pressure continuous-flow system for drug entrapment in liposomes. *Methods in Enzymology*. Vol. 149. *Drug and Enzyme Targeting*, Part B (R. Green and K. J. Widder, eds.). San Diego: Academic Press, pp. 64–77.
56. Talsma, H., Özer, A. Y., Van Bloois, L., and Crommelin, D. J. A. (1989). The size reduction of liposomes with a high pressure homogenizer (Microfluidizer™). Characterization of prepared dispersions and comparison with conventional methods. *Drug. Dev. Ind. Pharm. 15*: 197–207.
57. Mentrup, E., and Stricker, H. (1988). *Acta Pharm. Technol. 34*: 16S.
58. Gamble, R. C. (1985). patent Vestar.
59. Brandl, M., Bachmann, D., Drechsler, M., and Bauer, K. H. (1990). Liposome preparation by a new high pressure homogenizer Gaulin micron lab 40. *Drug Dev. Ind. Pharm. 16*: 2167–2191.
60. Talsma, H., Van Steenbergen, M. J., Borchert, J. C. H., and Crommelin, D. J. A. (1994). A novel technique for the one-step preparation of liposomes and non-ionic surfactant vesicles without the use of organic solvents. Liposome formation in a continuous gas stream: The 'bubble' method. *J. Pharm. Sci.*, in press.
61. Deamer, D. W., and Bangham, A. D. (1976). Large volume liposomes by an ether injection method. *Biochim. Biophys. Acta 443*: 629–634.
62. Martin, F. J. High encapsulation liposome processing method. U.S. pat;ent 4,752,425.
63. McGurk, J. G. (1986). A method for the preparation of liposomes, a pack for use therein and a composition for use in preparing a liposomal material. European Patent 0 190 926.
64. Kim, S., and Martin, G. M. (1981). Preparation of cell size unilamellar liposomes with high captured volume and defined size distribution. *Biochim. Biophys. Acta 646*: 1–9.
65. Kim, S., Turker, M. S., Chi, E. Y., Sela, S., and Martin, G. M. (1983). Preparation of multivesicular liposomes. *Biochim. Biophys. Acta 728*: 339–348.
66. Szoka, F., and Papahadjopoulos, D. (1978). Procedure for preparation of liposomes with large aqueous space and high capture by reverse-phase evaporation. *Proc. Natl. Acad. Sci. USA 75*: 4194–4198.
67. Szoka, F., and Papahadjopoulos, D. (1981). Liposomes: Preparation and characterization. *Liposomes: From Physical Structure to Therapeutic Applications* (C. G. Knight, ed.). Amsterdam: Elsevier/North Holland, pp. 51–82.
68. Handa, T., Takeuchi, H., Ohokubo, Y., and Kawashima, Y. (1987). Lyophilized liposomes prepared by a modified reverse-phase evaporation method. *Chem. Pharm. Bull. 35*: 748–755.
69. Pidgeon, C. (1985). European Patent 0 179 660.
70. Pidgeon, C., Hung, A. H., and Dittrich, K. (1986). *Pharm. Res. 3*: 23–34.

71. Gruner, S. M., Lenk, R. P., Janoff, A. S., and Orton, M. J. (1985). Novel multilayered lipid vesicles: Comparison of physical characteristics of multilamellar liposomes and stable plurilamellar vesicles. *Biochemistry 24*: 2833–2842.
72. Batzri, S., and Korn, E. D. (1973). Single bilayer liposomes prepared without sonication. *Biochim. Biophys. Acta 298*: 1015–1019.
73. Kremer, J. M. M., van der Esker, M. W. J., Pathmamanoharan, C., and Wiersema, P. H. (1977). *Biochemistry 16*: 3932–3935.
74. Perrett, S., Golding, M., and Williams, P. (1991). A simple method for the preparation of liposomes for pharmaceutical applications: Characterization of the liposomes. *J. Pharm. Pharmacol. 43*: 154–161.
75. Kagawa, Y., and Racker, E. (1972). Partial resolution of the enzymes catalysing oxidative phosphorylation. XXV Reconstitution of vesicles catalysing ^{32}Pi-adenosine triphosphate exchange. *J. Biol. Chem. 246*: 5477–5488.
76. Racker, E. (1972). Reconstitution of a calcium pump with phospholipids and a purified Ca^{++}-adenosine triphosphatase from sarcoplasmic reticulum. *J. Biol. Chem. 247*: 8198–8200.
77. Loyter, A., and Volsky, D. J. (1982). Reconstituted Sendai virus envelopes as carriers for the introduction of biological material into animal cells. *Membrane Reconstitution* (G. Poste and G. L. Nicholson, eds.). Elsevier Biomedical Press, pp. 215–266.
78. Bergers, J. J., Den Otter, W., De Groot, J. W., De Blois, A. W., Dullens, H. F. J., Steerenberg, P. A., and Crommelin, D. J. A. (1992). Reconstituted membranes of tumor cells (proteoliposomes) induce specific protection to tumor lymphoma cells. *Cancer Immunology and Immunotherapy 34*: 233–240.
79. Chander, R., and Schreier, H. (1992). Artificial viral envelopes containing recombinant human immunodeficiency virus (HIV) gp 160. *Life Sciences 50*: 481–489.
80. Allen, T. M. (1984). Removal of detergent and solvent traces from liposomes. *Liposome Technology*, Vol. 1 (G. Gregoriadis, edc.). Boca Raton, Fla.: CRC Press, pp. 109–122.
81. Teelmann, K., Schlaeppi, B., Schuepbach, M., and Kistler, A. (1984). Preclinical safety evaluation of intravenously administered mixed micelles. *Arzneim.-Forsch. 34*: 1517–1523.
82. Jiskott, W., Teerlink, T., Beuvery, E. C., and Crommelin, D. J. A. (1986). Preparation of liposomes via detergent removal from mixed micelles by dilution. The effect of bilayer composition and process parameters on liposome characteristics. *Pharm. Weekbl. Sci. Ed. 8*: 259–265.
83. Weder, H. G., and Zumbühl, O. (1984). The preparation of variably sized homogeneous liposomes for laboratory, clinical and industrial use by controlled detergent dialysis. *Liposome Technology*, Vol. 1 (G. Gregoriadis, ed.). Boca Raton, Fla.: CRC Press, pp. 79–107.
84. Huang, A., Tsao, Y. S., Kennel, S. J., and Huang, L. (1982). *Biochim. Biophys. Acta 716*: 140–150.
85. Harsh, M., Walther, P., Hengartner, H., and Weder, H. G. (1981). Targeting of monoclonal antibody-coated liposomes to sheep red blood cells. *Biochem. Biophys. Res. Commun. 103*: 1069–1076.

86. Hughes, B. J., Kennel, S., Lee, R., and Huang, L. (1989). Monoclonal antibody targeting of liposomes to mouse lung in vivo. *Cancer Research 49*: 6214-6220.
87. Philippot, J. R., Mutaftschiev, S., and Liautard, J. P. (1985). Extemporaneous preparation of large unilamellar liposomes. *Biochim. Biophys. Acta 821*: 79-84.
88. Son, K., and Alkan, H. (1989). Liposomes prepared dynamically by interactions between bile salts and phospholipid molecules. *Biochim. Biophys. Acta 981*: 288-294.
89. Son, K., and Alkan, H. (1990). Novel approach for intravenous delivery of insoluble drugs. *Pharm. Res. 7*: S-101.
90. Lawaczek, R., Kainosho, M., and Chan, S. I. (1976). *Biochim. Biophys. Acta 443*: 313-330.
91. Strauss, G. (1984). Freezing and thawing of liposome suspensions. *Liposome Technology*, Vol. 1 (G. Gregoriadis, ed.). Boca Raton, Fla.: CRC Press, pp. 197-219.
92. Fransen, G. J., Salemink, P. J. M., and Crommelin, D. J. A. (1986). Critical parameters in freezing of liposomes. *Int. J. Pharm. 33*: 37-35.
93. Mayer, L. D., Hope, M. J., Cullis, P. R., and Janoff, A. S. (1985). Solute distributions and trapping efficiencies observed in freeze-thawed multilamellar vesicles. *Biochim. Biophys. Acta 817*: 193-196.
94. Oku, N., and MacDonald, R. C. (1983). *Biochemistry 22*: 855-863.
95. Kirby, C. J., and Gregoriadis, G. (1984). A simple procedure for preparing liposomes capable of high encapsulation efficiency under mild conditions. *Liposome Technology*, Vol. 1 (G. Gregoriadis, ed.). Boca Raton, Fla.: CRC Press, pp. 19-27.
96. Ohsawa, T., Miura, H., and Harada, K. (1984). A novel method for preparing liposomes with a high capacity to encapsulate proteinous drugs: Freeze-drying method. *Chem. Pharm. Bull 32*: 2442-2445.
97. Seltzer, S. E., Gregoriadis, G., and Dick, R. (1988). Evaluation of the dehydration-rehydration method for production of contrast-carrying liposomes. *Investigative Radiology 23*: 131-138.
98. Gregoriadis, G., Da Silva, H., and Florence, A. T. (1990). A procedure for the efficient entrapment of drugs in dehydration-rehydration liposomes (DRVs). *Int. J. Pharm. 65*: 235-242.
99. Shew, R. L., and Deamer, D. W. (1985). A novel method for encapsulation of macromolecules in liposomes. *Biochim. Biophys. Acta 816*: 1-8.
100. Li, W., and Haines, T. H. (1986). Uniform preparations of large unilamellar vesicles containing anionic lipids. *Biochemistry 25*: 7477-7483.
101. Senior, J. (1987). Fate and behaviour of liposomes in vivo: A review of controlling factors. *CRC Crit. Reviews Therapeut. Drug Carrier Systems 3*: 123-193.
102. Talsma, H., Van Steenbergen, M. J., and Crommelin, D. J. A. (1991). The cryopreservation of liposomes: 3. Almost complete retention of a water-soluble marker in small liposomes in a cryoprotectant containing dispersion after a freezing/thawing cycle. *Int. J. Pharm. 77*: 119-126.

103. Jiskoot, W., Teerlink, T., Beuvery, E. C., and Crommelin, D. J. A. (1986). Preparation of liposomes via detergent removal from mixed micelles by dilution. The effect of bilayer composition and process parameters on liposome characteristics. *Pharm. Weekbl. Sci. Ed. 8*: 259–265.
104. Talsma, H., Gooris, G., v. Steenbergen, M., Salomons, M. A., Bouwstra, J., and Crommelin, D. J. A. (1992). The influence of the molar ratio of cholesteryl hemisuccinate/dipalmitoylphosphatidylcholine on 'liposome' formation after lipid film hydration. *Chem. Phys. Lipids 62*: 105–112.
105. Saunders, L., Perrin, J., and Gammack, D. (1962). Ultrasonic radiation of some phospholipid sols. *J. Pharm. Pharmacol. 14*: 567–572.
106. Lelkes, P. I. (1984). The use of French pressed vesicles for efficient incorporation of bioactive macromolecules and as drug carriers in vitro and in vivo. *Liposome Technology*, Vol. 1 (G. Gregoriadis, ed.). Boca Raton, Fla.: CRC Press, pp. 51–65.
107. Olson, F., Hunt, C. A., Szoka, F., Vail, W. J., and Papahadjopoulos, D. (1979). Preparation of liposomes of defined size distribution by extrusion through polycarbonate membranes. *Biochim. Biophys. Acta 557*: 9–23.
108. Szoka, F., Olson, F., Heath, T., Vail, W., Mayhew, E., and Papahadjopoulos, D. (1980). Preparation of unilamellar liposomes of intermediate size (0.1–0.2 μm) by a combination of reverse phase evaporation and extrusion through polycarbonate membranes. *Biochim. Biophys. Acta 601*: 559–571.
109. Jousma, H., Talsma, T., Spies, F., Joosten, J. G. H., Junginger, H. E., and Crommelin, D. J. A. (1987). Characterization of liposomes. The influence of extrusion of multilamellar vesicles through polycarbonate membranes on particle size, particle size distribution and number of bilayers. *Int. J. Pharm. 35*: 263–274.
110. Hope, M. J., Bally, M. B., Webb, G., and Cullis, P. R. (1985). Production of large unilamellar vesicles by a rapid extrusion procedure. Characterization of size distribution, trapped volume and ability to maintain a membrane potential. *Biochim. Biophys. Acta 812*: 55–65.
111. Amselem, S., Gabizon, A., and Barenholz, Y. (1993). A large scale method for preparation of sterile and non-pyrogenic liposomal formulations for clinical use. *Liposome Technology*, 2d ed. (G. Gregoriadis, ed.). Boca Raton, Fla.: CRC Press, 501–525.
112. Martin, F. J., and Morano, J. K. (1988). U.S. Patent No. 4,737,323, issued April 12.
113. Nicolay, K., Van der Neut, R., Fok, J. J., and DeKruyff, B. (1985). *Biochim. Biophys. Acta 819*: 55–65.
114. Barenholz, Y., Gibbs, D., Litman, B. J., Goll, J., Thompson, T. E., and Carlson, F. D. (1977). A simple method for the preparation of homogeneous phospholipid vesicles. *Biochemistry 16*: 2806–2810.
115. Bergers, J. J., Den Otter, W., Dullens, H. F. J., De Groot, J. W., Steerenberg, P. A., Mimpen, M. W. H., and Crommelin, D. J. A. (1993). Critical factors for liposome incorporated tumor-associated antigens to induce protective tumour immunity to SL2 tumours in mice. *Cancer Immunol. Immunother. 37*: 271–279.
116. Storm, G., Van Bloois, L., Brouwer, M., and Crommelin, D. J. A. (1985). The interaction of cytostatic drugs with adsorbents in aqueous media. The po-

tential implications for liposome preparation. *Biochim. Biophys. Acta 818*: 343–351.

117. Moro, L. C. V., Neri, G., and Rigamonti, A. (1980). Verfahren zum Reinigen von Liposomensuspensionen, patent DE 30 01 842 A1.
118. Storm, G., Steerenberg, P. A., Roerdink, F. H., De Jong, W. H., and Crommelin, D. J. A. (1987). Influence of lipid composition on the antitumor activity exerted by doxorubicin-containing liposomes in a rat solid tumor model. *Cancer Research 47*: 3366–3372.
119. Rogers, J. A., and Davis, S. S. (1980). Functional group contributions to the partitioning of phenols between liposomes and water. *Biochim. Biophys. Acta 598*: 392–404.
120. Choi, Y. W., and Rogers, J. A. (1991). Characterization of distribution behaviour of 2-imidazolines into multilamellar liposomes. *J. Pharm. Sci. 80*: 757–760.
121. Ma, L., Ramachandran, C., and Weiner, N. D. (1991). Partitioning of a homologous series of alkyl *p*-aminobenzoates in dipalmitoylphosphatidylcholine liposomes: Effect of liposome type. *Int. J. Pharm. 77*: 127–140.
122. Storm, G., Van Bloois, L., Steerenberg, P. A., Van Etten, E., De Groot, G., and Crommelin, D. J. A. (1989). Liposome encapsulation of doxorubicin: Pharmaceutical and therapeutic aspects. *J. Control. Rel. 9*: 215–229.
123. Barenholz, Y., and Amselem, S. (1993). Quality control in the development and clinical use of liposome based formulations. *Liposome Technology*, 2d ed. (G. Gregoriadis, ed.). Boca Raton, Fla.: CRC Press, pp.
124. Van Bloois, L., Dekker, D. D., and Crommelin, D. J. A. (1987). Solubilization of lipophilic drugs by amphiphiles: Improvement of the apparent solubility of almitrine bismesylate by liposomes, mixed micelles and O/W emulsions. *Acta Pharm. Technol. 33*: 136–139.
125. Lidgate, D. D., Felgner, P. L., Fleitman, J. S., Whatley, J., and Fu, R. C. (1988). *Pharm. Res. 5*: 759–764.
126. Kimelberg, H. K. (1976). Protein-liposome interactions and their relevance to the structure and function of cell membranes. *Mol. Cell Biochem. 10*: 171–190.
127. Jain, M. K., and Zakim, D. (1987). The spontaneous incorporation of proteins into preformed bilayers. *Biochim. Biophys. Acta 9062*: 33–68.
128. Pidgeon, C., Williard, R. L., and Schroeder, S. (1989). Amino acid bracketing the predicted transmembrane domains of membrane proteins. *Pharm. Res. 6*: 779–786.
129. Bergers, J., Vingerhoeds, M. H., Van Bloois, L., Herron, J. N., Janssen, L. H. M., Fischer, M., and Crommelin, D. J. A. (1993). The role of protein charge in protein-lipid interactions, pH dependent changes of the electrophoretic mobility of liposomes through adsorption of water-soluble, globular proteins. *Biochemistry 32*: 4641–4649.
130. Bergers, J. J., Den Otter, W., Dullens, H. F. J., Kerkvliet, C. T. M., and Crommelin, D. J. A. (1994). Interleukin-2 containing liposomes: Interaction of interleukin-2 with liposomal bilayers and preliminary studies on application in cancer vaccines. *Pharm. Res.,* in press.

131. Bergers, J. J., van Bloois, L., Barenholz, Y., and Crommelin, D. J. A. Conformational changes of myoglobin upon interaction with liposomes. *Biochim. Biophys. Acta* submitted.
132. Reddy, R., Zhou, F., Huang, L., Carbone, F., Bevan, M., and Rouse, B. T. (1991). pH sensitive liposomes provide an efficient means of sensitizing target cells to class I restricted CTL recognition of a soluble protein. *J. Immunol. Methods 141*: 157–163.
133. Nair, S., Zhou, F., Reddy, R., Huang, L., and Rouse, B. T. (1992). Soluble proteins delivered to dendritic cells via pH-sensitive liposomes induce primary cytotoxic T lymphocyte responses in vitro. *J. Exp. Med. 175*: 609–612.
134. Chu, C.-J., Dijkstra, J., Lai, M.-Z., Hong, K., and Szoka, F. C. (1990). Efficiency of cytoplasmic delivery by pH sensitive liposomes to cells in culture. *Pharm. Res. 7*: 824–834.
135. Yatvin, M. B., Kreutz, W., Horwitz, B. A., and Shinitzky, M. (1980). pH sensitive liposomes: Possible clinical implications. *Science 210*: 1253–1255.
136. Connor, J., and Huang, L. (1985). Efficient cytoplasmic delivery of a fluorescent dye by pH sensitive immunoliposomes. *J. Cell. Biol. 101*: 582–589.
137. Nayar, R., and Schroit, A. J. (1985). Generation of pH-sensitive liposomes: Use of large unilamellar vesicles containing *N*-succinyldioleylphosphatidylethanolamine. *Biochemistry 24*: 5967–5971.
138. Ellens, H., Bentz, J., and Szoka, F. C. (1984). pH-induced destabilization of phosphatidylethanolamine-containing liposomes: Role of bilayer contact. *Biochemistry 23*: 1532–1538.
139. Parente, R. A., Nir, S., and Szoka, F. C. (1990). Mechanism of leakage of phospholipid vesicle contents induced by the peptide GALA. *Biochemistry 29*: 8720–8728.
140. Scherphof, G., Damen, J., and Wilschut, J. Interactions of liposomes with plasma proteins. *Liposome Technology*, Vol. 3 (G. Gregoriadis, ed.). Boca Raton, Fla.: CRC Press, pp. 205–224.
141. Uremura, K., and Kinsky, S. C. (1972). Active vs. passive sensitization of liposomes toward antibody and complement by dinitrophenylated derivatives of phosphatidylethanolamine. *Biochemistry 11*: 4085–4094.
142. Ligler, F. S., Bredehorst, R., Talebian, A., Shriver, L. C., Hammer, C. F., Sheridan, J. P., Vogel, C. W., and Gaber, B. P. (1987). A homogeneous immunoassay for the Mycotoxin T-2 utilizing liposomes, monoclonal antibodies and complement. *Anal. Biochem. 163*: 369–375.
143. Masaka, T., Okada, N., Yasuda, R., and Okada, H. (1989). Assay of complement activity in human serum using large unilamellar liposomes. *J. Immunol. Methods 123*: 19–24.
144. Magin, R. L., and Weinstein, J. (1984). The design and characterization of temperature-sensitive liposomes. *Liposome Technology*, Vol. 3 (G. Gregoriadis, ed.). Boca Raton, Fla.: CRC Press, pp. 137–155.
145. Perkins, W. R., Minchey, S. R., Ostro, M. J., Taraschi, T. F., and Janoff, A. S. (1988). The captured volume of multilamellar vesicles. *Biochim. Biophys. Acta 943*: 103–107.
146. Chapman, C. J., Erdahl, W. E., Taylor, R. W., and Pfeiffer, D. R. (1991). *Chem. Phys. Lipids 60*: 201–208.

147. Chapman, C. J., Erdahl, W. E., Taylor, R. W., and Pfeiffer, D. R. (1990). *Chem. Phys. Lipids 55*: 73–83.
148. Nichols, J. W., and Deamer, D. W. (1976). Catecholamine uptake and concentration by liposomes maintaining pH gradients. *Biochim. Biophys. Acta 455*: 269–271.
149. Mayer, L. D., Bally, M. B., Hope, M. J., and Cullis, P. R. (1985). Uptake of neoplastic agents into large unilamellar vesicles in response to a membrane potential. *Biochim. Biophys. Acta 816*: 802–808.
150. Mayer, L. D., Bally, M. B., and Cullis, P. R. (1986). Uptake of adriamycin into large unilamellar vesicles in response to a pH gradient. *Biochim. Biophys. Acta 857*: 123–126.
151. Mayer, L. D., Bally, M. B., and Cullis, P. R. (1990). Strategies for optimizing liposomal doxorubicin. *J. Liposome Research 1*: 463–480.
152. Mayer, L. D., Tai, L. C. L., Bally, M. B., Mitilenes, G. N., Ginsberg, R. S., and Cullis, P. R. (1990). Characterization of liposomal systems containing doxorubicin entrapped in response to pH gradients. *Biochim. Biophys. Acta 1025*: 143–151.
153. Madden, T. D., Harrigan, P. R., Tai, L. C. L., Bally, M. B., Mayer, L. D., Redelemaier, T. E., Loughrey, H. C., Tilcock, C. P. S., Reinish, L. W., and Cullis, P. R. (1990). The accumulation of drugs within large unilamellar vesicles exhibiting a proton gradient: A survey. *Chem. Phys. Lipids 53*: 37–46.
154. Haran, G., Cohen, R., Bar, L. K., and Barenholz, Y. (1990). Ammonium ion gradients in liposomes: A method to obtain efficient entrapment and controlled release of amphipathic molecules. *Liposomes in Drug Delivery—21 Years On* (G. Gregoriadis, ed.). Abstract book. London.
155. Grit, M., De Smidt, J. H., Struijke, A., and Crommelin, D. J. A. (1989). Hydrolysis of phosphatidylcholine in aqueous liposome dispersions. *Int. J. Pharm. 50*: 1–6.
156. Grit, M., Underberg, W. J. M., and Crommelin, D. J. A. (1993). Hydrolysis of saturated soybean phosphatidylcholine in aqueous liposome dispersions. *J. Pharm. Sci. 82*: 362–366.
157. Grit, M., Zuidam, N. J., and Crommelin, D. J. A. (1993). Analysis and hydrolysis kinetics of phospholipids in aqueous dispersions. *Liposome Technology*, 2d ed. (G. Gregoriadis, ed.). Boca Raton, Fla.: CRC Press, pp. 455–486.
158. Hwang, K. J., Merriam, J. E., Beaumier, P. L., and Luk, K. S. (1982). Encapsulation with high efficiency of radioactive metal ions in liposomes. *Biochim. Biophys. Acta 716*: 101–109.
159. Hwang, K. J. (1984). The use of gamma ray perturbed angular correlation technique for the study of liposomal integrity in vitro and in vivo. *Liposome Technology*, Vol. 3 (G. Gregoriadis, ed.). pp. 247–262.
160. Gabizon, A., Huberty, J., Straubinger, R. M., Price, D. C., and Papahadjopoulos, D. (1988/89). An improved method for in vivo tracing and imaging of liposomes using a Gallium 67-deferoxamine complex. *J. Liposome Research 1*: 123–135.
161. Kemps, J. M. A., and Crommelin, D. J. A. (1988). Chemische stabiliteit van fosfolipiden in farmaceutische preparaten. I. Hydrolyse van fosfolipiden in waterig milieu. *Pharm. Weekbl. 123*: 355–363.

162. Kemps, J. M. A., and Crommelin, D. J. A. (1988). De chemische stabiliteit van fosfolipiden in waterig milieu. II. De peroxidatie van fosfolipiden in waterig milieu. *Pharm. Weekbl. 123*: 457–469.
163. Frøkjaer, S., Hjorth, E. L., and Wørts, O. (1984). Stability and storage of liposomes. *Optimization of Drug Delivery* (H. Bundgaard, A. Bagger Hansen, and H. Kofod, eds.). Copenhagen: Munksgaard, pp. 384–404.
164. Grit, M., Zuidam, N. J., Underberg, W. J. M., and Crommelin, D. J. A. (1993). Hydrolysis of partially saturated egg phosphatidylcholine in aqueous liposome dispersions and the effect of cholesterol incorporation on its hydrolysis kinetics. *J. Pharm. Pharmacol. 45*: 490–495.
165. Grit, M., and Crommelin, D. J. A. (1993). The effect of surface charge on the hydrolysis kinetics of partially hydrogenated egtg phosphatidylcholine and egg-phosphatidylglycerol in aqueous liposome dispersions. *Biochim. Biophys. Acta, 1167*: 49–55.
166. Grit, M., and Crommelin, D. J. A. (1992). The effect of aging on the physical stability of liposome dispersions. *Chem. Phys. Lipids 62*: 113–122.
167. Konings, A. W. T. (1984). Lipid peroxidation in liposomes. *Liposome Technology*, 1st ed. (G. Gregoriadis, ed.). Boca Raton, Fla.: CRC Press, pp. 139–161.
168. Pajean, M., Huc, A., and Herbage, D. (1991). Stabilization of liposomes with collagen. *Int. J. Pharmaceutics 77*: 31–40.
169. Zuidam, N. J., Lee, S. L. L., and Crommelin, D. J. A. Gamma-irradiation of liposomes. Effect of bilayer composition and irradiation dose on chemical degradation and physical destabilization, in preparation.
170. Hunt, C. A., and Tsang, S. (1981). Alpha-tocopherol retards autoxidation and prolongs shelf life of liposomes. *Int. J. Pharm. 8*: 101–110.
171. Crommelin, D. J. A., and Van Bommel, L. M. G. (1984). Stability of liposomes on storage: Freeze dried, frozen or as an aqueous dispersion. *Pharm. Res. 1*: 159–163.
172. Crowe, J. H., Crowe, L. M., Carpenter, J. F., and Wistrom, C. A. (1987). Stabilization of dry phospholipid bilayers and proteins by sugars. *Biochem. J. 242*: 1–10.
173. Özer, Y., Talsma, H., Crommelin, D. J. A., and Hincal, A. (1988). Influence of freezing and freeze-drying on the stability of liposomes dispersed in aqueous media. *Acta Pharm. Technol. 34*: 129–139.
174. Ausborn, M., Nuhn, P., and Schreier, H. (1992). Stabilization of liposomes by freeze-thaw and lyophilization techniques: Problems and opportunities. *Europ. J. Pharmaceutics and Biopharm. 38*: 133–139.
175. Hauser, H., and Strauss, G. (1987). Stabilization of small unilamellar phospholipid vesicles during spray drying. *Biochim. Biophys. Acta 897*: 1045–1052.
176. Van Bommel, E. M. G., and Crommelin, D. J. A. (1984). Stability of doxorubicin containing liposomes: As an aqueous dispersion, frozen or freeze dried. *Int. J. Pharm. 22*: 299–310.
177. Strauss, G. (1984). Freezing and thawing of liposome suspensions. *Liposome Technology*, Vol. 1 (G. Gregoriadis, ed.). Boca Raton, Fla.: CRC Press, pp. 197–219.

178. Crowe, L. M., Crowe, J. H., Rudolph, A., Womersley, C., and Appel, L. (1985). Preservation of freeze dried liposomes by trehalose. *Arch. Biochem. Biophys. 242*: 240–247.
179. Talsma, H., and Crommelin, D. J. A. (1993). Liposomes as drug delivery systems, Part 3: Stabilization. *Pharm. Technol.* 48–59.
180. Talsma, H., Van Steenbergen, M. J., and Crommelin, D. J. A. (1991). The cryopreservation of liposomes: 3. Almost complete retention of a water-soluble marker in small liposomes in a cryoprotectant containing dispersion after a freezing/thawing cycle. *Int. J. Pharm. 77*: 119–126.
181. Talsma, H., Van Steenbergen, M. J., Salemink, P. J. M., and Crommelin, D. J. A. (1991). The cryopreservation of liposomes. 1. A differential scanning calorimetry study of the thermal behavior of a liposome dispersion containing mannitol during freezing/thawing. *Pharm. Res. 8*: 1021–106.
182. Talsma, H., Van Steenbergen, M. J., and Crommelin, D. J. A. (1992). The cryopreservation of liposomes. Part 2. Effect of particle size on crystallization behaviour and marker retention. *Cryobiology 29*: 80–86.
183. Goren, D., Gabizon, A., and Barenholz, Y. (1990). The influence of physical characteristics of liposomes on their pharmacological behaviour. *Biochim. Biophys. Acta 1029*: 285–294.
184. Lautenschläger, H., Röding, J., and Ghyczy, M. (1988). Über die Verwendung von Liposomen aus Soja-Phospholipiden in der Kosmetik. *Seifen, Öle, Fette, Wachse 114*: 531–534.
185. Rose, P. E. (1980). Improved tables for the evaluation of sphere size distributions including the effect of section thickness. *J. Micros. 118*: 135–141.
186. Fowler, K., Bottomley, L. A., and Schreier, H. (1992). Surface topography of phospholipid bilayers and vesicles (liposomes) by scanning tunneling microscopy (STM). *J. Controlled Release 22*: 283–292.
187. Ruf, H., Georgalis, Y., and Grell, E. (1989). Dynamic laser light scattering to determine size distributions of vesicles. *Methods Enzym. 172*: 364–390.
188. Dos Ramos, J. G., and Silebi, C. A. (1990). The determination of particle size distribution of submicrometer particles by capillary hydrodynamic fractionation. *J. Colloid Interface Sci. 135*: 165–177.
189. Mason, J. T., and Huang, C. (1978). Hydrodynamic analysis of egg phosphatidylcholine vesicles. *Liposomes and Their Uses in Biology and Medicine* (D. Papahadjopoulos, ed.). *Ann. N.Y. Acad. Sci.*, pp. 29–49.
190. Talsma, H., Jousma, H., Nicolay, K., and Crommelin, D. J. A. (1987). Multilamellar or multivesicular vesicles? *Int. J. Pharm. 37*: 171–173.
191. Hiemenz, P. C. (1986). *Principles of Colloid and Surface Chemistry*, 2d ed. New York: Marcel Dekker.
192. Cevc, G. (1990). Membrane electrostatics. *Biochim. Biophys. Acta 1031-3*: 311–382.
193. Shinitzky, M., and Barenholz, Y. (1978). Fluidity parameters of lipid regions determined by fluorescence polarization. *Biochim. Biophys. Acta 515*: 367–394.
194. Ben-Yashar, V., and Barenholz, Y. (1989). The interaction of cholesterol and cholest-4-en-3-one with dipalmitoylphosphatidylcholine. Comparison based on the use of three fluorophores. *Biochim. Biophys. Acta 985*: 271–278.

195. Jones, G. R., and Cossins, A. R. (1990). Physical methods of study. *Liposomes: A Practical Approach* (R. R. C. New, ed.). Oxford: IRL Press, pp. 183-220.
196. Kikuchi, H., Carlsson, A., Yaki, K., and Hirota, S. (1985). Possibility of heat sterilization of liposomes. *Chem. Pharm. Bull. 39*: 1018-1022.
197. Cherian, M., Lenk, R. P., and Jedrusiak, J. A. (1990). Heat treating liposomes. *PCT Int. Appl. WO* 90:03808.
198. Zuidam, N. J., Lee, S. S. L., and Crommelin, D. J. A. (1993). Sterilization of liposomes by heat treatment. *Pharm. Res. 10*: 1591-1596.
199. Ianzini, F., Guidoni, L., Indovina, P. L., Viti, V., Erriu, G., Onnis, S., and Randaccio, P. (1984). Gamma irradiation effects on phosphatidylcholine multilayer liposomes: Calorimetric, NMR and spectrofluorimetric studies. *Radiat. Res. 98*: 154-166.
200. Albertini, G., Fanelli, E., Guidoni, L., Ianzini, F., Mariani, P., Masella, R., Rustichelli, F., and Viti, V. (1987). Studies on structural modifications induced by gamma irradiation in distearoylphosphatidylcholine liposomes. *Int. J. Radiat. Biol. 52*: 145-156.
201. Martin, F. J. (1990). Pharmaceutical manufacturing of liposomes. *Specialized Drug Delivery Systems: Manufacturing and Production Technology* (P. Tyle, ed.). New York: Marcel Dekker, pp. 267-316.
202. Pearson, F. C. (1985). Pyrogens. *Advances in Parenteral Sciences*. New York: Marcel Dekker.
203. Kriftner, R. W. (1992). Liposome production. The ethanol injection technique and the development of the first approved liposome dermatic. *Liposome Dermatics* (O. Braun-Falco, H. C. Korting, and H. I. Maibach, eds.). Berlin: Springer-Verlag, pp. 91-100.
204. Weder, H. G. (1992). Liposome production: The sizing-up technology starting from mixed micelles and the scaling-up procedure for the topical glucocorticoid betamethasone dipropionate and betamethasone. *Liposome Dermatics* (O. Braun-Falco, H. C. Korting, and H. I. Maibach, eds.). Berlin: Springer-Verlag, pp. 101-109.
205. Schwenderer, R. A. (1986). The preparation of large volumes of homogeneous, sterile liposomes containing various lipophilic cytostatic drugs by the use of a capillary dialyzer. *Cancer Drug Delivery 3*: 123-129.
206. Amselem, S., Gabizon, A., and Barenholz, Y. (1990). Optimization and upscaling of doxorubicin-containing liposomes for clinical use. *J. Pharm. Sci. 79*: 1045-1052.
207. Klimchak, R. J., and Lenk, R. P. (1988). Scale-up of liposome products. *Bio Pharm* 18-21.
208. Ohsawa, T., Matsukawa, Y., Takakura, Y., Hashida, M., and Sezaki, H. (1985). Fate of lipid and encapsulated drug after intramuscular administration of liposomes prepared by the freeze-thawing method in rats. *Chem. Pharm. Bull. 33*: 5013-5022.
209. Schreier, H., Levy, M., and Mihalko, P. (1987). Sustained release of liposome-encapsulated gentamicin and fate of phospholipid following intramuscular injection in mice. *J. Control. Release 5*: 187-192.

210. Kadir, F., Eling, W. M. C., Crommelin, D. J. A., and Zuidema, J. (1991). Influence of injection volume on the release kinetics of liposomal chloroquine administered subcutaneously or intramuscularly to mice. *J. Control. Release 17*: 277–284.
211. Kadir, F., Eling, W. M. C., Crommelin, D. J. A., and Zuidema, J. (1992). Kinetics and prophylactic efficacy of increasing dosages of liposome-encapsulated chloroquine after intramuscular injection into mice. *J. Control. Release 20*: 47–54.
212. Kadir, F., Seijsener, C. B. J., and Zuidema, J. (1992). Influence of the injection volume on the release pattern of intramuscularly administered propranolol to rats. *Int. J. Pharmaceut. 81*: 193–198.
213. Patel, H. M. (1988). Fate of liposomes in the lymphatics. *Liposomes as Drug Carriers* (G. Gregoriadis, ed.). New York: John Wiley.
214. Kadir, F., Zuidema, J., and Crommelin, D. J. A. (1993). Liposomes as drug delivery systems for intramuscular and subcutaneous injection. *Particulate Drug Carriers in Medical Applications* (Rolland, ed.). New York: Marcel Dekker, pp. 165–198.
215. Illum, L., Jacobsen, L. O., Müller, R. H., Mak, E., and Davis, S. S. (1987). Surface characteristics and the interaction of colloidal particles with mouse peritoneal macrophages. *Biomaterials 8*: 113–117.
216. Allen, T. M., and Chonn, A. (1987). Large unilamellar liposomes with low uptake into the reticuloendothelial system. *FEBS Lett. 223*: 42–46.
217. Gabizon, A., and Papahadjopoulos, D. (1988). Liposome formulations with prolonged circulation time in blood and enhanced uptake by tumors. *Proc. Natl. Acad. Sci. U.S.A. 85*: 6949–6953.
218. Senior, J. H., Trimble, K. R., and Maskiewicz, R. (1991). Interaction of positively-charged liposomes with blood: Implications for their application *in vivo*. *Biochim. Biophys. Acta. 1070*: 173–179.
219. Woodle, M. C., and Lasic, D. D. (1992). Sterically stabilized liposomes. *Biochim. Biophys. Acta 1113*: 171–199.
220. Allen, T. M. (1989). Stealth liposomes: Avoiding reticuloendothelial uptake in liposomes. *Liposomes in the Therapy of Infectious Diseases and Cancer* (G. Lopez-Berestein and I. J. Fidler, eds.). New York: Alan R. Liss, pp. 405–415.
221. Allen, T. M., Mehra, T., Hansen, C., and Chin, Y. C. (1992). Stealth liposomes: An improved sustained release system for 1-β-D-arabinofuranosylcytosine. *Cancer Res. 52*: 2431–2439.
222. Lasic, D. D., Martin, F. J., Gabizon, A., Huang, S. K., and Papahadjopoulos, D. (1991). Sterically stabilized liposomes: A hypothesis on the molecular origin of the extended circulation times. *Biochim. Biophys. Acta 1070*: 187–192.
223. Needham, D., McIntosh, T. J., and Lasic, D. D. (1992). Repulsive interactions and mechanical stability of polymer-grafted lipid membranes. *Biochim. Biophys. Acta 1108*: 40–48.
224. Papahadjopoulos, D., Allen, T. M., Gabizon, A., Mayhew, E., Matthay, K., Huang, S. K., Lee, K. D., Woodle, M. C., Lasic, D. D., Redemann, C., and Martin, F. J. (1991). Sterically stabilized liposomes: Improvements in pharma-

cokinetics and antitumor therapeutic efficacy. *Proc. Natl. Acad. Sci. U.S.A. 88*: 11460–11464.
225. Huang, S. K., Hong, K. L., Lee, K. D., Papahadjopoulos, D., and Friend, D. S. (1991). Light microscopic localization of silver enhanced liposome entrapped colloidal gold in mouse tissues. *Biochim. Biophys. Acta 1069*: 117–121.
226. Vaage, J., Mayhew, E., Lasic, D., and Martin, F. (1992). Therapy of primary and metastatic mouse mammary carcinomas with doxorubicin encapsulated in long circulating liposomes. *Int. J. Cancer 51*: 942–948.
227. Mayhew, E. G., Lasic, D., Babbar, S., and Martin, F. J. (1992). Pharmacokinetics and antitumor activity of epirubicin encapsulated in long-circulating liposomes incorporating a polyethylene glycol-derivatized phospholipid. *Int. J. Cancer 51*: 302–309.
228. Bakker-Woudenberg, I. A. J. M., Lokerse, A. F., ten Kate, M. T., and Storm, G. (1992). Enhanced localization of liposomes with prolonged blood circulation time in infected lung tissue. *Biochim. Biophys. Acta 1138*: 318–326.
229. Woodle, M. C., Storm, G., Newman, M. S., Jekot, J. J., Collins, L. R., Martin, F. J., and Szoka, F. C., Jr. (1992). Prolonged systemic delivery of peptide drugs by long-circulating liposomes: Illustration with vasopressin in the Brattleboro rat. *Pharm. Res. 9*: 260–265.
230. Maruyama, K., Kennel, S. J., and Huang, L. (1990). Lipid composition is important for highly efficient target binding and retention of immunoliposomes. *Proc. Natl. Acad. Sci. U.S.A. 87*: 5744–5748.
231. Storm, G., Wilms, H. P., and Crommelin, D. J. A. (1991). Liposomes and biotherapeutics. *Biotherapy 3*: 25–42.
232. Surewicz, W. K., Epand, R. M., Orlowski, R. C., and Mantsch, H. H. (1987). Structural properties of acidic phospholipids in complexes with calcitonin: A Fourier transform infrared spectroscopic investigation. *Biochim. Biophys. Acta 899*: 307–310.
233. Hui, S. W., Epand, R. M., Dell, K. R., Epand, R. F., and Orlowski, R. C. (1984). Effects of lipid structure on peptide-lipid interactions. Complexes of salmon calcitonin with phosphatidylglycerol and with phosphatidic acid. *Biochim. Biophys. Acta 772*: 264–272.
234. Epand, R. M., Epand, R. F., Orlowski, R. C., Schlueter, R. J., Boni, L. T., and Hui, S. W. (1983). Amphipathic helix and its relationship to the interaction of calcitonin with phospholipids. *Biochemistry 22*: 5074–5084.
235. Williams, N. A., and Weiner, N. D. (1989). Interaction of recombinant leukocyte alpha-interferon with lipid monolayers. *Int. J. Pharmaceut. 51*: 241–244.
236. Allison, A. C., and Gregoriadis, G. (1974). Liposomes as immunological adjuvants. *Nature 252*: 252.
237. van Rooijen, N., and van Nieuwmegen, R. (1982). Immunoadjuvant properties of liposomes. *Targeting of Drugs* (G. Gregoriadis, J. Senior, and A. Trouet, eds.). New York: Plenum Press, pp. 301–326.
238. van Rooijen, N. (1988). Liposomes as immunological adjuvants: Recent developments. *Liposomes as Drug Carriers* (G. Gregoriadis, ed.). New York: John Wiley, pp. 159–165.

239. Gregoriadis, G. (1990). Immunological adjuvants: A role for liposomes. *Immunol. Today 11*: 89–97.
240. Alving, C. R. (1991). Liposomes as carriers of antigens and adjuvants. *J. Immunol. Meth. 140*: 1–13.
241. Buiting, A. M. J., van Rooijen, N., and Claassen, E. (1992). Liposomes as antigen carriers and adjuvants in vivo. *Res. Immunol. 143*: 541–548.
242. Fidler, I. J., Sone, S., Fogler, W. E., and Barnes, Z. L. (1981). Eradication of spontaneous metastases and activation of alveolar macrophages by intravenous injection of liposomes containing muramyl dipeptide. *Proc. Natl. Acad. Sci. U.S.A. 78*: 1680–1684.
243. Schroit, A. J., and Fidler, I. J. (1982). Effects of liposome structure and lipid composition on the activation of the tumoricidal properties of macrophages by liposomes containing muramyl dipeptide. *Cancer Res. 42*: 161–167.
244. Phillips, N. C., Moras, M. L., Chedid, L., Petit, J. F., Tenu, J. P., Lederer, E., Bernard, J. M., and Lefrancier, P. (1985). Activation of macrophage cytostatic and cytotoxic activity *in vitro* by liposomes containing a new lipophilic murmyl peptide derivative, MDP-L-alanyl-cholesterol (MTP-CHOL). *J. Biol. Response Modif. 4*: 464–474.
245. Phillips, N. C., and Tsao, M. S. (1989). Inhibition of experimental liver tumor growth in mice by liposomes containing a lipophilic muramyl dipeptide derivative. *Cancer Res. 49*: 936–939.
246. Galelli, A., Charlot, B., Phillips, N. C., and Chedid, L. (1989). Induction of colony-stimulating activity in mice by injection of liposomes containing lipophilic muramyl peptide derivatives. *Cancer Res. 49*: 810–815.
247. Vosika, G. J., Cornelius, D. A., Gilbert, C. W., Sadlik, J. R., Bennek, J. A., Doyle, A., and Hertsgaard, D. (1991). Phase I trial of Immther, a new liposome-incorporated lipophilic disaccharide tripeptide. *J. Immunother. 10*: 256–266.
248. Kleinerman, E. S., Erickson, K. I., Schroit, A. J., Fogler, W. E., and Fidler, I. (1983). Activation of tumoricidal properties in human blood monocytes by liposomes containing lipophilic muramyl tripeptide. *Cancer Res. 43*: 2010–2014.
249. Sone, D., Utsugi, T., Tandon, P., and Ogawara, M. (1988). A dried preparation of liposomes containing muramyl tripeptide phosphatidylethanolamine as a potent activator of human blood monocytes to the anticancer state. *Cancer Immunol. Immunother. 22*: 191–196.
250. MacEwen, E. G., Kurzman, I. D., Rosenthal, R. C., Smith, B. W., Manley, P. A., Roush, J. K., and Howard, P. E. (1989). Therapy for osteosarcoma in dogs with intravenous injection of liposome-encapsulated muramyl tripeptide. *J. Natl. Cancer Inst. 81*: 935–938.
251. Szoka, F. C., Jr. (1992). The macrophage as the principal antigen-presenting cell for liposome-encapsulated antigen. *Res. Immunol. 143*: 186–188.
252. Torchilin, V. P. (1985). Liposomes as targetable drug carriers. *CRC Crit. Rev. Ther. Drug Carrier Syst. 2*: 65–115.
253. Weissig, V., Lasch, J., Klibanov, A. L., and Torchilin, V. P. (1986). A new hydrophobic anchor for the attachment of proteins to liposomal membranes. *FEBS Lett. 202*: 86–90.

254. Bogdanov, A. A., Klibanov, A. L., and Torchilin, V. P. (1988). Protein immobilization on the surface of liposomes via carbodiimide activation in the presence of *N*-hydroxysulfosuccinimide. *FEBS Lett. 231*: 381–384.
255. Schott, H., Hess, W., Hengartner, H., and Schwendener, R. A. (1988). Synthesis of lipophilic amino and carboxyl components for the functionalization of liposomes. *Biochim. Biophys. Acta 943*: 53–62.
256. Weissig, V., Lasch, J., and Gregoriadis, G. (1989). Covalent coupling of sugars to liposomes. *Biochim. Biophys. Acta 1003*: 54–57.
257. Schwendener, R. A., Trüb, T., Schott, H., Langhals, H., Barth, R. F., Groscurth, P., and Hengartner, H. (1990). Comparative studies of the preparation of immunoliposomes with the use of two bifunctional coupling agents and investigation of *in vitro* immunoliposome-target cell binding by cytofluorometry and electron microscopy. *Biochim. Biophys. Acta 1026*: 69–79.
258. Weissig, V., and Gregoriadis, G. (1993). Coupling of aminogroup-bearing ligands to liposomes. *Liposome Technology*, Vol. 3. 2d ed. (G. Gregoriadis, ed.). Boca Raton, Fla.: CRC Press, pp. 231–248.
259. Gould-Fogerite, S., Mazurkiewicz, J. E., Bhisitkul, D., and Mannino, R. J. (1988). The reconstitution of biologically active glycoproteins into large liposomes: Use as a delivery vehicle to animal cells. *Advances in Membrane Biochemistry and Bioenergetics* (C. H. Kim, H. Tedeschi, J. J. Diwan, and J. C. Salerno, eds.). New York: Plenum Press, pp. 569–586.
260. Ran, S., Nussbaum, O., Loyter, A., Marikowsky, Y., and Rivnay, B. (1988). Reconstitution of fusogenic Sendai virus envelopes by the use of the detergent CHAPS. *Arch. Biochem. Biophys. 261*: 437–446.
261. Stegmann, T., Morselt, H. W. M., Booy, F. P., van Breemen, J. F., Scherphof, G., and Wildschut, J. (1987). Functional reconstitution of influenza virus envelopes. *EMBO J. 6*: 2651–2659.
262. Schneider, J., Falk, H., and Hunsmann, G. (1983). Virosomes constructed from lipid and purified Friend leukemia virus glycoprotein. *J. Gen. Virol. 64*: 559–565.
263. Johnson, D. C., Wittels, M., and Spear, P. G. (1984). Binding to cells of virosomes containing *herpes simplex* virus type 1 glycoproteins and evidence for fusion. *J. Virol. 52*: 238–247.
264. Ho, R. J. Y., Burke, R. L., and Merigan, T. C. (1989). Antigen-presenting liposomes are effective in treatment of recurrent herpes simplex virus genitalis in guinea pigs. *J. Virol. 63*: 2951–2958.
265. Rubas, W., Banerjea, A. C., Gallati, H., Speiser, P. P., and Joklik, W. K. (1990). Incorporation of the reovirus M cell attachment protein into small unilamellar vesicles: Incorporation efficiency and binding capability to L929 cell *in vitro*. *J. Microencapsulation 7*: 385–395.
266. Russell, D. G., and Alexander, J. (1988). Effective immunization against cutaneous leishmaniasis with defined membrane antigens reconstituted into liposomes. *J. Immunol. 140*: 1274–1279.
267. Richards, R. L., Hayre, M. D., Hockmeyer, W. T., and Alving, C. R. (1988). Liposomes, Lipid A, and aluminum hydroxide enhance the immune response to a synthetic malaria sporozoite antigen. *Infect. Immun. 56*: 682–686.

268. Garcon, N., Gregoriadis, G., Taylor, M., and Summerfield, J. (1988). Mannose-mediated targeted immunoadjuvant action of liposomes. *Immunology 64*: 743–745.

269. Davis, D., and Gregoriadis, G. (1987). Liposomes as adjuvants with immunopurified tetanus toxoid: Influence of liposomal characteristics. *Immunology 61*: 229–234.

270. el Guink, N., Kris, R. M., Goodman-Snitkoff, G., Small, P. A., Jr., and Mannino, R. J. (1989). Intranasal immunization with proteoliposomes protects against influenza. *Vaccine 7*: 147–151.

271. Gregoriadis, G., Tan, L., Ben-Ahmedia, E. T. S., and Jennings, R. (1992). Liposomes enhance the immunogenicity of reconstituted influenza virus A/PR/8 envelopes and the formation of protective antibody by influenza virus A/Sichuan/87 (H3N2) surface antigen. *Vaccine 10*: 747–753.

272. Xiao, Q., Gregoriadis, G., and Ferguson, M. (1989). Immunoadjuvant action of liposomes for entrapped poliovirus peptide. *Biochem. Soc. Trans. 17*: 695.

273. Just, M., Berger, R., Drichsler, H., Brantschen, S., and Gluck, R. (1992). A single vaccination with an inactivated hepatitis A liposome vaccine induces protective antibodies after only two weeks. *Vaccine 10*: 737–739.

274. Perrin, P., Thibodeau, L., and Sureau, P. (1985). Rabies immunosome (subunit vaccine) structure and immunogenicity. Pre- and post-exposure protection studies. *Vaccine 3*: 325–332.

275. Oth, D., Mercier, G., Perrin, P., Joffret, M. L., Sureau, P., and Thibodeau, L. (1987). The association of the rabies glycoprotein with liposome (immunosome) induces an *in vitro* specific release of interleukin-2. *Cell. Immunol. 108*: 220–226.

276. Mansour, S., Thibodeau, L., Perrin, P., Sureau, P., Mercier, G., Joffret, M. L., and Oth, D. (1988). Enhancement of antigen-specific interleukin 2 production by adding liposomes to rabies antigens for priming. *Immunol. Lett. 18*: 33–36.

277. Thibodeau, L., Chagnon, M., Flamand, L., Oth, D., Lachappelle, L., Tremblay, C., and Montagnier, L. (1989). Rôle des liposomes dans la présentation de la glycoproteine de l'enveloppe du VIH et la réponse immunitaire chez la souris. *C.R. Acad. Sci. Paris 309* (III): 741–747.

278. Bergers, J. J., Storm, G., and den Otter, W. (1992). Liposomes as vehicles for the presentation of tumor-associated antigens to the immune system. *Liposome Technology*, Vol. 2. 2d ed. (G. Gregoriadis, ed.). Boca Raton, Fla.: CRC Press, pp. 141–166.

279. Bergers, J. J., den Otter, W., Dullens, H. F. J., de Groot, J. W., Steerenberg, P. A., Filius, P. M. G., and Crommelin, D. J. A. (1994). Specific tumor immunity induced by immunization with liposome incorporated tumor associated antigens combined with immunostimulants. *Cancer Immunol. Immunother.*, in press.

280. Reddy, R., Zhou, F., Nair, S., Huang, L., and Rouse, B. T. (1992). *In vivo* cytotoxic T lymphocyte induction with soluble proteins administered in liposomes. *J. Immunol. 148*: 1585–1589.

281. Zhou, F., Rouse, B. T., and Huang, L. (1992). Induction of cytotoxic T lymphocytes *in vivo* with protein antigen entrapped in membranous vehicles. *J. Immunol. 149*: 1599–1604.

282. Huang, L., Reddy, R., Nair, S. K., Zhou, F., and Rouse, B. T. (1992). Liposomal delivery of soluble protein antigens for class I MHC-mediated antigen presentation. *Res. Immunol. 143*: 192–196.
283. Michalek, S. M., Childers, N. K., Katz, J., Denys, F. R., Berry, A. K., Eldridge, J. H., McGhee, J. R., and Curtiss, R. (1989). Liposomes as oral adjuvants. *Curr. Top. Microbiol. Immunol. 146*: 51–58.
284. Childers, N. K., Denys, F. R., McGee, N. F., and Michalek, S. M. (1990). Ultrastructural study of liposome uptake by M cells of rat Peyer's patch: An oral vaccine system for delivery of purified antigen. *Reg. Immunol. 3*: 8–16.
285. Jackson, S., Mestecky, J., Childers, N. K., and Michalek, S. M. (1990). Liposomes containing anti-idiotypic antibodies: An oral vaccine to induce protective secretory immune responses specific for pathogens of mucosal surfaces. *Infect. Immun. 58*: 1932–1936.
286. Thompson, L. (1992). At age 2, gene therapy enters a growth phase. *Science 258*: 744–746.
287. Miller, A. D., and Rosman, G. J. (1989). Improved retroviral vectors for gene transfer and expression. *BioTechniques 7*: 980–990.
288. Rosenfeld, M. A., Yoshimura, K., Trapnell, B. C., Yoneyama, K., Rosenthal, E. R., Dalemans, W., Fukayama, M., Bargon, J., Stier, L. E., Stratford-Perricaudet, L., Perricaudet, M., Guggino, W. B., Pavirani, A., Lecocq, J. P., and Crystal, R. G. (1992). In vivo transfer of the human cystic fibrosis transmembrane conductance regulator gene to the airway epithelium. *Cell 68*: 143–155.
289. Curiel, D. T., Agarwal, S., Romer, M. U., Wagner, E., Cotten, M., Birnstiel, M. L., and Boucher, R. C. (1992). Gene transfer to respiratory epithelial cells via the receptor-mediated endocytosis pathway. *Am. J. Respir. Cell Mol. Biol. 6*: 247–252.
290. Ostro, M. J., Giacomoni, D., and Dray, S. (1977). Incorporation of high molecular weight RNA into large artificial lipid vesicles. *Biochem. Biophys. Res. Comm. 76*: 836–842.
291. Dimitriadis, G. J. (1979). Entrapment of plasmid DNA in liposomes. *Nucleic Acids Res. 6*: 2697–2705.
292. Fraley, R. T., Fornari, C. S., and Kaplan, S. (1979). Entrapment of a bacterial plasmid in phospholipid vesicles: Potential for gene transfer. *Proc. Natl. Acad. Sci. U.S.A. 76*: 3348–3352.
293. Wilson, T., Papahadjopoulos, D., and Taber, R. (1979). The introduction of poliovirus RNA into cells via lipid vesicles-liposomes. *Cell 17*: 77–84.
294. Fraley, R., Subramani, S., Berg, P., and Papahadjopoulos, D. (1980). Introduction of liposome-encapsulated SV40 DNA into cells. *J. Biol. Chem. 255*: 10431–10435.
295. Wong, T. K., Nicolau, C., and Hofschneider, P. H. (1980). Appearance of beta-lactamase activity in animal cells upon liposome-mediated gene transfer. *Gene 10*: 87–94.
296. Nicolau, C., and Rottem, S. (1982). Expression of a beta-lactamase activity in *Mycoplasma capriodolum* transfected with the liposome-encapsulated E. coli pBR322 plasmid. *Biochem. Biophys. Res. Comm. 108*: 982–986.

297. Schaefer-Ridder, M., Wang, Y., and Hofschneider, P. H. (1982). Liposomes as gene carriers: Efficient transformation of mouse L cells by thymidine kinase gene. *Science 215*: 166–168.
298. Nicolau, C., Lepape, A., Soriano, P., Fargette, F., Muh, J. P., and Juhel, M. F. (1982). *In vivo* transfer and expression of the liposome encapsulated rat insulin gene. *Progr. Clin. Biol. Res. 102(A)*: 321–330.
299. Mannino, R. J., and Gould-Fogerite, S. (1988). Liposome mediated gene transfer. *BioTechniques 6*: 682–690.
300. Nicolau, C., and Cudd, A. (1989). Liposomes as carriers of DNA. *Crit. Rev. Therap. Drug Carrier Syst. 6*: 239–271.
301. Hug, P., and Sleight, R. G. (1991). Liposomes for the transformation of eukaryotic cells. *Biochim. Biophys. Acta 1097*: 1–17.
302. Felgner, P. L., and Rhodes, G. (1991). Gene therapeutics. *Nature 349*: 351–352.
303. Fraley, R., Straubinger, R. M., Rule, G., Springer, E. L., and Papahadjopoulos, D. (1981). Liposome-mediated delivery of deoxyribonucleic acid to cells: Enhanced efficiency of delivery related to lipid composition and incubation conditions. *Biochemistry 20*: 6978–6987.
304. Itani, T., Ariga, H., Yamaguchi, N., Tadakuma, T., and Yasuda, T. (1987). A simple and efficient liposome method for transfection of DNA into mammalian cells grown in suspension. *Gene 56*: 256–276.
305. Gould-Fogerite, S., Mazurkiewicz, J. E., Raska, K., Jr., Voelkerding, K., Lehman, J. M., and Mannino, R. J. (1989). Chimerasome-mediated gene transfer in vitro and in vivo. *Gene 84*: 429–438.
306. Felgner, P. L., Gadek, T. R., Holm, M., Roman, R., Chan, H. W., Wenz, M., Northrop, J. P., Ringold, G. M., and Danielsen, M. (1987). Lipofection: A highly efficient, lipid-mediated DNA-transfection procedure. *Proc. Natl. Acad. Sci. U.S.A. 84*: 7413–7417.
307. Stamatatos, L., Levantis, R., Zuckerman, M. J., and Silvius, J. R. (1988). Interactions of cationic lipid vesicles with negatively charged phospholipid vesicles and biological membranes. *Biochemistry 27*: 3917–3925.
308. Leventis, R., and Silvius, J. R. (1990). Interactions of mammalian cells with lipid dispersions containing novel metabolizable cationic amphiphiles. *Biochim. Biophys. Acta 1023*: 124–132.
309. Gao, X., and Huang, L. (1991). A novel cationic liposome reagent for efficient transfection of mammalian cells. *Biochem. Biophys. Res. Comm. 179*: 280–285.
310. Debs, R., Pian, M., Gaensler, K., Clements, J., Friend, D. S., and Dobbs, L. (1992). Prolonged transgene expression in rodent lung cells. *Am. J. Respir. Cell Mol. Biol. 7*: 406–413.
311. Trubetskoy, V. S., Torchilin, V. P., Kennel, S., and Huang, L. (1992). Cationic liposomes ehnance targeted delivery and expression of exogenous DNA mediated by *N*-terminal modified poly(L-lysine)-antibody conjugate in mouse lung endothelial cells. *Biochim. Biophys. Acta 1131*: 311–313.
312. Legendre, J. Y., and Szoka, F. C., Jr. (1993). Cyclic amphipathic peptide-DNA complexes mediate high efficiency transfection of adherent mammalian cells. *Proc. Natl. Acad. Sci. U.S.A. 90*: 893–897.

313. Düzgünes, N., Goldstein, J. A., Friend, D. S., and Felgner, P. L. (1989). Fusion of liposomes containing a novel cationic lipid, *N*-[2,3-(dioleyloxy)propyl]-*N,N,N*-trimethylammonium: Induction by multivalent anions and asymmetric fusion with acidic phospholipid vesicles. *Biochemistry 28*: 9179–9184.
314. Malone, R. W., Felgner, P. L., and Verma, I. M. (1989). Cationic liposome-mediated RNA transfection. *Proc. Natl. Acad. Sci. U.S.A. 86*: 6077–6081.
315. Brigham, K. L., Meyrick, B., Christman, B., Berry, L. C., Jr., and King, G. (1989). Expression of a prokaryotic gene in cultured lung endothelial cells after lipofection with a plasmid vector. *Am. J. Resp. Cell Mol. Biol. 1*: 95–100.
316. Brigham, K. L., Meyrick, B. O., Christman, B., Magnuson, M., King, G., and Berry, L. C., Jr. (1989). In vivo transfection of murine lungs with a functioning prokaryotic gene using a liposome vehicle. *Am. J. Med. Sci. 298*: 278–281.
317. Plitman, J. D., Canonico, A. E., Conary, J. T., Meyrick, B. O., and Brigham, K. L. (1992). The effects of inhaled and intravenous DNA/liposomes on lung function and histology in the rabbit. *Am. Rev. Resp. Dis. 145*: A588.
318. Stribling, R., Brunette, E., Liggitt, D., Gaensler, K., and Debs, R. (1992). Aerosol gene delivery *in vivo*. *Proc. Natl. Acad. Sci. U.S.A. 89*: 11277–11281.
319. Lim, C. S., Chapman, G. D., Gammon, R. S., Mühlestein, J. B., Bauman, R. P., Stack, R. S., and Swain, J. L. (1991). Direct *in vivo* gene transfer into the coronary and peripheral vasculatures of the intact dog. *Circulation 83*: 2007–2011.
320. Nabel, E. G., Plautz, G., and Nabel, G. J. (1990). Site-specific gene expression *in vivo* by direct gene transfer into the arterial wall. *Science 249*: 1285–1288.
321. Stewart, M. J., Plautz, G. E., del Buono, L., Yang, Z. Y., Xu, L., Gao, X., Huang, L., Nabel, E. G., and Nabel, G. J. (1992). Gene transfer *in vivo* with DNA-liposome complexes: Safety and acute toxicity in mice. *Human Gene Ther. 3*: 267–275.
322. Hazinski, T. A., Ladd, P. A., and DeMatteo, C. A. (1991). Localization and induced expression of fusion genes in the rat lung. *Am. J. Resp. Cell Mol. Biol. 4*: 206–209.
323. Canonico, A. E., Conary, J. T., Meyrick, B. O., and Brigham, K. L. (1992). In vivo expression of a CMV promoter driven human alpha-1 antitrypsin (a1AT) gene after intravenous or airway administration of DNA/liposome complex. *Am. Rev. Resp. Dis. 145*: A200.
324. Hyde, S. C., Gill, D. R., Higgins, C. F., Trezise, A. E. O., MacVinish, L. J., Cuthbert, A. W., Ratcliff, R., Evans, M. J., and Colledge, W. H. (1993). Correction of the ion transport defect in cystic fibrosis transgenic mice by gene therapy. *Nature 362*: 250–255.
325. Soriano, P., Dijkstra, J., Legrand, A., Spanjer, H., Londos-Gagliardi, D., Roerdink, F., Scherhoff, G., and Nicolau, C. (1983). Targeted and nontargeted liposomes for *in vivo* transfer to rat liver cells of a plasmid containing the preproinsulin 1 gene. *Proc. Natl. Acad. Sci. U.S.A. 80*: 7128–7131.
326. Stavridis, J. C., Deliconstantinos, G., Psallidopoulos, M. C., Armenakis, N. A., Hadjiminas, D. J., and Hadjiminas, J. (1986). Construction of trans-

ferrin-coated liposomes for *in vivo* transport of exogenous DNA to bone marrow erythroblasts in rabbits. *Exp. Cell Res. 164*: 568–572.

327. Wang, C. Y., and Huang, L. (1987). pH-sensitive immunoliposomes mediated target-cell-specific delivery and controlled expression of a foreign gene in mouse. *Proc. Natl. Acad. Sci. U.S.A. 84*: 7851–7855.
328. Subbarao, N. K., Parente, R. A., Szoka, F. C., Jr., Nadshi, L., and Pongracz, K. (1987). pH-Dependent bilayer destabilization by an amphipathic peptide. *Biochemistry 26*: 2964–2972.
329. Legendre, J. Y., and Szoka, F. C., Jr. (1992). Delivery of plasmid DNA into mammalian cell lines using pH-sensitive liposomes: Comparison with cationic liposomes. *Pharm. Res. 9*: 1235–1242.
330. Akhtar, S., and Juliano, R. L. (1992). Liposome delivery of antisense oligonucleotides: Adsorption and efflux characteristics of phosphorothioate oligodeoxynucleotides. *J. Control. Release 22*: 47–56.
331. Bennett, C. F., Chiang, M. Y., Chan, H., and Grimm, S. (1993). Use of cationic lipids to enhance the biological activity of antisense oligonucleotides. *J. Liposome Res. 3*: 85–102.
332. Renneisen, K., Leserman, L., Matthes, E., Schröder, H. C., and Müller, W. E. G. (1990). Inhibition of expression of human immunodeficiency virus-1 *in vitro* by antibody-targeted liposomes containing antisense RNA to the env region. *J. Biol. Chem. 265*: 16337–16342.
333. Leonetti, J. P., Machy, P., Degols, G., Lebleu, B., and Leserman, L. (1990). Antibody-targeted liposomes containing oligodeoxyribonucleotides complementary to viral RNA selectively inhibit viral replication. *Proc. Natl. Acad. Sci. U.S.A. 87*: 2448–2451.
334. Thierry, A. R., Rahman, A., and Dritschilo, A. (1993). Overcoming multidrug resistance in human tumor cells using free and liposomally encapsulated antisense oligodeoxynucleotides. *Biochem. Biophys. Res. Comm. 190*: 952–960.
335. Thierry, A. L., and Dritschilo, A. (1993). Intracellular availability of unmodified, phosphorothioated and liposomally encapsulated oligodexynucleotides for antisense activity. *Nucleic Acids Res. 20*: 5691–5698.
336. Lee, V. H. L., Urrea, P. T., Smith, R. E., and Schanzlin, D. J. (1985). Ocular drug bioavailability from topically applied liposomes. *Survey Ophthalmol. 29*: 335–348.
337. Niesman, M. R. (1992). The use of liposomes as drug carriers in ophthalmology. *Crit. Rev. Therap. Drug Carrier Syst. 9*: 1–38.
338. Smolin, G., Okumoto, M., Feller, S., and Condon, D. (1981). Idoxuridine-liposome therapy for herpes simplex keratitis. *Am. J. Opthalmol. 91*: 220–225.
339. Shaeffer, H. E., and Krohn, D. L. (1982). Liposomes in topical drug delivery. *Invest. Ophthalmol. Vis. Sci. 21*: 220–227.
340. Singh, K., and Mezei, M. (1983). Liposomal ophthalmic drug delivery systems I. Triamcinolone acetonide. *Int. J. Pharmaceut. 16*: 339–344.
341. Singh, K., and Mezei, M. (1984). Liposomal ophthalmic delivery system. II. Dihydrostreptomycin sulfate. *Int. J. Pharmaceut. 19*: 263–269.

342. Stratford, R. E., Jr., Yang, D. C., Redell, M. A., and Lee, V. H. L. (1983). Effects of topically applied liposomes on disposition of epinephrine and inulin in the albino rabbit eye. *Int. J. Pharmaceut. 13*: 263–272.
343. Benita, S., Plenecassagne, J. D., Caves, G., Drouin, D., Dong, P. L. H., and Sincholle, D. (1984). Pilocarpine hydrochloride liposomes: Characterization *in vitro* and preliminary evaluation *in vivo* in rabbit eye. *J. Microencapsulation 1*: 203–216.
344. Barza, M., Baum, J., and Szoka, F. (1984). Pharmacokinetics of subconjunctival liposome-encapsulated gentamicin in normal rabbit eyes. *Invest. Ophthalmol. Vis. Sci. 25*: 486–490.
345. Lee, V. H. L. (1984). Precorneal factors influencing the ocular distribution of topically applied liposomal inulin. *Curr. Eye Res. 3*: 585–591.
346. Shaeffer, H. E., Brietfeller, J. M., and Krohn, D. L. (1982). Lectin-mediated attachment of liposomes to cornea: Influence of transcorneal drug flux. *Invest. Ophthalmol. Vis. Sci. 23*: 530–533.
347. Norley, S. G., Huang, L., and Rouse, B. T. (1986). Targeting of drug loaded immunoliposomes to *herpes simplex* virus infected corneal cells: An effective means of inhibiting virus replication *in vitro*. *J. Immunol. 136*: 681–685.
348. Norley, S. G., Sendele, D., Huang, L., and Rouse, B. T. (1987). Inhibition of HSV replication in the mouse cornea by drug containing immunoliposomes. *Invest. Ophthalmol. Vis. Sci. 28*: 591–595.
349. Fitzgerald, P., Hadgraft, J., and Wilson, C. G. (1987). A gamma scintigraphic evaluation of the precorneal residence of liposomal formulations in the rabbit. *J. Pharm. Pharmacol. 39*: 487–490.
350. Fitzgerald, P., Hadgraft, J., Kreuter, J., and Wilson, C. G. (1987). A gamma scintigraphic evaluation of microparticulate ophthalmic delivery systems: Liposomes and nanoparticles. *Int. J. Pharmaceut. 40*: 81–84.
351. Taniguchi, K., Yamamoto, Y., Itakura, K., Miichi, H., and Hayashi, S. (1988). Assessment of ocular irritability of liposome preparations. *J. Pharmaco-biodyn. 11*: 607–611.
352. Guo, L. S. S., Redemann, C. T., and Radhakrishnan, R. (1987). Bioadhesive liposomes in ophthalmic drug delivery. *Invest. Ophthalmol. Vis. Sci.* (Suppl.) *28*: 72.
353. Guo, L. S. S., Sarris, A. M., and Levy, M. D. (1989). A safe bioadhesive liposomal formulation for ophthalmic applications. *Invest. Ophthalmol. Vis. Sci.* (Suppl.) *29*: 439.
354. McCalden, T. A., and Levy, M. (1990). Retention of topical liposomal formulation on the cornea. *Experientia 40*: 713–715.
355. Davies, N. M., Farr, S. J., Hadgraft, J., and Kellaway, I. W. (1992). Evaluation of mucoadhesive polymers in ocular drug delivery. II. Polymer-coated vesicles. *Pharm. Res. 9*: 1137–1142.
356. Schäfer-Korting, M., Korting, H. C., and Braun-Falco, O. (1989). Liposome preparations: A step forward in topical drug therapy for skin disease? *J. Am. Acad. Dermatol. 21*: 1271–1275.
357. Egbaria, K., and Weiner, N. (1990). Liposomes as a topical drug delivery system. *Adv. Drug Deliv. Rev. 5*: 287–300.

358. Korting, H. C., Blecher, P., Schäfer-Korting, M., and Wendel, A. (1991). Topical liposome drugs to come: What the patent literature tells us. *J. Am. Acad. Dermatol. 25*: 1068–1071.
359. Schreier, H., and Bouwstra, J. (1994). Liposomes as topical drug carriers: Dermal and transdermal drug delivery. *J. Control. Release*, in press.
360. Mezei, M., and Gulasekharam, V. (1980). Liposomes—A selective drug delivery system for the topical route of administration. I. Lotion dosage form. *Life Sci. 26*: 1473–1477.
361. Mezei, M., and Gulasekharam, V. (1982). Liposomes—A selective drug delivery system for the topical route of administration: Gel dosage form. *J. Pharm. Pharmacol. 34*: 473–474.
362. Ganesan, M. G., Weiner, N. D., Flynn, G. L., and Ho, N. F. H. (1984). Influence of liposomal drug entrapment on percutaneous absorption. *Int. J. Pharmaceut. 20*: 139–154.
363. Ho, N. F. H., Ganesan, M. G., Weiner, N. D., and Flynn, G. L. (1985). Mechanisms of topical delivery of liposomally entrapped drugs. *J. Control. Release 2*: 61–65.
364. Knepp, V. M., Hinz, R. S., Szoka, F. C., Jr., and Guy, R. H. (1988). Controlled drug delivery from a novel liposomal delivery system. I. Investigation of transdermal potential. *J. Control. Release 5*: 211–221.
365. Knepp, V. M., Szoka, F. C., Jr., and Guy, R. H. (1990). Controlled drug release from a novel liposomal delivery system. II. Transdermal delivery characteristics. J. Control. Release 12: 25–37.
366. Lasch, J., Laub, R., and Wohlrab, W. (1991). How deep do intact liposomes penetrate into human skin? *J. Control. Release 18*: 55–58.
367. Hofland, H. E. J., Bouwstra, J. A., Ponec, M., Bodde, H. E., Spies, F., Coos Verhoef, J., and Junginger, H. E. (1991). Interactions of nonionic surfactant vesicles with cultured keratinocytes and human skin in vitro: A survey of toxicological aspects and ultrastructural changes in stratum corneum. *J. Control. Release 16*: 155–168.
368. Abraham, W., Wertz, P. W., Landmann, L., and Downing, D. T. (1987). Stratum corneum lipid liposomes: Calcium-induced transformation into lamellar sheets. *J. Invest. Dermatol. 88*: 212–214.
369. Golden, G. M., Guzek, D. B., Kennedy, A. H., McKie, J. E., and Potts, R. O. (1987). Stratum corneum lipid phase transitions and water barrier properties. *Biochemistry 26*: 2382–2388.
370. Goundalkar, A., and Mezei, M. (1984). Chemical modification of triamcinolone acetonide to improve liposomal encapsulation. *J. Pharm. Sci. 73*: 834–835.
371. Wohlrab, W., and Lasch, J. (1987). Penetration kinetics of liposomal hydrocortisone in human skin. *Dermatologica 174*: 18–22.
372. Korting, H. C., Zienicke, H., Schäfer-Korting, M., and Braun-Falco, O. (1990). Liposome encapsulation improves efficacy of betamethasone dipropionate in atopic eczema but not in psoriasis vulgaris. *Eur. J. Clin. Pharmacol. 39*: 349–351.
373. Lasch, J., and Wohlrab, W. (1986). Liposome-bound cortisol: A new approach to cutaneous therapy. *Biomed. Biochim. Acta 10*: 1295–1299.

374. Vermorken, A. J. M., Hukkelhoven, M. W. A. C., Vermeesch-Markslag, A. M. G., Goos, C. M. A. A., Wirtz, P., and Ziegenmeyer, J. (1984). The use of liposomes in the topical application of steroids. *J. Pharm. Pharmacol. 36*: 334–336.
375. Kato, A., Ishibashi, Y., and Miyake, Y. (1987). Effect of egg yolk lecithin on transdermal delivery of bunazosin hydrochloride. *J. Pharm. Pharmacol. 39*: 399–400.
376. Kimura, T., Nagahara, N., Hirabayashi, K., Kurosaki, Y., and Nakayama, T. (1989). Enhanced percutaneous penetration of flufenamic acid using lipid disperse systems containing glycosylceramides. *Chem. Pharm. Bull. 37*: 454–457.
377. Michel, C., Purmann, T., Mentrup, E., Seiller, E., and Kreuter, J. (1992). Effect of liposomes on percutaneous penetration of lipophilic materials. *Int. J. Pharmaceut. 84*: 93–105.
378. Cevc, G., and Blume, G. (1992). Lipid vesicles penetrate into intact skin owing to the transdermal osmotic gradients and hydration force. *Biochim. Biophys. Acta 1104*: 226–232.
379. Planas, M. E., Gonzelez, P., Rodriguez, L., Sanchez, S., and Cevc, G. (1992). Noninvasive percutaneous induction of topical analgesia by a new type of drug carrier, and prolongation of local pain insensitivity by anesthetic liposomes. *Anesth. Analg. 75*: 615–621.
380. Miyachi, Y., Imamura, S., and Niwa, Y. (1987). Decreased skin superoxide dismutase activity by a single exposure of ultraviolet radiation is reduced by liposomal superoxide dismutase pretreatment. *J. Invest. Dermatol. 89*: 111–112.
381. Brown, G. L., Curtsinger, L. J., White, M., Mitchell, R. O., Pietsch, J., Nordquist, R., von Fraunhofer, A., and Schultz, G. S. (1988). Acceleration of tensile strength of incisions treated with EGF and TGF-β. *Ann. Surg. 208*: 788–794.
382. Egbaria, K., Ramachandran, C., Kittayanond, D., and Weiner, N. (1990). Topical delivery of liposomally encapsulated interferon evaluated by *in vitro* diffusion studies. *Antimicrob. Agents Chemother. 34*: 107–110.
383. Mihalko, P. J., Schreier, H., and Abra, R. M. (1988). Liposomes: A pulmonary perspective. *Liposomes as Drug Carriers* (G. Gregoriadis, ed.). New York: John Wiley, pp. 679–694.
384. Kellaway, I. W., and Farr, S. J. (1990). Liposomes as drug delivery systems to the lung. *Adv. Drug Deliv. Rev. 5*: 149–161.
385. Schreier, H. (1992). Liposome aerosols. *J. Liposome Res. 2*: 145–184.
386. Schreier, H., Gonzalez-Rothi, R. J., and Stecenko, A. A. (1993). Pulmonary delivery of liposomes. *J. Control. Release 24*: 209–223.
387. Niven, R. W., and Schreier, H. (1990). Nebulization of liposomes. I. Effects of lipid composition. *Pharm. Res. 7*: 1127–1133.
388. Niven, R. W., Speer, M., and Schreier, H. (1991). Nebulization of liposomes. II. The effects of size and modeling of solute release profiles. *Pharm. Res. 8*: 217–221.
389. Niven, R. W., Carvajal, M. T., and Schreier, H. (1992). Nebulization of liposomes. III. Effect of operating conditions. *Pharm. Res. 9*: 515–520.

390. Taylor, K. M. G., Taylor, G., Kellaway, I. W., and Stevens, J. (1990). The stability of liposomes to nebulization. *Int. J. Pharmaceut. 58*: 57–61.
391. Gilbert, B. E., Six, H. R., Wilson, S. Z., Wyde, P. R., and Knight, V. (1988). Small particle aerosols of enviroxime-containing liposomes. *Antiviral Res. 9*: 355–365.
392. Farr, S. J., Kellaway, I. W., Parry-Jones, D. R., and Woolfrey, S. G. (1985). 99mTechnetium as a marker of liposomal deposition and clearance in the human lung. *Int. J. Pharmaceut. 26*: 303–316.
393. Gonzalez-Rothi, R. J., Cacace, J., Straub, L., and Schreier, H. (1991). Liposomes and pulmonary alveolar macrophages: Functional and morphological interactions. *Exp. Lung Res. 17*: 687–705.
394. Myers, M. A., Thomas, D. A., Straub, L., Soucy, D. W., Niven, R. W., Kaltenbach, M., Hood, C. I., Schreier, H., and Gonzalez-Rothi, R. J. (1993). Pulmonary effects of chronic exposure to liposome aerosols in mice. *Exp. Lung Res. 19*: 1–19.
395. Schreier, H., McNicol, K. J., Ausborn, M., Soucy, D. W., Derendorf, H., Stecenko, A. A., and Gonzalez-Rothi, R. J. (1992). Pulmonary delivery of anikacin liposomes and acute liposome toxicity in the sheep. *Int. J. Pharmaceut. 87*: 183–193.
396. Thomas, D. A., Myers, M. A., Wichert, B. M., Schreier, H., and Gonzalez-Rothi, R. J. (1991). Acute effects of liposome aerosol inhalation on pulmonary function in healthy human volunteers. *Chest 99*: 1268–1270.
397. Jobe, A., and Ikegami, M. (1987). Surfactant for the treatment of respiratory distress syndrome. *Am. Rev. Respir. Dis. 136*: 1256–1275.
398. Forsgren, P. E., Modig, J. A., Dahlbäck, C. M. O., and Axelsson, B. I. (1990). Prophylactic treatment with an aerosolized corticosteroid liposome in a porcine model of early ARDS induced by endotoxaemia. *Acta Chir. Scand. 156*: 423–431.
399. Forsgren, P., Modig, J., Gerdin, B., Axelsson, B., and Dahlbäck, M. (1990). Intrapulmonary deposition of aerosolized Evans Blue dye and liposomes in an experimental porcine model of early ARDS. *Upsala J. Med. Sci. 95*: 117–136.
400. McCullogh, H. N., and Juliano, R. L. (1979). Organ-selective action of an antitumor drug: Pharmacologic studies of liposome-encapsulated beta-cytosine arabinoside administered via the respiratory system of rats. *J. Natl. Cancer Inst. 63*: 727–731.
401. Juliano, R. L., and McCullough, H. N. (1980). Controlled delivery of an antitumor drug: Localized action of liposome-encapsulated cytosine arabinoside administered via the respiratory system. *J. Pharmacol. Exp. Therap. 214*: 381–387.
402. Padmanabhan, R. V., Gudapaty, R., Liener, I. E., Schwartz, B. A., and Hoidal, J. R. (1985). Protection against pulmonary oxygen toxicity in rats by the intratracheal administration of liposome-encapsulated superoxide dismutase or catalase. *Am. Rev. Respir. Dis. 132*: 164–167.
403. Shek, P. N., Jurima-Romet, M., Barber, R. F., and Demeester, J. (1988). Liposomes: Potential for inhalation prophylaxis and therapy. *J. Aerosol Med. 1*: 257–258.

404. Jurima-Romet, M., Barber, R. F., Demeester, J., and Shek, P. N. (1990). Distribution studies of liposome-encapsulated glutathione administered to the lung. *Int. J. Pharmaceut. 63*: 227–235.
405. Debs, R. J., Straubinger, R. M., Brunette, E. N., Lin, J. M., Lin, E. J., Montgomery, B., Friend, D. S., and Papahadjopoulos, D. (1987). Selective enhancement of pentamidine uptake in the lung by aerosolization and delivery in liposomes. *Am. Rev. Respir. Dis. 135*: 731–737.
406. Wyde, P. R., Six, H. R., Wilson, S. Z., Gilbert, B. E., and Knight, V. (1988). Activity against rhinoviruses, toxicity, and delivery in aerosol of enviroxime in liposomes. *Antimicrob. Agents Chemother. 32*: 890–895.
407. Gilbert, B. E., Wyde, P. R., and Wilson, S. Z. (1992). Aerosolized liposomal amphtericin B for treatment of pulmonary and systemic *Cryptococcus neoformans* infections in mice. *Antimicrob. Agents Chemother. 36*: 1466–1471.
408. Pettenazzo, A., Jobe, A., Ikegami, M., Abra, R., Hogue, E., and Mihalko, P. (1989). Clearance of phosphatidylcholine and cholesterol from liposomes, liposomes loaded with metaproterenol, and rabbit surfactant from adult rabbit lungs. *Am. Rev. Respir. Dis. 138*: 752–758.
409. McCalden, T. A., Fielding, R. M., Mihalko, P. J., and Kaplan, S. A. (1989). Sustained bronchodilator therapy using inhaled liposomal formulations of beta-2 adrenergic agonists. *Novel Drug Delivery and Its Therapeutic Application* (L. F. Prescott and W. S. Nimmo, eds.). New York: John Wiley, pp. 297–303.
410. Taylor, K. M. G., Taylor, G., Kellaway, I. W., and Stevens, J. (1989). The influence of liposomal encapsulation on sodium cromoglycate pharmacokinetics in man. *Pharm. Res. 6*: 633–636.
411. Sone, S., Tachibana, K., Shono, M., Ogushi, F., and Tsubura, E. (1984). Potential value of liposomes containing muramyl dipeptide for augmenting the tumoricidal activity of human alveolar macrophages. *J. Biol. Resp. Modif. 3*: 185–194.
412. Wichert, B. V., Gonzelez-Rothi, R. J., Straub, L. E., Wichert, B. M., and Schreier, H. (1992). Amikacin liposomes: Characterization, aerosolization, and *in vitro* activity against *Mycobacterium avium-intracellulare* in alveolar macrophages. *Int. J. Pharmaceut. 78*: 227–235.
413. Woodley, J. F. (1985). Liposomes for oral administration of drugs. *CRC Crit. Rev. Therap. Drug Carrier Syst. 2*: 1–18.
414. Schmidt, P. C., and Michaelis, J. (1990). Liposomen zur oralen Anwendung. 1. Mitteilung: Literaturüberblick zur Stabilität und Resorption von Liposomen *in vitro/in vivo* und zu Ergebnissen mit *in vivo* getesteten Arzneistoffen. *Pharm. Ztg. Wiss. 135*: 125–134.
415. Rowland, R. N., and Woodley, J. F. (1980). The stability of liposomes *in vitro* to pH, bile salts and pancreatic lipase. *Biochim. Biophys. Acta 620*: 400–409.
416. Rowland, R. N., and Woodley, J. F. (1981). The uptake of distearoyl phosphatidylcholine/cholesterol liposomes by rat intestinal sacs *in vitro*. *Biochim. Biophys. Acta 673*: 217–223.
417. Kimura, T., Nishimura, H., Kurosaki, Y., and Nakayama, T. (1990). Use of liposomal dosage form of flufenamic acid for treatment of oral ulcer. *Pharm. Res. 7*: 149S.

418. Sveinsson, S. J., and Mezei, M. (1992). *In vitro* oral mucosal absorption of liposomal triamcinolone acetonide. *Pharm. Res. 9*: 1359–1361.
419. Gregoriadis, G., Swain, C. P., Wills, E. J., and Tavill, A. S. (1974). Drug-carrier potential of liposomes in cancer chemotherapy. *Lancet i*: 1313–1316.
420. Segal, A. W., Gregoriadis, G., Lavender, J. P., Tarin, D., and Peters, T. J. (1976). Tissue and hepatic subcellular distribution of liposomes containing bleomycin after intravenous administration to patients with neoplasms. *Clin. Sci. Mol. Med. 51*: 421–425.
421. Osborne, M. P., Richardson, V. J., Jeyasingh, K., and Ryman, B. E. (1979). Radionuclide-labelled liposomes: A new lymph node imaging agent. *Int. J. Nucl. Biol. Med. 6*: 75–83.
422. Osborne, M. P., Payne, J. H., Richardson, V. J., McCready, V. R., and Ryman, B. E. (1983). The preoperative detection of axillary lymph node metastases in breast cancer by isotope imaging. *Br. J. Surg. 70*: 141–144.
423. Lopez-Berestein, G., Kasi, L., Rosenblum, M. G., Haynie, T., Jahns, M., Glenn, H., Mehta, R., Mavligit, G. M., and Hersh, E. M. (1984). Clinical pharmacology of 99mTc labelled liposomes in patients with cancer. *Cancer Res. 44*: 375–378.
424. Perez-Soler, R., Lopez-Berestein, G., Kasi, L. P., Cabanillas, F., Jahns, M., Glenn, H., Hersh, E. M., and Haynie, T. (1985). Distribution of technetium-99m-labelled multilamellar liposomes in patients with Hodgkin's disease. *J. Nucl. Med. 26*: 72–75.
425. Frühling, J., Coune, A., Ghanem, G., Sculier, J. P., Verbist, A., Brassinne, C., Laduron, C., and Hildebrand, J. (1984). Distribution in man of 111-In-labelled liposomes containing a water-insoluble antimitotic agent. *Nucl. Med. Comm. 5*: 205–208.
426. Patel, H. M., Boodle, K. M., and Vaughan-Jones, R. (1984). Assessment of the potential uses of liposomes in lymphoscintigraphy and lymphatic drug delivery. *Biochim. Biophys. Acta 801*: 76–86.
427. Begent, R. H. J., Green, A. J., Bagshawe, K. D., Jones, B. E., Keep, P. A., Searle, F., Jewkes, R. F., Barratt, G. M., and Ryman, B. E. (1982). Liposomally entrapped second antibody improves tumour imaging with radiolabelled (first) antitumour antibody. *Lancet 2*: 739–742.
428. Barratt, G. M., Ryman, B. E., Chester, K. A., and Begent, R. H. J. (1984). Liposomes as aids to tumour detection. *Biochem. Soc. Trans. 12*: 348–349.
429. Presant, C. A., Proffitt, R. T., Turner, F., Williams, L. E., Winsor, D., Werner, J. L., Kennedy, P., Wisemann, C., Gala, K., McKenna, R. J., Smith, J. D., Bouzaglou, S. A., Callahan, R. A., Baldeschwieler, J., and Crossley, R. J. (1988). Successful imaging of human cancer with indium-111-labeled phospholipid vesicles. *Cancer 62*: 905–911.
430. Presant, C. A., Blayney, D., Profitt, R. T., Turner, A. F., Williams, L. E., Nadel, H. I., Kennedy, P., Wiseman, C., Gala, K., Crossley, R. J., Preiss, S. J., Ksionski, G. E., and Presant, S. L. (1990). Preliminary report: Imaging of Kaposi sarcoma and lymphoma in AIDS with indium-111-labeled liposomes. *Lancet 335*: 1307–1309.

431. Turner, A. F., Presant, C. A., Proffitt, R. T., Williams, L. E., Winsor, D. W., and Werner, J. L. (1988). In-111-labeled liposomes: Dosimetry and tumor depiction. *Radiology 166*: 761–765.
432. Tyrrell, D. A., Ryman, B. E., Keeton, B. R., and Dubowitz, V. (1976). Use of liposomes in treating type II glycogenesis. *Br. Med. J. 2*: 88–89.
433. Belchetz, P. E., Braidman, J. P., Crawley, J. C. W., and Gregoriadis, G. (1977). Treatment of Gaucher's disease with liposome entrapped glucocerebroside:β-glucosidase. *Lancet ii*: 116–117.
434. Gregoriadis, G., Neerunjun, D., Meade, T. W., Goolamali, S. K., Weereratne, H., and Bull, G. (1980). Experiences after long term treatment of a type I Gaucher disease patient with liposome entrapped glucocerebroside:β-glucosidase. *Birth Defects* (Desnick, R. J., ed.). New York: Alan R. Liss, pp. 383–392.
435. Gregoriadis, G., Weereratne, H., Blair, H., and Bull, G. M. (1982). Liposomes in Gaucher type I disease: Use in enzyme therapy and the creation of an animal model. *Prog. Clin. Biol. Res. 85*: 681–701.
436. Lopez-Berestein, G., Mehta, R., Hopfer, R. L., Mills, K., Kasi, L., Mehta, K., Fainstein, V., Luna, M., Hersh, E. M., and Juliano, R. (1983). Treatment and prophylaxis of disseminated infection due to *Candida albicans* in mice with liposome-encapsulated amphotericin B. *J. Infect. Dis. 47*: 939–945.
437. Lopez-Berestein, G., Hopfer, R. L., Mehta, R., Mehta, K., Hersch, E. M., and Juliano, R. L. (1984). Liposome-encapsulated amphotericin B for treatment of disseminated candidiasis in neutropenic mice. *J. Infect. Dis. 150*: 278–283.
438. Tremblay, C., Barza, M., Fiore, C., and Szoka, F. (1984). Efficacy of liposome-intercalated amphotericin B in treatment of systemic candidiasis in mice. *Antimicrob. Agents Chemother. 216*: 170–173.
439. Szoka, F. C., Jr., Millholland, D., and Barza, M. (1987). Effect of lipid composition and liposome size on toxicity and in vitro fungicidal activity of liposome-intercalated amphotericin B. *Antimicrob. Agents Chemother. 31*: 421–429.
440. Juliano, R. L., Grant, C. W. M., Barber, K. R., and Kalp, M. A. (1987). Mechanism of the selective toxicity of amphotericin B incorporated into liposomes. *Mol. Pharmacol 21*: 1–11.
441. Lopez-Berestein, G., Fainstein, V., Hopfer, R., Mehta, K., Sullivan, M. P., Keating, R. L., and Bodey, G. P. (1985). Liposomal amphotericin B for the treatment of systemic fungal infections in patients with cancer. A preliminary study. *J. Infect. Dis. 151*: 704–709.
442. Shirkoda, A., Lopez-Berestein, G., Holbert, J. M., and Luna, M. A. (1986). Hepatosplenic fungal infection: CT and pathologic evaluation after treatment with liposomal amphotericin B. *Radiology 159*: 349–353.
443. Sculier, J. P., Coune, A., Meunier, F., Brassinne, C., Laduron, C., Hollaert, C., Collette, N., Heymans, C., and Klastersky, J. (1988). Pilot study of amphotericin B entrapped in sonicated liposomes in cancer patients with fungal infections. *Eur. J. Cancer Clin. Oncol. 24*: 527–538.
444. Meunier, F., Sculier, J. P., Coune, A., Brassinne, C., Heymans, C., Laduron, C., Collette, N., Hollaert, C., Bron, D., and Klastersky, J. (1988). Amphoteri-

cin B encapsulated in liposomes administered to cancer patients. *Ann. N.Y. Acad. Sci. 544*: 598–610.

445. Sculier, J. P., Delcroix, C., Brassinne, C., Laduron, C., Hollaert, C., and Coune, A. (1989). Pharmacokinetics of amphotericin B in patients receiving repeated i.v. high doses of amphotericin B entrapped into sonicated liposomes. *J. Liposome Res. 1*: 151–166.
446. Sculier, J. P., Bron, D., Coune, A., and Meunier, F. (1989). Successful treatment with liposomal amphotericin B in two patients with persisting fungemia. *Eur. J. Clin. Microb. Infect. Dis. 8*: 903–907.
447. Sculier, J. P., Klastersky, J., Libert, P., Ravez, P., Brohee, D., Vandermoten, G., Michel, J., Thiriaux, J., Bureau, G., Schmerber, J., Sergysels, R., and Coune, A. (1990). Cyclophosphamide, doxorubicin and vincristine with amphtericin B in sonicated liposomes as salvage therapy for small cell lung cancer. *Eur. J. Cancer. 26*: 919–921.
448. Tollemar, J., Ringden, O., and Tyden, G. (1990). Liposomal amphotericin-B (AmBisome) treatment in solid organ and bone marrow transplant recipients. Efficacy and safety evaluation. *Clin. Transplantation 4*: 167–175.
449. Katz, N. M., Pierce, P. F., Anzeck, R. A., Visner, M. S., Canter, H. G., Foegh, M. L., Pearle, D. L., Tracy, C., and Rahman, A. (1990). Liposomal amphotericin B for treatment of pulmonary aspergillosis in a heart transplant patient. *J. Heart Transplant. 9*: 14–17.
450. Viviani, M. A., Cofrancesco, E., Boschetti, C., Tortorano, A. M., and Cortellaro, M. (1991). Eradication of *fusarium* infection in a leukopenic patient treated with liposomal amphotericin B. *Mycoses 34*: 255–256.
451. Schurmann, D., de Matos Marques, B., Grunewald, T., Pohle, H. D., Hahn, H., and Ruf, B. (1991). Safety and efficacy of liposomal amphotericin B in treating AIDS-associated disseminated cryptococcosis. *J. Infect. Dis. 164*: 620–622.
452. Ringden, O., Meunier, F., Tollemar, J., Ricci, P., Tura, S., Kuse, E., Viviani, M. A., Gorin, N. C., Klastersky, J., Fenaux, P. P., Prentice, H. G., and Ksionski, G. (1991). Efficacy of amphotericin B encapsulated in liposomes (AmBisome) in the treatment of invasive fungal infections in immunocompromised patients. *J. Antimicrob. Chemother. 28* (Suppl. B): 73–82.
453. Meunier, F., Prentice, H. G., and Ringden, O. (1991). Liposomal amphotericin B (AmBisome): Safety data from a phase II/III clinical trial. *J. Antimicrob. Chemother. 28* (Suppl. B): 83–91.
454. Hudson, J., Scott, G. L., and Warnock, D. W. (1991). Treatment of hepatic candidosis with liposomal amphotericin B in patient with acute leukemia. *Lancet 338*: 1534.
455. Chopra, R., Blair, S., Strang, J., Cervi, P., Patterson, K. G., and Goldstone, A. H. (1991). Liposomal amphotericin B (AmBisome) in the treatment of fungal infections in neutropenic patients. *J. Antimicrob. Chemother. 28* (Suppl. B): 93–104.
456. Coker, R. J., Murphy, S. M., and Harris, J. R. W. (1991). Experience with liposomal amphotericin B (AmBisome) in cryptococcal meningitis in AIDS. *J. Antimicrob. Chemother. 28* (Suppl. B): 105–109.

457. Ringden, O., Tollemar, J., and Tyden, G. (1992). Liposomal amphotericin B. *Lancet 339*: 374.
458. Lackner, H., Schwinger, W., Urban, C., Müller, W., Ritschl, E., Reiterer, F., Kuttnig-Haim, M., Urlesberger, B., and Hauer, C. (1992). Liposomal amphotericin-B (AmBisome) for treatment of disseminated fungal infections in two infants of very low birth weight. *Pediatrics 89*: 1259-1261.
459. Coker, R. J., and Horner, P. J. (1992). Short-course treatment and response to liposomal amphotericin B in AIDS-associated cryptococcosis. *J. Infect. Dis. 165*: 593.
460. Lopez-Berestein, G., Bodey, G. P., Frankel, L. S., and Mehta, K. (1987). Treatment of hepatosplenic candidiasis with liposomal amphotericin B. *J. Clin. Oncol. 5*: 310-317.
461. Lopez-Berestein, G., Bodey, G. P., Fainstein, V., Keating, M., Frankel, L. S., Zeluff, B., Gentry, L., and Mehta, K. (1989). Treatment of systemic fungal infections with liposomal amphotericin B. *Arch. Int. Med. 149*: 2533-2536.
462. Lopez-Berestein, G. (1990). Therapeutic activity of liposomal amphotericin B in systemic mycoses. *J. Liposome Res. 1*: 511-516.
463. Gabizon, A., Daga, A., Goren, D., Barenholz, Y., and Fuks, Z. (1982). Liposomes as *in vivo* carrier of adriamycin: Reduced cardiac uptake and preserved antitumor activity in mice. *Cancer Res. 42*: 4734-4739.
464. Forssen, E. A., and Tokes, Z. A. (1983). Improved therapeutic benefits of doxorubicin by entrapment in anionic liposomes. *Cancer Res. 43*: 546-550.
465. van Hoesel, Q. G. C. M., Steerenberg, P. A., Crommelin, D. J. A., van Dijk, A., van Oort, W., Klein, S., Douze, J. M. C., de Wildt, D. J., and Hillen, F. C. (1984). Reduced cardiotoxicity and nephrotoxicity with preservation of antitumor activity of doxorubicin entrapped in stable liposomes in the Lou/M Wsl rat. *Cancer Res. 44*: 3698-3705.
466. Rahman, A., White, G., More, N., and Schein, P. (1985). Pharmacological, toxicological and therapeutic evaluation in mice of doxorubicin entrapped in cardiolipin liposomes. *Cancer Res. 45*: 796-803.
467. Gabizon, A., Meshorer, A., and Barenholz, Y. (1986). Comparative long-term study of the toxicities of free and liposome-associated doxorubicin in mice after intravenous administration. *J. Natl. Cancer Inst. 77*: 459-469.
468. van Hoesel, Q. G. C. M., Steerenberg, P. A., Dormans, J. A. M. A., de Jong, W. H., de Wildt, D. J., and Vos, J. G. (1986). Time-course study of doxorubicin-induced nephropathy and cardiomyopathy in male and female Lou/M Wsl rats: Lack of evidence for a causal relationship. *J. Natl. Cancer Inst. 76*: 299-307.
469. Storm, G., van Hoesel, Q. G. C. M., de Groot, G., Kop, W., Steerenberg, P. A., and Hillen, F. C. (1989). A comparative study on the antitumor effect, cardiotoxicity and nephrotoxicity of doxorubicin given as a bolus, continuous infusion or entrapped in liposomes in the Lou/M Wsl rat. *Cancer Chemother. Pharmacol. 24*: 341-348.
470. Storm, G., Regts, J., Beijnen, J. H., and Roerdink, F. H. (1989). Processing of doxorubicin-containing liposomes by liver macrophages *in vitro*. *J. Liposome Res. 1*: 195-210.

471. Crommelin, D. J. A., Nässander, U. K., Peeters, P. A. M., Steerenberg, P. A., de Jong, W. H., Eling, W. M. C., and Storm, G. (1990). Drug-laden liposomes in antitumor therapy and in the treatment of parasitic diseases. *J. Control. Release 11*: 233–243.
472. Gabizon, A., Chisin, R., Amselem, S., Druckmann, S., Cohen, R., Goren, D., Fromer, I., Peretz, T., Sulkes, A., and Barenholz, Y. (1990). Pharmacokinetic and imaging studies in patients receiving a formulation of liposome-associated adriamycin. *Br. J. Cancer 64*: 1125–1132.
473. Crommelin, D. J. A., and van Bloois, L. (1983). Preparation and characterization of doxorubicin-containing liposomes. II. Loading capacity, long-term stability and doxorubicin-bilayer interaction mechanisms. *Int. J. Pharmaceut. 17*: 135–144.
474. Crommelin, D. J. A., Slaats, N., and van Bloois, L. (1983). Preparation and characterization of doxorubicin-containing liposomes. I. Influence of liposome charge and pH of hydration medium on loading capacity and particle size. *Int. J. Pharmaceut. 16*: 79–92.
475. Janssen, M. J. H., Crommelin, D. J. A., Storm, G., and Hulshoff, A. (1985). Doxorubicin decomposition on storage. Effect of pH, type of buffer and liposome encapsulation. *Int. J. Pharmaceut. 23*: 1–11.
476. Amselem, S., Cohen, R., Druckmann, S., Gabizon, A., Goren, D., Abra, R. M., Huang, A., New, R., and Barenholz, Y. (1992). Preparation and characterization of liposomal doxorubicin for human use. *J. Liposome Res. 2*: 93–123.
477. Kumai, K., Takahashi, T., Tsubouchi, T., Yoshino, K., Ishibiki, K., and Abe, O. (1985). Selective hepatic arterial infusion of liposomes containing antitumour agents. *Jpn. J. Cancer Chemother. 12*: 1946–1948.
478. Gabizon, A., Peretz, T., Ben-Yosef, R., Catane, R., Biran, S., and Barenholz, Y. (1986). Phase I study with liposome-associated adriamycin: Preliminary report. *Proc. Am. Soc. Clin. Oncol. 5*: 43.
479. Gabizon, A., Peretz, T., Sulkes, A., Anselem, S., Ben-Yosef, R., Ben-Baruch, N., Catane, R., Biran, S., and Barenholz, Y. (1989). Systemic administration of doxorubicin-containing liposomes in cancer patients: A Phase I study. *Eur. J. Cancer Clin. Oncol. 25*: 1795–1803.
480. Delgado, G., Potkul, R. K., Treat, J. A., Lewandowski, G. S., Barter, J. F., Forst, D., and Rahman, A. (1989). Phase I/II study of intraperitoneally administered doxorubicin entrapped in cardiolipin liposomes in patients with ovarian cancer. *Am. J. Obstet. Gynecol. 160*: 812–819.
481. Rahman, A., Treat, J., Roh, J. K., Potkul, L. A., Alvord, W. G., Forst, D., and Woolley, P. V. (1990). A Phase I clinical trial and pharmacokinetic evaluation of liposome-encapsulated doxorubicin. *J. Clin. Oncol. 8*: 1093–1100.
482. Owen, R. R., Sells, R. A., Gilmore, I. T., New, R. R. C., and Stringer, R. E. (1992). A Phase I clinical evaluation of liposome-entrapped doxorubicin (lipdox) in patients with primary and metastatic hepatic malignancy. *Anti-Cancer Drug 3*: 101–107.
483. Kleinerman, E. S., Murray, J. L., Snyder, J. S., Cunningham, J. E., and Fidler, I. J. (1989). Activation of tumoricidal properties in monocytes from can-

cer patients following intravenous administration of liposomes containing muramyl tripeptide phosphatidylethanolamine. *Cancer Res.* *49*: 4665–4670.

484. Hanagan, J. R., Trunet, P., LeSher, D., Andrejcio, K., and Frost, H. (1989). Phase I development of CGP 19835A lipid (MTP-PE encapsulated in liposomes). *Liposomes in the Therapy of Infectious Diseases and Cancer* (G. Lopez-Berestein and I. J. Fidler, eds.). New York: Alan R. Liss, pp. 305–315.
485. Creaven, P. J., Brenner, D. E., Cowens, J. W., Huben, R., Karakousis, C., Han, T., Dadey, B., Andrejcio, K., and Cushman, M. K. (1989). Initial clinical trial of muramyl tripeptide derivative (MTP-PE) encapsulated in liposomes: An interim report. *Liposomes in the Therapy of Infectious Diseases and Cancer* (G. Lopez-Berestein and I. J. Fidler, eds.). New York: Alan R. Liss, pp. 297–303.
486. Urba, W. J., Hartmann, L. C., Longo, D. L., Steis, R. G., Smith, J. W., Kedar, I., Creekmore, S., Sznol, M., Conlon, K., Kopp, W. C., Huber, C., Herold, M., Alvord, W. G., Snow, S., and Clark, J. W. (1990). Phase I and immunomodulatory study of a muramyl peptide, muramyl tripeptide phosphatidylethanolamine. *Cancer Res.* *50*: 2979–2986.
487. Kleinerman, E. S., Jia, S. F., Griffin, J., Seibel, N. L., Benjamine, R. S., and Jaffe, N. (1992). Phase II study of liposomal muramyl tripeptide in osteosarcoma: The cytokine cascade and monocyte activation following administration. *J. Clin. Oncol.* *10*: 1310–1316.
488. Murray, J. L., Kleinerman, E. S., Tatom, J. R., Cunningham, J. E., Lepe-Zuniga, J., Gutterman, J. U., Andrejcio, K., Fidler, I. J., and Krakoff, I. H. (1989). A pilot Phase I trial of liposomal *N*-acetyl-muramyl-L-alanyl-D-iso-glutaminyl-L-alanyl-phosphatidylethanolamine [MTP-PE (CGP 19835A)] in cancer patients. *Liposomes in the Therapy of Infectious Diseases and Cancer* (G. Lopez-Berestein and I. J. Fidler, eds.). New York: Alan R. Liss, pp. 329–342.
489. Coune, A., Sculier, J. P., Frühling, J., Strykmans, P. B., Brasinne, Ch., Ghanem, G., Laduron, C., Atassi, G., Ruysschaert, J. M., and Hildebrand, J. (1983). I.v. administration of a water insoluble antimitotic compound entrapped in liposomes. Preliminary report on the infusion of large volumes of liposomes to man. *Cancer Treat. Rep.* *67*: 1031–1033.
490. Sculier, J. P., Coune, A., Brassinne, C., Laduron, C., Atassi, G. R., Ruysschaert, J. M., and Frühling, J. (1986). Intravenous infusion of high doses of liposomes containing NSC 251635, a water-insoluble cytostatic agent. A pilot study with pharmacokinetic data. *J. Clin. Oncol.* *4*: 789–797.
491. Pestalozzi, B., Schwendener, R., and Sauter, C. (1992). Phase I/II study of liposome-complexed mitoxantrone in patients with advanced breast cancer. *Ann. Oncol.* *3*: 445–449.
492. Fries, L. F., Gordon, D. M., Richards, R. L., Egan, J. E., Hollingdale, M. R., Gross, M., Silverman, C., and Alving, C. R. (1992). Liposomal malaria vaccine in humans: A safe and potent adjuvant strategy. *Proc. Natl. Acad. Sci. U.S.A.* *89*: 358–362.
493. de Silva, M., Hazleman, B. L., Page Thomas, D. P., and Wraight, P. (1979). Liposomes in arthritis: A new approach. *Lancet i*: 1320–1322.

494. Robillard, E., Alarie, Y., Dagenais-Perusse, P., Baril, E., and Guilbeault, A. (1964). Microaerosol administration of synthetic dipalmitoyl lecithin in the respiratory distress syndrome: A preliminary report. *Can. Med. Assoc. J. 90*: 55–57.
495. Martin, F. (1990). Development of liposomes for aerosol delivery: Recent pre-clinical and clinical results. *J. Liposome Res. 1*: 407–429.

4

Niosomes

J. A. Bouwstra and H. E. J. Hofland

Leiden/Amsterdam Center for Drug Research, Leiden University, Leiden, The Netherlands

I. INTRODUCTION

Niosomes are vesicles mainly consisting of nonionic surfactants. This class of vesicles was introduced by Handjani-Vila et al. (1) in 1979. One of the reasons for preparing niosomes is the assumed higher chemical stability of the surfactants than that of phospholipids, which are used in the preparation of liposomes. Due to the presence of ester bonds, phospholipids are easily hydrolyzed (2). This can lead to phosphoryl migration at low pH. Another type of degradation is the peroxidation of unsaturated phospholipids (3). In order to avoid peroxidation processes, vesicles are often stored under nitrogen atmosphere. Unreliable reproducibility arising from the use of lecithins in liposomes leads to additional problems and has led scientists to search for vesicles prepared from other material, such as nonionic surfactants.

Niosomes have been prepared from several classes of nonionic surfactants, e.g., polyglycerol alkylethers (1,4), glucosyl dialkylethers (5), crown ethers (6), and polyoxyethylene alkyl ethers and esters (7). Often a charged surfactant is intercalated in the bilayers in order to introduce electrostatic

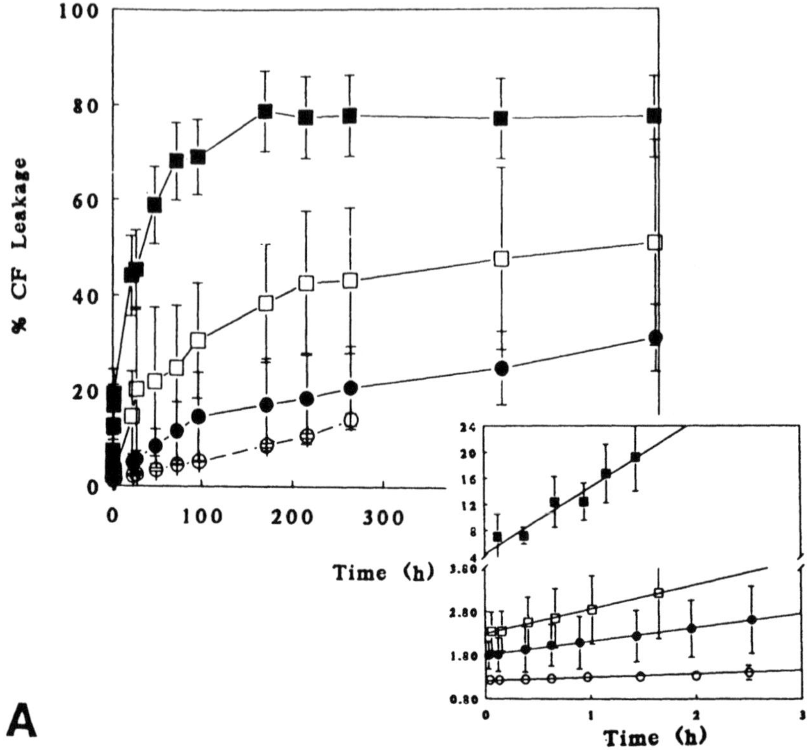

Figure 1 The release profile of CF is plotted against time. The relation between CF release and cholesterol content based on $C_{9=9}EO_{10}$ was investigated. ■ 20; □ 40; ● 60; ○ 80% cholesterol. Data are presented as means SD, $n = 3$. The dotted lines indicate that the suspension is inhomogeneous, which means that a part of the material does not form vesicles. The interpretation of these release profiles is questionable. A, niosomes prepared by sonication leading to a mean diameter of approximately 150 nm as determined by dynamic light scattering; B, niosomes prepared by the film method leading to diameters larger than 1 μm. Niosomes prepared by the film method lead to niosomes with a mean number of bilayers larger than 5 (determined by x-ray diffraction techniques).

repulsion between the vesicles, thus increasing their stability. This is especially recommended at higher surfactant concentrations.

One of the aims in developing delivery systems is controlling the release of drugs from the carrier system in order to achieve a controlled uptake in the body. The release can be controlled by various parameters. In Fig. 1 the release of a model substance, 5,6 carboxyfluorescein, is shown. The niosomes were prepared using various molar ratios of Brij 96 and cholesterol (8). From

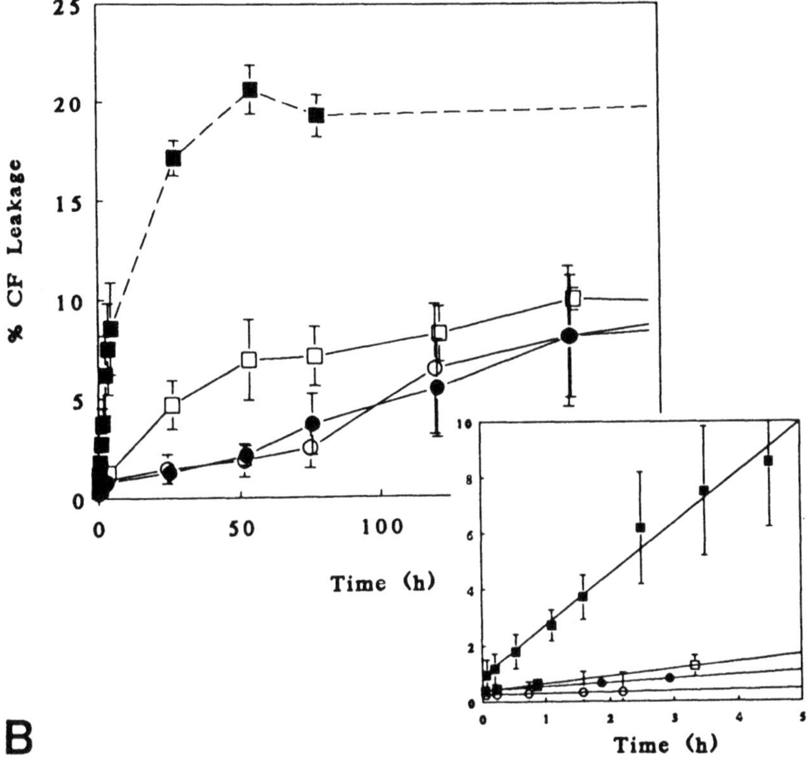

this figure it is clear that the release has been influenced by the composition, the size, and the number of bilayers of the vesicles. This shows that these parameters are important for the use of vesicles in pharmaceutical applications.

In the first part of this chapter the ability of nonionic surfactants in forming vesicles will be discussed. We will mainly focus on polyoxyethylene alkyl ethers (C_nEO_m, in which n is the number of carbon atoms in the alkyl chain, while m is the number of oxyethylene units), since the corresponding binary H_2O + C_nEO_m T–X phase diagrams have been studied extensively. The various preparation methods and the obtained entrapment efficiencies will be described, and the toxicity of niosomes in relation to gel-liquid transition temperatures will be discussed. Finally, the administration of drugs encapsulated in vesicles by means of various routes will be described.

II. THE ABILITY OF SURFACTANTS TO FORM NIOSOMES

A. Surfactants

Various research groups have used different types of surfactants. A summary is given in Table 1. Okahata et al. (9) reported that some of the members of the dialkyl polyoxyethylene surfactant series could form vesicles. Handjani-Vila et al. (1) reported that vesicular systems were formed when a mixture of cholesterol and a nonionic surfactant were hydrated. They used ether surfactants consisting of either a mono- or a dialkylether chain. The latter were similar in structure to phospholipids. Amphiphiles forming niosomes also might be of an ester type but may possibly degrade to triglycerides and fatty acids in the presence of esterases (10). It is expected that the ester type surfactants are chemically less stable than the ether type surfactants but also less toxic (7). Although various surfactants are used in preparing niosomes, general rules can be used for predicting whether or not niosomes may be formed from the various nonionic surfactants.

B. Geometric Considerations

Considering the geometry of models for the possible structures of micelles, the interfacial area A per surfactant molecule plays an important role. It is postulated that the hydrocarbon interior is a continuum. This means that the shortest distance between the center of the micelle and the interface should be smaller or equal to the maximally attainable length l_c of the hydrocarbon tail. With V_c being the hydrocarbon volume, simple geometric conditions can be derived for the forming of micelles and vesicles (11). These are

$$\text{spheric micelles:} \quad A > \frac{3V_c}{l_c}$$

$$\text{rod micelles:} \quad A > \frac{2V_c}{l_c}$$

$$\text{bilayer micelles, vesicles:} \quad A > \frac{V_c}{l_c}$$

A schematic presentation of these rules is given by Israelachvili (12); see Fig. 2. If, in accordance with Mitchell et al. (13), an increase in surfactant concentration results in an ordering of the surfactants, the following disorder-to-order transitions are expected: bilayer micelles will be packed into a lamellar phase, rod micelles will be packed into a hexagonal phase, and spherical micelles will be packed into a cubic or a hexagonal closed packed phase.

Table 1 Nonionic Surfactants Used for the Preparation of Vesicles

Nonionic surfactant	Company/laboratory	Structural formula	Reference
Hexadecylpoly(3)glycerol	L'Oreal	$C_{16}H_{33}O(CH_2CH_2O)_3H$ (with CH_2OH branch)	1, 4, 10, 27, 28, 30, 33, 36, 41, 42
Cholesterolpoly(24)oxyethylene ether			31, 33
Dialkylpoly(7)glycerol ether		$C_{16}H_{33}-CH-O(CH_2CHO)_7H$ (with CH_2 branch on CH, linked to $C_{12}H_{25}O$; CH_2OH branch on CHO)	41, 19
Alkylglucoside	Faculty of Pharmaceutical Science, University of Tokyo	$H(CH_2)_n-C_6H_{11}O_6$ n = 8, 10, 12, 14, 16, 18	20, 24, 34
Cetyl mannoside		$C_{16}H_{33}-C_6H_{11}O_6$	20
Cetyl lactoside		$C_{16}H_{33}-C_6H_{11}O_6$	20
Alkyl galactoside		$H(CH_2)_n-C_6H_{11}O_6$	20, 34
Polyoxyethylenealkyl ether	SERVO/ATLAS NIKKO	$H(CH_2)_n-(OC_2H_4)_mOH$ n = 10, 12, 14, 16, 18 m = 3, 4, 5, 6, 7, 8	7, 8, 26, 49, 51
Neutral crown ethers			6
Polyoxyethylene glycerol-α, α-diether	Department of Organic Synthesis, Faculty of Engineering, Kyoshu University	$(H(CH_2)_n-OCH_2)_2CHO(C_2H_4O)_xH$ n = 12, 14, 16, 18 x = 6 to 30	9
n-Decyloxyethyleneoctadecylmyricylamine		$H(CH_2)_{18}-N-(C_2H_4O)_{10}H$ (N bonded to) $H(CH_2)_{13}-C=O$	9
Cetyldiglycerolester	L'Oreal	$C_{15}H_{31}CO-(OCHCH_2)OCH_2CHOHCH_2OH$ (with CH_2OH branch on OCH)	10

Lipid	Critical packing parameter $v/a_0 l_c$	Critical packing shape	Structures formed
Single-chained lipids with large head-group areas: NaDS *in low salt* *some lysophospholipids*	$< \frac{1}{3}$	cone 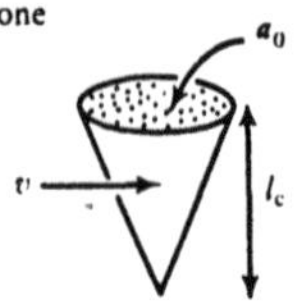 (a_0, v, l_c)	spherical micelles 
Single-chained lipids with small head-group areas: NaDS *in high salt* $C_{16}TAB$ *in high salt*, *lysolecithin* *nonionic surfactants*	$\frac{1}{3} - \frac{1}{2}$	truncated cone or wedge 	cylindrical micelles; 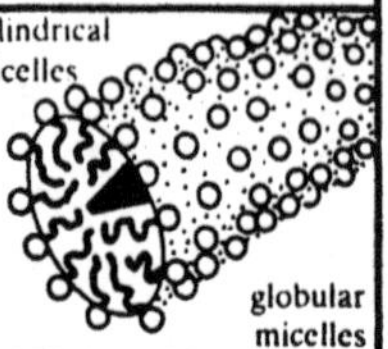 globular micelles
Double-chained lipids with large head-group areas, fluid chains: $(C_{12})_2DAB$, *lecithin, sphingomyelin, phosphatidylserine, phosphatidylglycerol phosphatidylinositol, phosphatidic acid disugardiglycerides*	$\frac{1}{2} - 1$	truncated cone 	vesicles, flexible bilayers 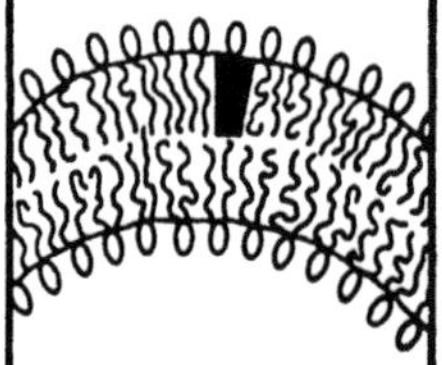
Double-chained lipids with small head-group areas: *anionic lipids in high salt saturated frozen chains, e.g. phosphatidylethalonamine phosphatidylserine* + Ca^{2+}	~1	cylinder 	planar bilayers 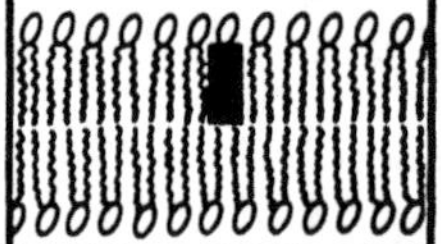
Double-chained lipids with small head-group areas, nonionic lipids, poly (cis) unsaturated chains, high *T* *unsaturated phosphatidylethanolamine Cardiolipin* + Ca^{2+} *phosphatidic acid* + Ca^{2+} *monosugardiglycerides, cholesterol (rigid)*	>1	inverted truncated cone 	inverted micelles 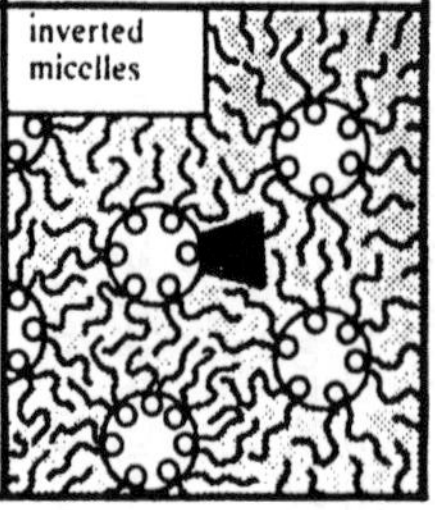

Figure 2 A schematic view of the geometric considerations for forming micelles and bilayers.

From this simplified theory it is expected that vesicles can only be prepared from those surfactants that form a lamellar phase upon concentrating their micellar solution, and that these surfactants exhibit a small interfacial area per molecule (see equations). Surfactants exhibiting this phase behavior possess a small hydrophilic head group in comparison with the alkyl chain length.

C. Formation of Vesicles

The ability of surfactants to form vesicles has been studied by freeze fracture electron microscopy and polarization microscopy. By using polarization microscopy large multilamellar vesicles are recognized by the so-called Maltese crosses. In Table 2 the ability of C_nEO_m surfactants in forming vesicles is presented. Surfactants with a small head group such as C_nEO_3 surfactants are indeed able to form vesicles, but when using surfactants that form gel state bilayers ($C_{16}EO_3$ and $C_{18}EO_3$) at room temperature, mainly aggregates of vesicles are found. First focussing on the $C_{12}EO_3$ surfactant, in accordance with the theory a lamellar phase is formed at the water-rich side of the binary $C_{12}EO_3 + H_2O$ T-X phase diagram (13). For clarity, the above-mentioned phase diagram studies have been carried out by using pure surfactants, that may differ in phase behavior from the technical grade surfactants used in preparing the niosomes. In previous studies (14), it was found that the phase diagram of the technical grade surfactant $C_{12}EO_7$ was similar to that of pure $C_{12}EO_6$. From these findings it was concluded that purity influences the phase behavior only to a small extent. Although the phase diagram of H_2O +

Table 2 Formation of Vesicles Consisting of 100% Polyoxyethylene Alkyl Ether Surfactant (30 mM) as Determined by Freeze Facture Electron Microscopy

Surfactant	EO_3	EO_5	EO_7
C_{10}	+	−	−
C_{12}	+	−	−
C_{14}	+	+	−
C_{16}	$+^{ps}$	$+^{sv}$	$+^{v}$
C_{18}	$+^{ps}$	$+^{ps}$	$-^{sv}$

+ indicates the formation of vesicles; ps means that phase separation occurred; sv indicates that the formulation was slightly viscous; v stands for viscous formulation.

$C_{14}EO_3$ is not known, again a lamellar phase is expected in the water-rich regions, because H_2O + $C_{12}EO_3$ shows such a phase behavior (13). It seems that all these surfactants obey the theory described above. However, as mentioned, $C_{18}EO_3$ and $C_{16}EO_3$ form many aggregating vesicles, which is probably due to the nature of the bilayers: gel state bilayers of uncharged vesicles are known to destabilize the suspension (15). The gel-liquid transition of hydrated $C_{16}EO_3$ and $C_{18}EO_3$ surfactants takes place in a temperature range from 40 to 48°C and 50 to 58°C respectively (16).

That the instability of the vesicle suspension is due to the gel state bilayers has been confirmed by the observation that aggregates are formed upon cooling the suspension and not during the hydration, which takes place at elevated temperatures (60–80°C). Another surfactant that forms niosomes without adding cholesterol or other additives is $C_{12}EO_4$. In the binary $C_{12}EO_4$ + H_2O T-X phase diagram a lamellar phase has been established also at the water-rich side of the phase diagram (17).

By using C_nEO_5 surfactants, niosomes only can be prepared from $C_{14}EO_5$, $C_{16}EO_5$ and $C_{18}EO_5$. Gel state niosomes prepared from $C_{16}EO_5$ and $C_{18}EO_5$ again are not stable and form large aggregates. It seems that the head groups of $C_{10}EO_5$ and $C_{12}EO_5$ are too bulky for forming bilayer structures. This is confirmed by the phase behavior of the binary mixtures H_2O + $C_{12}EO_5$ and H_2O + $C_{12}EO_6$; both surfactants form a hexagonal phase upon concentrating the micellar solution (13).

From the series of C_nEO_7 surfactants, niosomes can be prepared from $C_{16}EO_7$, although large aggregates are formed. This again points in the direction of a destabilized suspension. The gel-liquid transition of hydrated $C_{15}EO_7$ and $C_{18}EO_7$ is located between 30–40°C and 35–55°C respectively (16). In the case of 10 to 14 C atoms in the alkyl chain the hydrophilic head group is too bulky for forming bilayered structures.

From the observations given above one can conclude that the polyoxyethylene alkyl ethers behave in accordance with the theory. However, the niosomes consisting of longer alkyl chain surfactants are destabilized by the formation of gel state bilayers.

A possibility for forming vesicles from single chain surfactants with a bulky head group is the addition of cholesterol, which possesses only a hydroxyl group as a hydrophilic moiety. On one hand, cholesterol compensates for the large hydrophilic head groups of the surfactants. It is thus possible to prepare vesicles from C_nEO_m surfactant with relatively large head groups such as $C_{10}EO_5$, $C_{12}EO_5$, and $C_{12}EO_7$. On the other hand, cholesterol is known to decrease the chain order below T_c (17). The latter mechanism may explain the stabilization of vesicles prepared from $C_{16}EO_n$ and $C_{18}EO_n$ after adding cholesterol to the solution. It seems that cholesterol also increases the membrane undulations of lecithin liposomes, which results in larger repulsion forces between the vesicles (18). Using C_nEO_m and cholesterol in a molar ratio of 3/2 it is possible to prepare vesicles from the whole series of C_nEO_m surfactants. However, when using single chain surfactants that have the possibility to form micelles, it can never be excluded that after addition of cholesterol a part of these surfactants still may form micelles. Surfactants that might have this behavior are e.g. $C_{12}EO_5$ and $C_{12}EO_7$.

D. Gel-Liquid Transitions

At least two publications deal with the gel-liquid transitions of the bilayers of niosomes. Ribier et al. (19) studied the fluidity of the bilayers using fluorometry and showed that the fluidity was strongly dependent on the hydrophilic head group, alkyl chain length, and cholesterol content. Using this method, the fluidity may be characterized by the dynamic parameter D_w, which reflects the mobility of the hydrocarbon chains, and the static parameter Θ_c. This parameter accounts for the orientational degree of the chains. Ribier et al. (19) used polyoxyethylene alkyl ether surfactants as well as polyglycerol alkyl ether surfactants (C_nG_m, in which n is the number of carbon atoms in the alkyl chain and m the number of glycerol units). They observed a strong increase in Θ_c when passing through the gel-liquid transition, accounting for an increased disorientation of the surfactants in the bilayers of the vesicles. The gel-liquid transition of $C_{16}G_2$ appeared to be approximately 38°C, while the mobility of both the double alkyl chain surfactant $C_{12}/C_{16}G_7$ and $C_{12}G_2$ increased gradually as a function of temperature; see Fig. 3. Com-

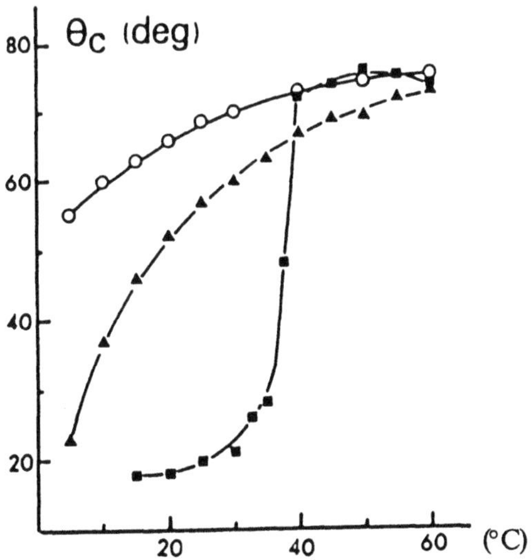

Figure 3 The influence of the hydrocarbon chain length of polyglycerol on the cone angle Θ_c: ○, $C_{12}G_2$; ■, $C_{16}G_2$; ▲, $C_{12/16}G_7$.

paring the fluidity of $C_{16}EO_5$ and $C_{16}G_2$, surfactants with approximately equal hydrophilic/lipophilic balance, Θ_c and D_ω are both higher in the case of vesicles prepared from $C_{16}EO_5$ compared to vesicles prepared from $C_{16}G_2$. This is in accordance with the higher interfacial area for the $C_{16}EO_5$ surfactant compared to the $C_{16}G_2$ surfactant. Buckton et al. (16) studied the gel-liquid transitions of dry and hydrated polyoxyethylene surfactants by high sensitivity differential scanning calorimetry. No single transition temperatures were observed, but the transitions occurred in a temperature range from 32 to 45°C for $C_{16}EO_3$ and between 20 and 40°C for $C_{16}EO_7$. These values are in agreement with the results obtained by Ribier et al. (19). The transition temperatures for $C_{18}EO_n$ were approximately 10°C higher. An increase in cholesterol content resulted in a decrease in enthalpy of transition, while no significant change was observed in the temperature range at which the transition occurred. This behavior is in fact in agreement with the observations of Ribier et al. (19), who found a decrease in fluidity between gel and liquid state with increasing cholesterol content. This behavior is also similar to lecithin/cholesterol mixtures, and it is therefore very likely that, by adding cholesterol to surfactants, an ordered liquid phase is formed comparable to that observed in lecithins. In an additional publication, Kiwada et al. (20) reported the gel-liquid transitions of glycosyl and galactoside surfactants that were also used in the preparation of vesicles. They observed very high transition temperatures varying between 75°C for cetyl glucoside and 105°C for stearyl galactoside surfactants. Probably the gel-liquid transition for the hydrated surfactants was not dramatically different, which means that in using these surfactants for the preparation of vesicles in all cases gel state bilayers are formed.

III. PREPARATION OF VESICLES

A. Preparation Methods

The preparation methods should be chosen according to the use of the niosomes, since the preparation methods influence the number of bilayers, size, size distribution, and entrapment efficiency of the aqueous phase and the membrane permeability of the vesicles (4,8). The preparation methods are similar to those used for the preparation of liposomes.

1. Ether injection method (4): The surfactant/cholesterol mixture is dissolved in diethyl ether and injected slowly through a needle into the aqueous phase at 60°C. Large unilamellar vesicles are formed during the evaporation of the ether. The disadvantage of this method is that a small amount of ether is often still present in the vesicle suspension and is often very difficult to remove.

2. Hand-shaking (film) method (4): The surfactant/cholesterol mixture is dissolved in diethyl ether in a round-bottomed flask, and the organic solvent is removed at room temperature under reduced pressure. The dried surfactant film is hydrated with an aqueous phase at 50–60°C during gentle agitation. Large multilamellar vesicles are prepared. Examples of vesicles prepared using hand-shaking, ether injection, and sonication are given in Fig. 1.

3. Sonication (4,8): An aqueous phase is added to the surfactant/cholesterol mixture in a glass vial. The mixture then is probe sonicated for a certain time period. The resulting vesicles are small and unilamellar. In the case of niosomes the resulting vesicle sizes are in general larger than liposomes, niosomes being no smaller than 100 nm in diameter.

4. Method described by Handjani-Vila (1): Equivalent amounts of lipid (or mixtures of lipids) and an aqueous solution of the active substance are mixed and agitated in order to get a homogeneous lamellar phase. The resulting mixture is homogenized at a controlled temperature by means of agitation or ultracentrifugation.

5. Reversed phase evaporation (20): Lipids are dissolved in chloroform and 1/4 volume of PBS (phosphate-buffer saline). The mixture is sonicated and evaporated under reduced pressure. The lipids form a gel, which is then hydrated. The evaporation is continued until the hydration is completed.

6. The size and number of bilayers of vesicles consisting of polyoxyethylene alkyl ethers and cholesterol can be changed in an alternative way (8). A temperature rise above 60°C transforms small unilamellar vesicles to large multilamellar vesicles (>1 μm), while vigorous shaking at room temperature results in the opposite effect by changing multilamellar vesicles into unilamellar ones. The transformation from unilamellar to multilamellar vesicles at higher temperatures might be characteristic for the polyoxyethelene alkylether (ester) surfactants, since it is known that polyethylene glycol (PEG) and water demixes at higher temperatures due to a breakdown of hydrogen bondings between water and PEG moieties (21).

Generally, free drug is removed from the encapsulated drug by gel permeation chromatography dialysis methods or by centrifugation. Often weight density differences between niosomes and the external phase are smaller than in the case of liposomes, which makes separation by centrifugation very difficult. A possibility is to add protamine to the vesicle suspension (22) in order to facilitate separation during centrifugation.

B. Entrapment Efficacies

As in the case of liposomes, entrapment efficacies of hydrophilic and lipophilic compounds depend on the preparation method. According to the results of Baillie et al. (4), niosomes prepared by ether injection resulted in entrapment efficacies of carboxy fluorescein that were significantly higher than those of vesicles prepared by hand-shaking.

Both Baillie et al. (4) and Hunter et al. (10) used glycerol surfactants and reported that the entrapment efficacy decreased as the amount of cholesterol added in the nonionic surfactant vesicles increased.

High entrapment efficacies were observed when incorporating an octapeptide, DGAVP, in niosomes prepared from polyoxyethylene alkyl ether surfactants (23). By using the freezing/thawing method, values of up to 40% entrapment efficacy were reported, while only a low surfactant concentration (90 mM) was used.

Entrapment efficacies of ^{14}C sucrose solution by alkyl glycoside vesicles prepared by the REV method were also measured (20). An increase in charge density in the bilayers resulted in a strong increase in encapsulation volume. It is even not unlikely that without addition of dicetylphosphate no proper vesicles could be prepared from the surfactants. Shorter alkyl chain glucoside surfactants resulted in lower entrapment efficacies than longer alkyl chain glucoside surfactants. The entrapment efficacy of galactoside vesicles is lower than that of glucoside vesicles. The authors argued that galactoside surfactants are difficult to hydrate, which might partly lead to nonvesicular structures.

In plasma the stability of alkyl glycoside vesicles (24) was poor when compared with that of liposomes. It was suggested that this was due to the single hydrocarbon chain, which leads to a less rigid bilayer structure than double alkyl chain surfactants. Small unilamellar vesicles were more stable than large multilamellar vesicles.

Moser et al. (25) reported that the entrapment efficacy of hemoglobin in niosomes prepared from glycerolalkyl ethers was high, being 0.3–0.5 g/g lipid. Hemoglobin niosomes are physically stable, whereas hemoglobin undergoes a progressive oxidation to methemoglobin, reaching 30% after 5 months stored at 4°C. Methemoglobin formation was due to the permeability of vesicles to oxygen.

IV. TOXICITY ASPECTS

There is an enormous amount of literature available about the toxicity of surfactant molecules, but only a few studies are available in which the toxicity of niosomes has been investigated. Hofland et al. (26) studied the toxicity of C_nEO_m surfactants with two different models. These models are (a) the ciliary beat frequency (CBF) of trachea, which is important for intranasal administration, and (b) the cell proliferation of keratinocytes, which is important for the transdermal application of vesicles. A decrease in CBF was considered to be a measure for the toxicity of the formulation.

The CBF of chicken embryo trachea was measured at 31°C, which is the physiological temperature of mucosa nasal tissue. The vesicles were prepared from 60 mol % surfactant and 40 mol % cholesterol. Application of gel state niosomes did not cause a decrease in CBF, whereas liquid state niosomes did result in a significant decrease in ciliary beat frequencies. Only in the case of niosomes prepared from $C_{12}EO_3$ was the decrease in CBF irreversible. No relationship was found between the potential of surfactants to form micelles in equilibrium with the niosomes (e.g., $C_{12}EO_7$, see Sec. II) and the decrease in ciliary beat frequency.

As a toxicity model for topical application of niosomes, the cell proliferation of keratinocytes was measured after 3 days incubation with niosomes. In this study, ester and ether type surfactants were compared. It appeared that ester linked surfactants were significantly less toxic than the ether linked surfactants, probably due to the degradation of the ester bond by enzymes. no difference in toxicity was observed between gel and liquid state vesicles, which is different from the results obtained by measuring the CBF. The authors suggest that this is probably due to a difference in incubation times. In the case of CBF measurements the incubation was only 1 hour, while 3 days' incubation time was used in the cell proliferation experiments. In the latter case the kinetic aspects, such as exchange of surfactants from the bilayers to the environment, does not affect the toxicity, while in the case of the CBF the kinetic aspects might be very important.

V. THERAPEUTIC APPLICATIONS

A. Administration of Drugs by the Intravenous Route

Several drugs encapsulated in niosomes were studied after intravenous administration, which will be described in this section.

1. Doxorubicin (Adriamycin)

Niosomes prepared from $C_{16}G_3$ (27) with and without cholesterol were administered intravenously in S180 tumor-bearing mice. The vesicles were approximately 800–1000 nm in diameter. Doxorubicin was encapsulated in the niosomes by the film method, the final encapsulated concentration being 1 mg/mL. After a bolus injection in the tail vein of the rat, the concentration of doxorubicin in serum and the accumulation in lungs, liver, heart, and spleen was determined.

In serum a significant increase in doxorubicin concentration was observed, as shown in Fig. 4, when compared to the bolus injection of the free drug. No difference in doxorubicin level was obtained when using vesicles prepared from only $C_{16}G_3$ or vesicles prepared from $C_{16}G_3$:cholesterol in a ratio of 1:1. This was not expected, since in an earlier paper (28) a significant

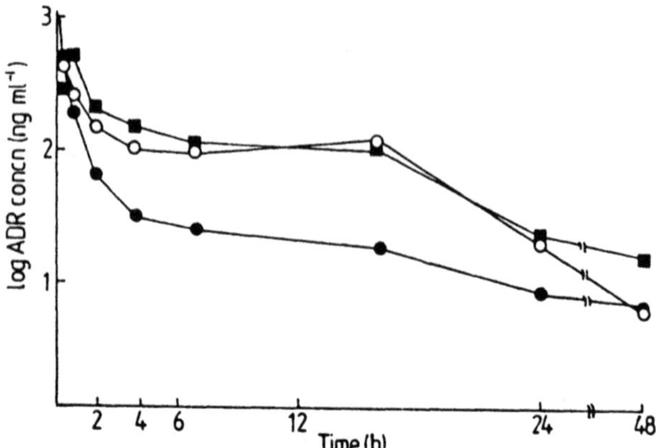

Figure 4 Serum concentration of doxorubicin as a function of time after administration of 5 mg kg^{-1} as a bolus injection into the tail vein of male NMR1 mice. ●, solution of drug; ■, niosomes prepared from $C_{16}G_3$; and ○, niosomes prepared from $C_{16}G_3$ and cholesterol, 1:1 ratio.

difference in release of doxorubicin was reported depending on the composition of the vesicles.

Contrary to the findings, intravenous administration of methotrexate-loaded niosomes (see Sec. V.B below), prepared from the same surfactants, did not lead to increased accumulation of the drug in the liver compared to administration of free drug. This may be due to the difference in size of the vesicles used in the two studies or to a modification of the vesicles by the encapsulated drug, or to the presence of dicetylphosphate in the bilayers of the methotrexate niosomes. It is known that size, charge, and hydrophilicity of the vesicles can change the distribution of the encapsulated drug when administered intravenously (29). The findings of drug accumulation in the spleen were similar to those of the liver; no significant differences between the different formulations were obtained.

In the heart, there was a very small decrease in accumulation of doxorubicin when encapsulated in $C_{16}G_3$ vesicles without cholesterol, although the differences were very small. This might favor the use of niosomes, since cardiotoxicity is one of the problems in doxorubicin therapy.

In the lungs, the level of doxorubicin was increased over the first 6 hours when administered in vesicles compared to free drug. Following the authors, this may be due to entrapment of the large vesicles in the alveoli or to the phagocytic capacity of the rapidly proliferating alveolar phagocytic cells within the basement epithelium.

Finally, drug accumulation in the tumor was increased when administered in cholesterol-containing vesicles. In addition, the tumor growth was decreased. In a very extensive follow-up study, the antitumor activities of doxorubicin encapsulated in niosomes was investigated in more detail (30) by using a molar ratio cholesterol/surfactant of 1:1. In this study, human squamous lung tumor xenografts were transplanted into mice. Doxorubicin solution or doxorubicin encapsulated in niosomes was administered intravenously, after which the mice were sacrificed at various time intervals in order to measure plasma concentrations of the doxorubicin and the accumulation in various organs. In the liver the concentration of two metabolites also was measured. Moreover, tumor growth was determined over a period of 20 days. As in the first study, the concentration of doxorubicin was decreased in cardiac tissue when encapsulated in niosomes, with a shorter tissue half-life, which might decrease its toxic effects. Metabolite concentrations in the liver were lower when the doxorubicin was encapsulated. On the other hand, in the second study the tumor growth was similar after doxorubicin was encapsulated in niosomes or administered as free drug. From these studies the authors concluded that an increase in doxorubicin dose might be possible when the drug is encapsulated in niosomes, since cardiotoxicity is one of the major problems in doxorubicin therapy, and since the concentrations of doxorubicin were reduced in the heart when administered in niosomes.

Cable et al. (31) encapsulated doxorubicin in vesicles prepared from $C_{16}G_3$ and solulan C24, which contains 24 oxyethylene units. It was thought that increasing the hydrophilicity of the surface of the vesicles results in an increase in circulation time of the vesicles. However, Cable et al. found that encapsulation of doxorubicin into vesicles resulted in a higher accumulation of doxorubicin in the liver in addition to a higher plasma concentration, but in a lower accumulation in the kidney than when it was administered as a free drug. It was suggested that an increase in liver uptake was caused by the penetration-enhancing activity of solulan C24. From these studies it was clear that an increase in hydrophilicity of the surface of the vesicles did not result in a lower uptake into the liver and in longer circulation times as shown in the case of hydrophobic microspheres coated by polaxomers or polaxamines (32). This different behavior cannot be due to simple increase in size when adding solulan C24, since size was similar to the vesicles without solulan C24 [prepared by sonication (33)].

2. Inulin

The tissue distribution of ^{3}H inulin encapsulated in stearyl galactoside vesicles, phosphatidylcholine vesicles, or stearyl glucoside vesicles was determined (34). Vesicles were prepared from either surfactants or phospholipids to which cholesterol and dicetylphosphate were added. The glucoside vesicles did not lead to significant differences in drug tissue distribution compared to drug encapsulated in phosphatidylcholine liposomes. The galactoside vesicles showed higher accumulation in the liver compared with glucoside and phosphatidylcholine vesicles. In this study it was not clear whether the galactoside vescicles were able to bind to the galactoside receptors or whether the vesicles were able to pass the fenestration of the liver sinusoids. With respect to the spleen, the accumulation of inulin was lower when encapsulated in galactoside and glucoside vesicles compared to the phosphatidylcholine vesicles. In lung and kidney the inulin accumulation was higher when administered in galactoside vesicles compared to the other types of vesicles. In an earlier publication the stability of the vesicles in plasma was assessed. Glucoside vesicles showed a rapid release of about 40% of the total encapsulated content in 24 hours, while phosphatidyl choline vesicles released 30% of their content. However, phosphatidyl choline vesicles disintegrated gradually, whereas glucoside vesicles remained stable for 48 hours. In a more recent publication (24), the stability phenomenon has been elucidated in more detail. In this study the stability of palmitoyl glucoside vesicles was compared with that of phosphatidyl choline vesicles. Palmitoyl glucoside vesicles appeared to release 70% of their content in plasma at 37°C, whereas phosphatidylcholine vesicles retained 65% of their content. The amount released by the glucoside vesicles depended on their size and type (glucoside vesicles prepared by reversed phase evaporation released 56% of their content; this preparation method was used in the tissue distribution studies). After reheating plasma to 56°C, no significant amount of inulin was released from phosphatidylcholine vesicles, whereas in the case of the glucoside vesicles 35% of the content was released. It seems that two factors are important for the destabilization of the glucoside vesicles, a heat stability dependent and a heat stability independent factor. With the knowledge of these data it is remarkable that no difference in tissue distribution was found between phosphatidylcholine vesicles and glucoside vesicles. It is difficult to give a convincing explanation, since the interpretations of these data are hampered by the lack of knowledge of drug distribution after administration of the free drug and the use of different surfactants in the stability and distribution studies. Stearylglucoside was used in the tissue distribution studies, while in the stability studies palmitoylglucoside was used (24).

3. Sodium Stiboglucinate

Sodium stiboglucinate is a drug used in therapy of visceral leishmaniasis, a protozoan infection of the reticuloendothelial system (RES). In the first of a series of studies (35), sodium stiboglucinate was administered intravenously as free drug, encapsulated in niosomes consisting of $C_{16}G_3$ and cholesterol, and encapsulated in liposomes (DPPC, cholesterol, and DCP). Two experiments were carried out. (a) Drug distribution in the tissue of uninfected mice was determined, and (b) mice were infected with *L. donovani* and treated with free drug and drug encapsulated in niosomes. After two weeks the mice were sacrificed and the parasite concentration in liver cell nuclei were counted.

The distribution of the antimony drug was extremely affected by encapsulation into vesicles. Although the free drug was still detectable after 4 hours of administration, administration of the drug in vesicles resulted in negligible amounts of drug in serum. In the liver, the concentrations of the drug were significantly higher when encapsulated in vesicles than after administration of free drug. From these observations it was concluded that drug encapsulated into vesicles was rapidly cleared from the systemic circulation by the RES. There also was a clear difference in effect on the parasite burden in the liver cells between free drug and niosome-encapsulated drug. The niosomal drug appeared to be an order of magnitude more effective. From this study and also by comparison with other studies it was concluded that no significant differences in drug distribution and therapeutic effect were observed between drugs encapsulated in niosomes and those in liposomes.

In a second study (36), the differences in therapeutic effect were studied when sodium stibugluconate was encapsulated in different types of niosomes. The results confirmed those of the first study. The encapsulation of the drug resulted in a better therapeutic effect than the free drug, but no significant differences were found between the different liposomal and niosomal formulations. The authors also concluded that no therapeutic advantages for niosomes could be found over liposomes. The advantage of niosomes may rather be found in the lower cost and higher stability of these vesicular systems.

In a third study (37), the long-term effects of the therapy were studied. Sodium stiboglucinate, free or encapsulated in small liposomes (DPPC and cholesterol) and niosomes ($C_{16}G_3$ and cholesterol) was administered intravenously in BALB/c mice with *L. donovani* parasites. The effect of drug entrapment after administration was followed for a period of 84 days. It appeared that free drug administration was less effective in late stages of the infections and that drug encapsulated in either liposomes or niosomes was less sensitive to the length of the infections. From this it may be concluded that the benefit of using vesicles increased during the time course of the infection.

B. Peroral Administration of Drugs

The effect of encapsulation in niosomes on the metabolism and urinary secretion of methotrexate has already been studied in 1985. In the case of liposomes the main difficulty is the low stability of the vesicles in the gastrointestinal tract, particularly due to the influence of bile salts and enzymes (38,39,40). It seems that only liposomes prepared from distearylphosphatidylcholine are stable, at least in vitro. No in vivo data on the stability of liposomes has been reported. In the case of niosomes, the stability might be increased due to ether linkage of the alkyl chain to the head group moiety.

Azmin et al. (41) studied the encapsulation of methotrexate into niosomes in vitro. Nonionic $C_{16}G_3$ was used as surfactant, to which cholesterol and dicetylphosphate were added. Two molar ratios were used, 47.5/47.5/5 and 60/30/5 for surfactant, cholesterol, and dicetylphosphate respectively. The vesicles were prepared by the film method followed by sonication, the mean diameter being approximately 100 nm. Groups of mice obtained the preparations orally or intravenously. In one publication (41) it was reported that intravenous and peroral administration of methotrexate encapsulated in vesicles resulted in a higher uptake in liver, brain, and serum compared to free drug or free drug in the presence of polysorbate 80. No significant differences were found between administration of methotrexate in the two different types of niosomes.

In a follow-up study (42), the amount of methotrexate and its metabolite 7-hydroxymethotrexate was measured in urine and feces for 24 hours after oral administration. Oral administration in niosomes resulted in lower concentrations of methotrexate and its metabolite in feces compared to administration of the free drug, indicating a higher uptake of methotrexate (MTX) from the gastrointestinal tract when administered in niosomes. On the other hand, a smaller amount of MTX administered in niosomes was found in urine, which in fact contradicts the interpretations of the results. In addition, after intravenous administration the uptake of methotrexate and its metabolite was also measured in the liver; the respective results are given in Fig. 5. The ratio between 7-hydroxymethotrexate and methotrexate was plotted against time after intravenous administration. Administration in vesicles resulted in a smaller 7-hydroxymethotrexate/methotrexate ratio, while the absolute quantities of the drug and its metabolite were significantly higher compared to intravenous administration of methotrexate solution or a methotrexate solution to which polysorbate 80 was added. After intravenous administration, the amount of free drug in the urine was much higher when administered as solutions then when administered in vesicles. It seems that niosomes facilitate the uptake of methotrexate into the liver, while the excretion via the kidney is decreased. This was also found for liposomes (42). Whether or not niosomes have advantages over administration as free drug or in liposomes is not yet known, since the inhibition of tumor growth also should be studied. Tumor growth was investigated by using intravenous administration of another drug, doxorubicin (see previous section).

Yoshida et al. (23) studied the effect of niosomes on 9-desglycinamide 8-arginine vasopressin (DGAVP) absorption across rat jejunum in an in vitro model. Niosomes were prepared from $C_{18}EO_3$, cholesterol, and dicetylphos-

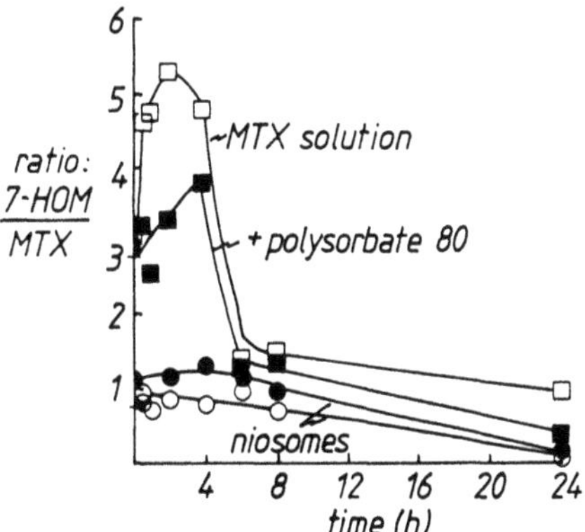

Figure 5 The ratio of 7-hydroxy-MTX (metabolite) to unchanged MTX, measured in the liver of mice after intravenous injection of 2.72 mg. □, MTX kg^{-1} as free solution; ■, MTX in 6 percent polysorbate solution; ○, MTX entrapped in niosomes, $C_{16}G_3$ and cholesterol ratio 1; ●, MTX entrapped in niosomes with a $C_{16}G_3$/cholesterol ratio of 2.

phate in a molar ratio of 57/38/5 by the freeze thawing method. This composition was chosen because it appeared to be the most stable vesicle formulation when exchanging the external PBS solution for a Krebs Ringer buffer solution. They administered free DGAVP, free DGAVP in the presence of niosomes, and DGAVP encapsulated in vesicles at the luminal side of the intestinal loop (donor phase). The time course of the DGAVP concentration at the donor and acceptor phase (serosal side) was measured. The presence of vesicles did not affect the stability of DGAVP at the luminal side of the small intestine, whereas encapsulation of DGAVP in vesicles increased its stability significantly; see Fig. 6. The transport of DGAVP across the intestinal mucosa was not affected by the presence of empty niosomes. The surfactants appeared not to act as penetration enhancers. However, encapsulation of DGAVP into vesicles increased its absorption across the small intestinal wall. This increase was observed after 2 hours of administration. The authors suggested that the most likely mechanism for this increased absorption was either an absorption of the vesicles on the luminal side, which reduced the DGAVP degradation during transport across the small intestinal mucosa, or intact penetration of the vesicles. Although the studies by Azmin et al. (41,42) were performed in vivo, the results of the two studies were very similar and strongly indicate a higher uptake of the drug by the small intestine when incorporated in vesicles. The vesicles used in both studies also were very similar, that is, of a similar composition, and

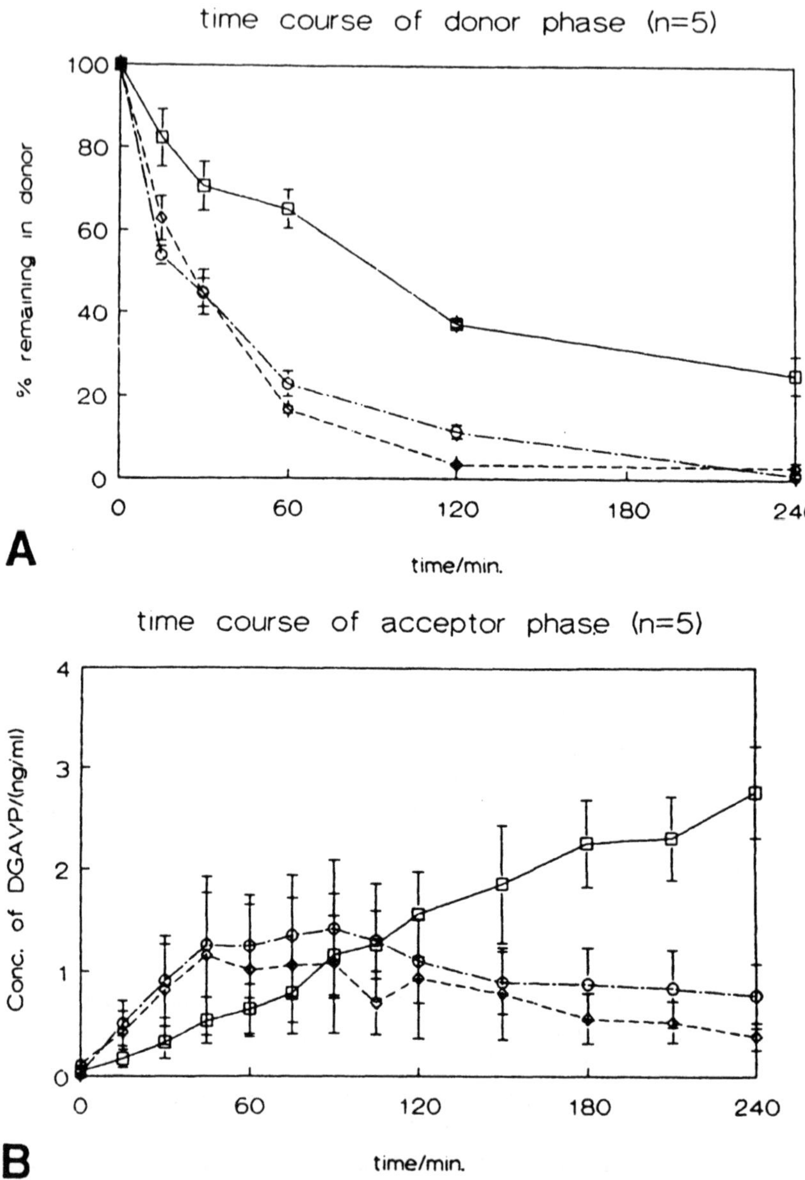

Figure 6 A, stability in rat intestinal lumen of DGAVP administered in solution ○, administered in the presence of niosomes ◇, or entrapped in niosomes □. B, concentration of DGAVP in acceptor phase of the rat intestinal lumen. Symbols as in A.

in both cases ether linked surfactants were used. Both studies show that administration of drugs in vesicles by the peroral route should be investigated in more detail.

C. Administration of Drugs by the Transdermal Route

Very recently studies have been performed on the transdermal administration of drugs in vitro. Although the administration of drugs by the transdermal route has advantages such as avoiding the first pass effect, it has one important drawback, the slow penetration rate of drugs through the skin. The barrier for most drugs has been found to be in the upper layer of the skin, the stratum corneum, which consists of corneocytes embedded in lipid lamellar regions. Transport of drugs is thought to take place mainly through these lipid bilayers (43).

In many publications it has been shown that liposomes do have an effect on the transport of drugs through the skin, but only a few studies deal with the mechanistic aspects (44,45,46). In general one can conclude from these studies that no large amounts of phospholipids are transported through the skin (44,47), except in one study (48).

With niosomes only in vitro studies were carried out (49). The niosomes were prepared from polyoxyethylene alkyl ethers and cholesterol. No charge inducers were added to these vesicles. Transport studies were carried out by using estradiol as a model drug, which is mainly intercalated in the bilayers of the vesicles. In all the experiments, estradiol saturated formulations were prepared. In this way the thermodynamic activity of the drug in the various formulations is equal to 1, which implies that any changes in estradiol transport rate are only due to changes in drug solubility, drug concentration gradient, or the drug diffusion coefficient in the skin. Changes in transport rate through the skin therefore are only based on changes in parameters related to the skin and not to differences in estradiol solubility in the formulations.

In the transport studies stratum corneum was used, since this is the main barrier for diffusion of estradiol through the skin. In the first studies the influence of gel state niosomes prepared from $C_{18}EO_3$ on estradiol transport was compared with two liquid state formulations prepared from $C_{12}EO_3$ and $C_{9=9}EO_{10}$. It appeared that estradiol intercalated in the bilayers of liquid state vesicles, resulting in high estradiol fluxes, whereas estradiol incorporated in gel state vesicles resulted in much lower drug transport rates through the stratum corneum.

The question that arose was, of course, whether this increase in flux was due to a penetration enhancement of the individual surfactants, or was it essential to encapsulate the drug into the vesicles. In additional experiments

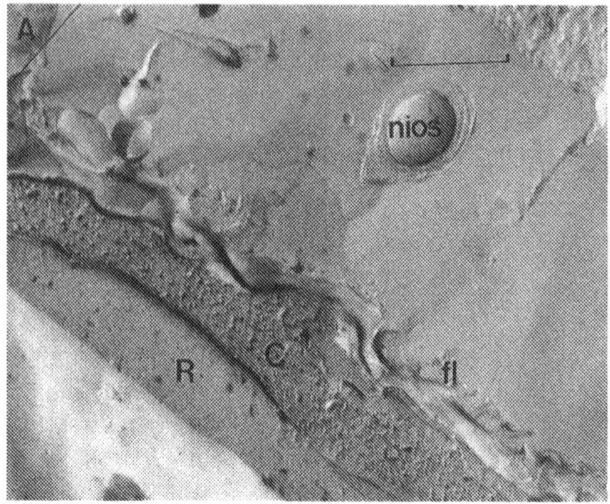

(A)

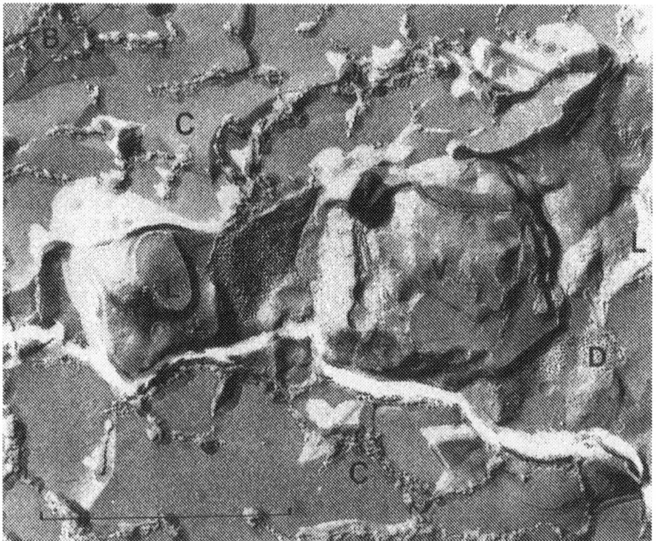

(B)

Figure 7 $C_{12}EO_3$ niosomes seem to fuse at the interface; see A. After treatment of $C_{12}EO_3$ vesicles, irregular structures B and water pools C were found in deeper regions of the stratum corneum, which are normally not found in stratum corneum samples. W, water pool; V, vesicular structures; C, corneocyte; R, rough fracture patterns; NIOS, niosomes; L, lamellar structures.

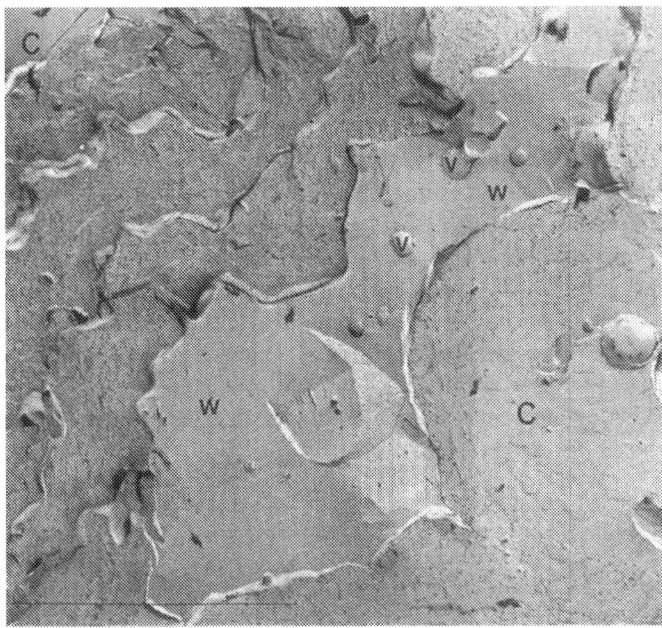

(C)

it appeared that pretreatment of stratum corneum followed by drug transport studies resulted in a significantly lower estradiol transport through stratum corneum then when estradiol was intercalated in the bilayers of niosomes. In fact a similar result was found by Komatsu et al. (50) using liposomes. From these studies it was concluded that it was essential to incorporate the drug in niosomes.

In a separate study (51), the interactions between vesicles and skin were investigated using freeze fracture electron microscopy. The employment of liquid state vesicles resulted in a fusion of the vesicles at the surface of the stratum corneum and in changes in the lipid bilayer structures in deeper layers of the stratum corneum (Fig. 7). Vesicular structures also were found. Two mechanisms were proposed: diffusion of intact niosomes through the stratum corneum, and diffusion of individual compounds through the stratum corneum forming new vesicular structures in certain regions of the stratum corneum. These vesicular structures often were found in clusters close to desmosomes. The second mechanism, reformation of vesicles, seems to be the more probable, because the diameter of the vesicles is much larger than the lipid lamellar spacings in the stratum corneum, which are only approximately

6 and 13 nm (52). Gel state vesicles did not lead to structural changes in the lipid bilayer regions in lower layers of the stratum corneum.

Returning to estradiol transport across the stratum corneum, the following mechanism has been proposed. Fusion of the vesicles at the surface of the stratum corneum can result in very high estradiol concentrations in the upper layers of the stratum corneum, since the bilayers of the vesicles are saturated with estradiol. The high activity gradient of estradiol in the stratum corneum may lead to a high transport rate of estradiol across the skin.

Whether or not the results of the above diffusion experiments can be explained by this mechanism has to be elucidated by additional experiments. One important conclusion can be drawn from these studies. The niosomes that have been used in these studies are promising vehicles at least for lipophilic drugs. Whether niosomes have advantages over liposomes, besides better stability, still has to be determined.

REFERENCES

1. Handjani-Vila, R. M., Ribier, A., Rondot, B., and Vanlerberghe, G. (1979). Dispersions of lamellar phases of non-ionic lipids in cosmetic products. *Int. J. Cos. Sci. 1*: 303–314.
2. Kemps, J. M. A., and Crommelin, D. J. A. (1988). Hydrolyse van fosfolipiden in waterig milieu. *Pharm. Weekbl. 123*: 355–363.
3. Kemps, J. M. A., and Crommelin, D. J. A. (1988). Peroxidatie van fosfolipiden. *Pharm. Weekbl. 123*: 457–469.
4. Baillie, A. J., Florence, A. T., Hume, L. R., Muirhead, G. T., and Rogerson, A. (1985). The preparation and properties of niosomes—Non-ionic surfactant vesicles. *J. Pharm. Pharmacol. 37*: 863–868.
5. van Hal, D., Bouwstra, J. A., and Junginger, H. E. (1992). Preparation and characterization of new dermal dosage-form for antipsoriatic drug dithranol, based on non-ionic surfactant vesicles. *Eur. J. Pharm. Biopharm. 38*: 47s.
6. Echoyen, L. E., Hernandez, J. C., Kaifer, A. E., Gokel, G. W., and Echoyen, L. (1988). Aggregates of steroidal lariat ethers: The first example of non-ionic liposomes (niosomes) formed from neutral crown ether compounds. *J. Chem. Soc., Chem. Commun. 8*: 836–837.
7. Hofland, H. E. J., Bouwstra, J. A., Ponec, M., Bodde, H. E., Spies, F., Verhoef, J. C., and Junginger, H. E. (1991). Interactions of non-ionic surfactant vesicles with cultured keratinocytes and human skin in vitro: A survey of toxicological aspects and ultrastructural changes in stratum corneum. *J. Contr. Rel. 16*: 155–168.
8. Hofland, H. E. J. (1992). Vesicles for transdermal drug delivery. Thesis, Leiden, The Netherlands.
9. Okahata, Y., Tanamachi, S. D., Nagai, M., and Kunitake, G. T. (1981). Synthetic bilayer membranes prepared from bialkyl amphiphiles with non-ionic zwitterionic head groups. *J. Colloid Interface Sci. 82*: 401–417.

10. Hunter, J. A., Dolan, T. F., Coombs, G. H., and Baillie, A. J. (1988). Vesicular systems (niosomes and liposomes) in experimental murine visceral leishmaniasis for delivery of sodium stibogluconate. *J. Pharm. Pharmacol. 40*: 161–165.
11. Mitchell, D. J., and Ninham, B. W. (1981). Micelles, vesicles and microemulsions. *J. Chem. Soc., Faraday Trans. 2, 77*: 601.
12. Israelachvili, J. N. (1983). *Physics of Amphiphiles: Micelles, Vesicles and Microemulsions* (V. Degorgio and M. Corti, eds.). Amsterdam: Elsevier, p. 31.
13. Mitchell, J. D., Tiddy, G. J. T., Warning, L., Bistock, T., and McDonlad, M. P. (1983). *J. Chem. Soc., Faraday Trans. 79*: 975–1000.
14. Bouwstra, J. A., Jousma, J., van der Meulen, M. A., Vijverberg, C. C., Spies, F., and Junginger, H. E. (1989). A structural study of the mesophases of two oxyethylene alkyl ethers-decane-water systems, a comparison between technical grade $C_{12}EO_7$ and pure $C_{12}EO_6$. *Colloid Polym. Sci. 267*: 531–538.
15. Ostrowsky, N., and Sornetti, D. (1983). *Physics of Amphiphiles: Micelles, Vesicles and Microemulsions* (V. Degorgio and M. Corti, eds.). Amsterdam: Elsevier.
16. Buckton, G., Chowdry, B. Z., Armstrong, J. K., Leharne, S. H., Bouwstra, J. A., and Hofland, H. E. J. (1992). The use of high sensitivity differential scanning calorimetry to characterize dilute aqueous dispersions of surfactants. *Int. J. Pharm. 83*: 115–121.
17. Silver, B. L. (1985). *The Physical Chemistry of Membranes* (B. L. Silver, ed.). New York: Alan and Unwin and The Salomon Press.
18. Michels, B., Fazel, N., and Cert, R. (1989). *Europ. Biophys. J. 17*: 187–189.
19. Ribier, A., Handjani-Vila, R. M., Bardez, R. M., and Valeur, E. (1984). Bilayer fluidity of non-ionic vesicles. An investigation by differential polarized phase fluorometry. *Colloids and Surfaces 10*: 155–161.
20. Kiwada, H., Nimura, H., Fujisali, Y., Yamada, S., and Kato, Y. (1985). Application of synthetic alkyl glycoside vesicles as drug carriers. I. Preparation and physical properties. *Chem. Pharm. Bull. 33*: 753–759.
21. Kaneshina, S., Shibata, O., and Nakamura, M. (1982). The effect of pressure on the mutual solubility of a nonionic surfactant-water system. *Bull. Chem. Soc. Japan. 55*: 951–952.
22. New, R. R. C., ed. (1990). Liposomes. A Practical Approach. New York: IRL Press, p. 127.
23. Yoshida, H., Lehr, C.-M., Kok, W., Junginger, H. E., Verhoef, J. C., and Bouwstra, J. A. (1992). Niosomes for oral delivery of peptide drugs. *J. Contr. Rel. 21*: 145–154.
24. Kowada, H., Nakajima, I., Matsuura, H., Tsuji, M., and Kato, Y. (1988). Application of synthetic alkyl glycoside vesicles as drug carrier systems. III. Plasma components affecting stability of the vesicles. *Chem. Pharm. Bull. 36*: 1841–1846.
25. Moser, P., Marchand-Arvier, M., Labrude, P., Handjani-Vila, R. M., and Vigneron, C. (1989). Hemoglobin niosomes. I. Preparation, functional and physicochemical properties, and stability. *Pharmaceutica Acta Helvetiae. 64*: 192–202.
26. Hofland, H. E. J., Bouwstra, J. A., Verhoef, J. C., Buckton, G., Chowdry, B. Z., Ponec, M., and Junginger, H. E. (1992). Safety aspects of non-ionic surfactant vesicles: A toxicity study related to the physicochemical characteristics of non-ionic surfactants. *J. Pharm. Pharmacol. 44*: 287–294.

27. Rogerson, A., Cummings, J., Willmott, N., and Florence, A. T. (1988). The distribution of doxorubicin in mice following administration of niosomes. *J. Pharm. Pharmacol. 40*: 337–342.
28. Rogerson, A., Baillie, A. J., and Florence, A. T. (1987). Adriamycin loaded niosomes: Drug entrapment, stability and release. *J. Microencap. 4*: 321–328.
29. Papahadjopoulos, D., and Gabizon, A. (1991). Liposome targetting to tumor cells in vivo. *Targetting of Drugs 2. Optimizing Strategies* (G. Gregoriadis, A. C. Allison, and G. Poste, eds.). *NATO ASI Series 199*: 95–103.
30. Kerr, D. J., Rogerson, A., Morrison, G. J., Florence, A. T., and Kay, S. B. (1988). Antitumour activity and pharmacokinetics of niosome encapsulated adriamycin in monolayer, spheroid and xenograft. *Br. J. Cancer 58*: 432–436.
31. Cable, C., Cassidy, J., Kaye, S. B., and Florence, A. T. (1988). Doxorubicin in cholesteryl polyoxyethylene modified niosomes: Evidence of enhancement of absorption in mice. *J. Pharm. Pharmacol.* (Suppl.) *40*: 31P.
32. Illum, L., Davis, S. S., Muller, R. H., Mak, E., and West, P. (1987). The organ distribution and circulation time of intravenous injected colloidal carriers sterically stabilized with block-copolymers—Polaxamine 908. *Life Sciences 40*: 367–374.
33. Cable, C., and Florence, A. T. (1988). Mixed polyglycerol-polyoxyethylene ether niosomes. *J. Pharm. Pharmacol.* (Suppl.) *40*: 30P.
34. Kiwada, H., Niimura, H., and Kato, Y. (1985). Tissue distribution and pharmacokinetic evaluation of the targetting efficiency of synthetic alkyl glycoside vesicles. *Chem. Pharm. Bull. 33*: 2475–2482.
35. Baillie, A. J., Coombs, G. H., Dolan, T. F., and Laurie, J. (1986). Non-ionic surfactant vesicles, niosomes, as a delivery system for anti-leishmanial drug, sodium stibogluconate. *J. Pharm. Pharmacol. 38*: 502–505.
36. Hunter, C. A., Dolan, T. F., Coombs, G. H., and Baillie, A. J. (1988). Vesicular systems (niosomes and liposomes) for delivery of sodium stibogluconate in experimental murine visceral Leishmaniasis. *J. Pharm. Pharmacol. 40*: 161–165.
37. Baillie, A. J., Dolan, T. F., Alexander, J., and Carter, K. C. (1989). Visceral Leishmaniasis in the BALB/c mouse: Sodium stibogluconate treatment during acute and chronic stages of infection. *Int. J. Pharm. 57*.
38. Woodley, J. F. (1980). Liposomes for oral administration of drugs. *CRC Critical Reviews in Therapeutic Drug Carrier Systems 2*: 1–18.
39. Patel, H. M., and Ryman, B. E. (1981). Systemic and oral administration of liposomes. *Liposomes: From Physical Structure to Therapeutic Applications* (K. Knight, ed.). Elsevier/North Holland Biomedical Press, pp. 410–441.
40. Roland, R. W., and Woodley, J. F. (1980). The stability of liposomes in vitro to pH, bile salts and pancreatic lipase. *Biochim. Biophys. Acta 620*: 400–409.
41. Azmin, M. N., Florence, A. T., Handjani-Vila, R. M., Stuart, J. B. F., Vanlerbeghe, G., and Wittaker, G. S. (1985). The effect of non-ionic surfactant vesicles (niosomes) entrapment on the absorption and distribution of methotrexate in mice. *J. Pharm. Pharmacol. 37*: 237–242.
42. Azmin, M. N., Florence, A. T., Handjani-Vila, R. M., Stuart, J. B. F., Vanlerberghe, G., and Wittaker, G. S. (1989). The effect of niosomes and poly-

sorbate80 on the metabolism and excretion of methotrexate in the mouse. *J. microencap. 3*: 95–100.
43. Bodde, H. E., Kruithof, N. A. M., Brussee, J., and Koerten, H. K. (1989). Visualization of normal and enhanced $HgCl_2$ transport through human skin in vitro. *Int. J. Pharm. 53*: 13-24.
44. Ganesan, M. G., Weiner, N. D., Flynn, G. L., and Ho, N. H. F. (1984). Influence of liposomal drug entrapment on percutaneous absorption. *Int. J. Pharm. 20*: 139-154.
45. Ho, N. H. F., Ganesan, M. G., Weiner, N. D., and Flynn, G. L. (1985). Mechanisms of topical delivery of liposomal entrapped drugs. *J. Contr. Rel. 2*: 61–65.
46. Knepp, V. M., Szoka, F. C., and Guy, R. H. (1990). Controlled release from a novel liposomal delivery system. II. Transdermal delivery characteristics. *J. Contr. Rel. 12*: 25-30.
47. Komatsu, H., Higaki, K., Okamoto, H., Miyagawa, K., Hashida, M., and Sezaki, H. (1986). Preservative activity and in vivo percutaneous penetration of butyl paraben entrapped in liposomes. *Chem. Pharm. Bull. 34*: 3415-3422.
48. Cevc, G., and Blume, G. (1991). Lipid vesicles penetrate into intact skin owing to transdermal osmotic gradients and hydration forces. *Biochim. Biophys. Acta 1104*: 226-232.
49. Hofland, H. E. J., Van der Geest, R., Bodde, H. E., Junginger, H. E., and Bouwstra, J. A. (1994). Estradiol permeation from non-ionic surfactant vesicles through human skin in vitro. *Pharm. Res.*, in press.
50. Komatsu, H., Okamoto, H., Miyagawa, K., Hashida, M., Sezaki, H. (1986). Percutaneous absorption of butyl paraben from liposomes in vitro. *Chem. Pharm. Bull. 34*: 3423-3430.
51. Hofland, H. E. J., Bouwstra, J. A., Bodde, H. E., Spies, F., Nagelkerke, J. F., Cullander, C., and Junginger, H. E. (1994). Interactions between non-ionic surfactant vesicles and human skin in vitro: Freeze fracture electron microscopy and confocal laser scanning microscopy. *J. Liposome Res.*, submitted.
52. Bouwstra, J. A., Gooris, G. S., van der Spek, J. A., and Bras, W. (1991). The structure of human stratum corneum studied by small angle X-ray scattering. *J. Invest. Derm. 97*: 1005-1012.

5

Nanoparticles

Jörg Kreuter

Institut für Pharmazeutische Technologie, Johann Wolfgang Goethe-Universität, Frankfurt am Main, Germany

I. INTRODUCTION

One of the primary objectives in the design of novel drug delivery systems is controlled delivery of the pharmacological agent to its site of action at a therapeutically optimal rate and dose regimen (1). This site-specific or targeted delivery combined with delivery at an optimal rate would not only improve the efficacy of a drug but would also reduce the possibility of unwanted toxic side effects of the drug, thus improving the therapeutic index (2). Among the most promising systems to achieve this goal are colloidal drug delivery systems (see the Preface of this book). Colloidal drug delivery systems include the drug carrier systems liposomes, niosomes, nanoparticles, and microemulsions. (Ointments, which also can be conceived of as colloidal drug delivery systems, are mainly used topically and therefore are very different in their way of action from the other four mentioned colloidal systems.)

Liposomes, niosomes, nanoparticles, and microemulsions are very similar in their size, shape, and mode of administration, and for this reason they may be used alternatively. Nevertheless, these systems have a number of different advantages and disadvantages. The great advantage of liposomes, for instance, is that their main components are lecithin and, in most cases, cholesterol. Similar lecithins as well as cholesterol are materials that also exist in the body in major amounts, and consequently a good bioacceptability may be expected. Nanoparticles, on the other hand, possess a better stability. This property, for instance, may be very important for many modes of targeting. As a consequence of the mentioned advantages and disadvantages, the most suitable system will have to be chosen depending on the drug and the therapeutic goal to be reached. As will be shown later, the same drug, e.g., mitoxantrone, can be more effective against one type of tumor, entrapped in liposomes, whereas the same drug bound to nanoparticles may be more efficient against another type of tumor.

Nanoparticles are made of artificial or natural polymers. The use of these polymers is often restricted by their bioacceptability. In addition to the properties of the carrier material, i.e., the polymer and the auxilliary substances, bioacceptability is also influenced by the particle size. A reduction of the particle size enables an intravenous injection. The diameter of the smallest blood capillaries is 4 μm (3); consequently, solid particles have to have a smaller diameter than this in order to be able to traverse all capillaries. A small particle size is also desirable for intramuscular and subcutaneous administration. Little and Parkhouse (4) have demonstrated that a reduction in particle size of polymeric particles minimizes possible irritant reactions at the injection site. Moreover, Nothdurft (5), Nothdurft and Mohr (6), and Stinson (7,8) showed that cancerogenic effects are not related to the chemical nature of materials like glass, steel, and a number of artificial polymers but rather to the size of these materials, when implanted into animals. In the mentioned experiments, discs of different diameters and shapes, fibers, powders, and small particles of sizes ranging between 1 μm and 76 μm were implanted or injected. In large animals, such as guinea pigs, no tumors were observable. In smaller animals, the occurrence of tumors was clearly size dependent, and no tumors could be found after administration of powders and small particles. During these experiments, Stinson (8) also observed that irritant tissue reactions only occurred around agglomerates of particles, not around single particles.

The final choice of the appropriate polymer, particle size, and manufacturing method will primarily depend on the bioacceptability of the polymer, secondarily on the physicochemical properties of the drug, and thirdly on the therapeutic goal to be reached.

II. DEFINITION OF NANOPARTICLES

Nanoparticles are solid colloidal particles ranging in size from 10 nm to 1000 nm (1 μm). They consist of macromolecular materials in which the active principle (drug or biologically active material) is dissolved, entrapped, or encapsulated, and/or to which the active principle is adsorbed or attached (9).

This definition includes not only particles with structures as described by Birrenbach and Speiser (10) by the term "nanopellets" but also "nanocapsules" with a shell-like wall as well as "microspheres," if they are below 1 μm in size (11–15). It also includes polymer latices such as the "molecular scale drug entrapment" products by Banker et al. (16–20). The above definition of nanoparticles was chosen because it is often very difficult to prove whether these particles have a continuous matrix or a shell-like wall, or whether their content is incorporated into or adsorbed onto the particles.

III. PREPARATION METHODS

A. Emulsion Polymerization

Emulsion polymerization is among the most frequently employed methods of producing nanoparticles. The term "emulsion polymerization" is somewhat misleading because this process can be carried out without any emulsifier. For this reason, this process will be shortly reviewed here.

The term "emulsion polymerization" was created because the monomer was emulsified in a nonsolvent by means of emulsifiers. After polymerization, a small polymer particle suspension was obtained. Initially it was assumed that these polymer particles were produced by polymerization of the monomer emulsion droplets. Later, however, it was realized that the resulting polymer particles were smaller than the initial emulsion droplets. For this reason, the theory of emulsion polymerization was revised in that the location of polymerization was shifted to the emulsifier micelles (9,21). These micelles coexisted with single emulsifier molecules that were present in solution and with the emulsifier molecules that were adsorbed at the emulsion/droplet interface, thus stabilizing the emulsion droplets. It was assumed that monomer molecules would diffuse from the emulsion droplets into the emulsifier micelles and that the solubilized monomer molecules in the micelles were then polymerized to form the polymer latex.

Later, Fitch and coworkers (22–25) observed that the emulsifier concentration did not affect the polymerization rate, and that formation of particles observed by Tyndall scattering was independent of the rate of polymerization. Particle formation in a specific solvent, in this case water, always occurred at a specific concentration that was characteristic for the polymer. In addition, the number of emulsifier micelles present did not affect the number of particles formed (26,27). At low monomer concentrations emulsion polymerization even could be carried out without any emulsifier molecules present. These observations led Fitch (25) to the conclusion that the location of the polymerization initiation is in the solvent phase. Initiation takes place in this phase when dissolved monomer molecules are hit by a starter molecule or by high-energy radiation. The polymerization and chain growth is maintained by further monomer molecules that diffuse to the growing polymer. Diffusion of these monomer molecules to the growing polymer particles is much faster than the polymerization process, thus providing sufficient monomer in the vincinity of the location of polymerization (28). The monomer droplets and the emulsifier micelles mainly act as reservoirs for further monomer and at later stages as reservoirs for emulsifier molecules that stabilize the polymer particles after phase separation and prevent coagulation. As mentioned above, some systems can be polymerized without any emulsifers present (9,29).

Initially, during emulsion polymerization the growing polymer molecules remain dissolved in the continuous surrounding phase. After a certain molecular weight is reached, the formed molecules become insoluble, so that phase separation and particle formation indicated by Tyndall scattering occurs. After phase separation, additional monomer and polymer molecules, including micro- and macroradicals, diffuse into these growing polymer particles, maintaining further particle growth. The termination of the polymerization by the reaction of two radicals can take place before or after particle formation (30). Consequently, a single polymer particle consists of a large number of macromolecules. The molecular weights of nanoparticles range between 10^3 Da in the case of poly(alkyl cyanoacrylate) nanoparticles (31) and 4×10^5 Da for poly(methyl methacrylate) nanoparticles (32). Assuming for instance a particle size of 100 nm and a density of about 1.0 g/cm^2 (real densities of nanoparticles range between 1.0 and 1.15 g/cm^2), one particle would consist of between about 10^3 and 5×10^5 single macromolecules, depending on the molecular weight of the final polymer.

B. Polymerization in a Continuous Aqueous Phase

1. Poly(methyl Methacrylate) Nanoparticles

Poly(methyl methacrylate) nanoparticles are very slowly biodegradable. For this reason, they are suitable as adjuvants for vaccines when the achievement of a very prolonged immune response is desired. They are also of great value for basic body distribution studies where the determination of the fate of intact particles has to be followed over extended time periods.

Monomeric methyl methacrylate is soluble in water in concentrations up to 1.5%. [In the presence of some materials such as certain viruses the solubility can even be increased up to 2% (33).] After dissolution of the monomer, the polymerization is initiated either by high-energy radiation (34) or chemically by addition of a polymerization initiator such as ammonium or potassium peroxodisulfate and heating to elevated temperatures (29).

Initiation with γ-rays produced by a ^{60}Co source (34) has the advantage that no auxilliary agents need to be used. The polymerization can be carried out in the pure system water-methyl methacrylate. Instead of pure water, a buffer solution or the solution of a drug or of another material to be bound to the nanoparticles can be used as the polymerization medium. The resulting molecular weights depend on the monomer concentration (30). A dose of 500 krad was found to be optimal for this polymerization process.

The other possibility for the polymerization of methyl methacrylate is chemical initiation. In this case, the system is heated to temperatures above 65°C; the initiator (potassium or ammonium peroxodisulfate) is added during heating at about 40 to 50°C. The molecular weights as well as the particle size of the resulting nanoparticles (Table 1) increase very significantly with increasing monomer concentration, and decrease very slightly with increasing temperature and with increasing initiator concentration (9,29). At a certain temperature and initiator concentration, the number of nucleating radicals is constant, resulting in a certain number of particles. Additional monomer, therefore, will not influence the number of radicals but will increase the molecular weight. An increase in initiator concentration at a constant temperature increases the number of radicals generated. At a constant monomer concentration, this will decrease the molecular weights of the resulting polymer molecules. An increase in the temperature has the same effect, because it increases the decay rate of the initiator molecules and hence the number of nucleating radicals, although the initiator molecule concentration is kept constant (Table 1).

Poly(methyl methacrylate) nanoparticles are generally produced without the addition of any emulsifiers. Nevertheless, hydrophilic macromolecules that may be present in the polymerization media for various reasons lead to a very homogeneous particle size distribution (9), because they do act as dispersing agents. These macromolecules can be part of the material to be bound to the nanoparticles or they may be added as auxilliary materials.

The mentioned macromolecules, drugs, antigens, or other biological agents to be bound to the nanoparticles may be present in the polymerization medium, or they may be added after polymerization. Polymerization in the presence of these materials, of course, can only be performed if the material is stable against the polymerization reaction.

2. Poly(alkyl Cyanoacrylate) Nanoparticles

Poly(alkyl cyanoacrylate) nanoparticles are rapidly biodegradable. For this reason, they are eliminated from the body in a few days (35). Their degradation rate can be monitored by combining cyanoacrylates with different side-chain esters simply by mixing the monomers prior to addition to the polymerization medium (36).

Table 1 Influence of Monomer Concentration, Initiator Concentration, and Temperature on Particle Size and Molecular Weight of Poly(methyl Methacrylate) Nanoparticles

Potassium peroxodisulfate concentration (mmol)	Particle size (nm)								Molecular weight ($\overline{M}_w$)		
	Methyl methacrylate concentration (mmol) at two temperatures										
	10		33.75		80		156.25		80		156
	65°C	85°C	65°C	85°C	65°C	85°C	65°C	85°C	65°C	85°C	85°C
0.3	85	72	129	128	181	170	256	262	–	434,000	–
1.65	98	88	151	169	212	193	248	248	–	–	–
3.0	92	72	135	149	223	177	250	258	289,000	220,500	400,000

Source: Adapted from Berg, U., Immunstimulation durch hochdisperse Polymersuspensionen, Diss. ETH Zürich No. 6481, Zürich, 1979, pp. 34, 67.

The cyanoacrylate monomers are added to the aqueous polymerization medium in concentrations between 0.05% and 7% (37). Because the solubilities of the cyanoacrylate monomers are exceeded at most of these concentrations, monomer droplets are formed and the system has to be stirred permanently.

The polymerization mechanism is an anionic process initiated by bases present in the polymerization medium. The cyanoacrylates are mainly initiated by the OH^- ions resulting from the dissociation of water, but some basic drugs also can act as initiators. This OH^--induced polymerization is very rapid. For this reason, the pH has to be kept below 3.5, with some drugs even below 1.0, to enable the formation of nanoparticles. The influence of temperature, pH, monomer, type and concentration of electrolyte, acidifying agent, and emulsifier (stabilizer) on the polymerization process employed is very complex. The reason for this is that the polymerization medium generates the initiating OH^- ions as well as simultaneously the H^+ ions that terminate the polymerization reaction (Fig. 1). As a result of this termination by H^+ ions, the molecular weights after polymerization are very low and decrease with decreasing pH (32,37,38). The influence of pH on the particle size is somewhat different: a particle size minimum exists at around a pH of 2, whereas the polydispersity falls with increasing pH until a plateau is reached at about a pH of 2.5 and above (40).

As a result of the low molecular weights, the growing particles are very soft and prone to agglomeration. For this reason, stabilizers have a significant influence on the particle size and the molecular weight (39,40). This influence of the stabilizers is also rather complex. Higher stabilizer concentrations generally produce larger particles. Table 2 gives a summary of the results of various investigations on stabilizers. High concentrations of above

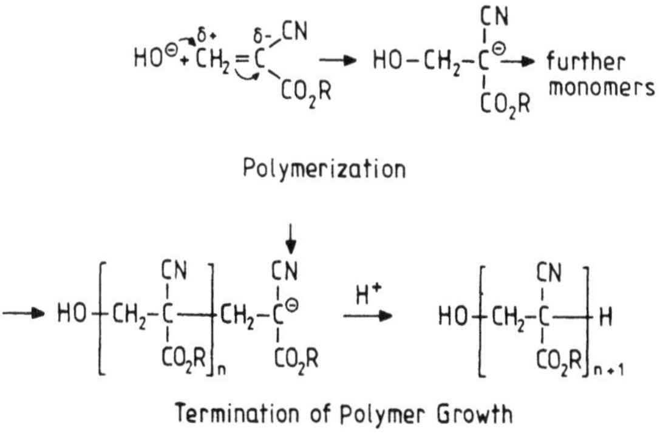

Figure 1 Polymerization mechanism of poly(alkyl cyanoacrylates) (R = alkyl function). (Reproduced from Ref. (1) with permission of the copyright holder.)

Table 2 Influence of Stabilizers and Acid Medium on the Particle Size of Poly(butyl Cyanoacrylate) Nanoparticles

Surfactant	Concentration (% w/v)	Type of acid (0.01 N, pH 2.25)	Particle size (nm)	Ref.
Dextran 70	0.05	HCl	212	40
	0.05	HCl	148	
	1.0	HCl	138	
	2.5	HCl	126	
Dextran 40	0.05	HCl	243	40
	0.5	HCl	145	
	1.0	HCl	134	
	2.5	HCl	131	
Dextran 10	0.05	HCl	770	40
	0.5	HCl	168	
	1.0	HCl	139	
	2.5	HCl	109	
β-Cyclodextrin	0.75	HCl	3450	40
	1.0	HCl	3000	
	1.75	HCl	2700	
Polysorbate 20	0.5	HCl	51,58	40,233
Polysorbate 40	0.5	HCl	46	40
Polysorbate 60	0.5	HCl	38	40
Poloxamer 184	0.5	HCl	254	40
Poloxamer 188	0.5	HCl	160	40
Poloxamer 188[a]	2.0	HCl	33	41
Poloxamer 237	0.5	HCl	118	40
Poloxamer 238	0.5	HCl	71	40
Poloxamer 338	0.5	HCl	73	40
Dextran 70	0.5	Citric	131	39
	0.5	Sulfuric	176	39
	0.5	Nitric	158	39
	0.5	Acetic	811	39
	0.5	Phosphoric	219	39

[a]Poly(isobutyl cyanoacrylate) nanoparticles.
Source: Reproduced from Ref. (1) with permission of the copyright holder.

2% of one of the most frequently used stabilizers, poloxamer 188, reduce the particle size of polyisobutylcyanoacrylate nanoparticles from around 200 nm without emulsifier down to about 31 to 56 nm (41).

Nanoparticles produced without stabilizers or by using polysorbates as surfactants have a monomodal molecular weight distribution with mean molecular weights of 1000–4000 Ka, whereas some stabilizers such as dextrans or poloxamers lead to a distinctive bimodal distribution of the molecular weights with peaks at 1000–4000 Da and 20,000–40,000 Da. This bimodal distribution is indicative of two separate polymerization reactions (37). One reaction probably occurs in the aqueous phase where polymerization termination by H^+ ions is rapid, leading to small molecular weights. After formation of primary particles, another pathway seems to be possible, namely polymerization of captured growing unterminated polymer molecules within these particles. Due to the reduced H^+ concentration in this environment, the termination frequency is reduced. Consequently, the resulting molecular weights would be much higher (37). However, it also may be possible that the increased molecular weight is caused by the incorporation of molecules such as dextrans or poloxamers into the polymer chain (40). It has to be mentioned that some drugs present during polymerization such as doxorubicin slightly increase the molecular weights of the lower peak and also can lead to the formation of a second peak at 50,000 Da (38).

Other factors contributing to the particle size of the nanoparticles include monomer concentration and stirring speed. Because the termination of the polymerization of the cyanoacrylates in water is governed by the H^+-induced termination, one would not expect a major contribution of the monomer concentration. Indeed, the monomer concentration influence is much less pronounced than with poly(methyl methacrylate). A slight particle size minimum is observed at a monomer concentration of about 2% (39). As mentioned above, the cyanoacrylate system has to be stirred during polymerization. Nevertheless, the particle size increases slightly with increasing stirring speed (39). This increase in particle size with increased agitation can be observed not only with the cyanoacrylates but also with other colloidal systems. It is caused by the higher kinetic energy of the system at higher agitation forces. This increased kinetic energy level probably enables some oligomers, small semisolid particles, and even larger particulates to surmount the interfacial energy barrier surrounding the particles, leading the coalescence with other particles (42,43). This particle size increase is coupled with an increase in polydispersity. Ultrasonication as well as the use of an ultraturrax significantly increases this effect (39).

Addition of sulfur dioxide by bubbling through the monomer shortly before its addition to the polymerization medium and start of the polymerization again induced very small particles down to diameters as small as 10 nm even without addition of further stabilizers. This method led to slightly larger particles (18 nm) after addition of 1% dextran 70 (44). The presence of sulfur dioxide gave the particles a high negative charge that probably inhibited particle growth.

Due to the initiation of the cyanoacrylate polymerization by bases, basic drugs also can act as polymerization starters. In this case, these drugs will be incorporated into the polymer chain. The increase in the molecular weight and the appearance of a second peak at much higher molecular weights observed during polymerization in the presence of doxorubicin (38) mentioned above is an indication that this interaction may have happened. The drug incorporation can be prevented in many cases by a reduction of the pH or by a change in the composition of the polymerization medium. Strong basic drugs, however, still may interact with the monomer. This interaction can be avoided by addition of the drug to the nanoparticles after polymerization.

Because of the complexity of the cyanoacrylate polymerization process, the resulting particle size in such multicomponent systems is difficult to predict. For this reason, the optimization of the polymerization conditions may require extensive experimental work. However, because this process is thermodynamically controlled, in our experience in most cases scaling-up is relatively simple (1,45).

3. Acrylic Copolymer Nanoparticles

Acrylic copolymer nanoparticles were produced by a number of authors including Rembaum et al. (46–48), Molday et al. (49), Yen et al. (50), Kreuter et al. (51,52), Lukowski et al. (53), and Scholsky and Fitch (54). Chemical initiation as well as γ-irradiation was employed for the polymerization. The following monomers were used during these studies: methyl methacrylate, 2-hydroxyethyl methacrylate, methacrylic acid, ethylene glycol dimethylacrylate, acrylamide, *N,N'*-bimethyleneacrylamide, and 2-dimethylamine. Some examples of copolymer compositions are given in Table 3. Reference 49 gives a good compilation of the production conditions.

An increase in the acrylic acid content of the poly(methyl methacrylate) copolymer nanoparticles neither increased the zeta-potential nor correlated with any of the observed slight variations in particle size (53). This is somewhat surprising, since the acrylic acid was expected to increase the surface charge of the particle and, as a result of this, to influence the particle growth and reduce the particle size. However, on the other hand, as expected the acrylic acid incorporation distinctly reduced the surface hydrophobicity and decreased the adsorptive properties for a model sorptive, rose bengal (53).

In a number of the above studies involving polymerization with γ-irradiation, polyethyleneglycols were used as stabilizers. These stabilizers were found to prevent pH-dependent aggregation and decrease the particle size (46). Antibodies, radioactive amino acids, and fluorescent molecules were coupled to the above nanoparticles by covalent binding using the cyanogen bromide, the carbodiimide, or the glutaraldehyde method (46,49).

4. Polystyrene Nanoparticles

Polystyrene nanoparticles can be produced by similar methods to polyacrylic nanoparticles (55,56). The water solubility of styrene, however, is lower than

Table 3 Acrylic Copolymer Nanoparticles

Composition	%	Diameter (nm)	Reference
Methyl methacrylate, and	66.6	360	51
2-hydroxyethyl methacrylate	33.3		
Methyl methacrylate, and	50	200	51
2-hydroxyethyl methacrylate	50		
Methyl methacrylate,	53	40	47
2-hydroxyethyl methacrylate,	30		
methacrylic acid, and	10		
ethyleneglycol dimethacrylate	7		
Methyl methacrylate,	33	80	47
2-hydroxyethyl methacrylate,	25		
methacrylic acid,	10		
acrylamide, and	25		
ethyleneglycol dimethacrylate	7		
2-Hydroxyethyl methacrylate,	70	150	47
methacrylic acid, and	20		
N,N'-bismethyleneacrylamide	10		
2-Hydroxyethyl methacrylate,	30	60	47
acrylamide,	30		
N,N'-bismethyleneacrylamide, and	30		
methacrylic acid	10		

Source: Reproduced from Ref. (1) with permission of the copyright holder.

that of most acrylates. For this reason, the use of surfactants is required. In addition, this material is not biodegradable. For this reason, it is mainly used as an immunosorbent (55) or for basic biodistribution and vaccination studies (see Sec. VI and Sec. VII.G).

5. Poly(vinyl Pyridine) Nanoparticles

Poly(4-vinyl pyridine) nanometer-sized particles also were prepared by Rembaum et al. (47,48). The polymerization was carried out in aqueous methanol or acetone solutions containing *N,N'*-bis-methyleneacrylamide as a cross-linking agent. Poly(ethylene oxide) was also present in amounts of 0.1 to 4% (w/v). Concentrations of poly(ethylene oxide) in water of 1% or higher yielded cross-linked gels when exposed to ionizing radiation. This suggests that the stabilization of the particles with this agent may be due partly to grafting and partly to physical entanglement of the poly(ethylene oxide) chains (47).

The size of the poly(4-vinyl pyridine) particles could be monitored by the monomer concentration and by the amount of methanol or acetone present in water. Especially, increasing amounts of the latter two solvents significantly increased the particle size.

Poly(4-vinyl pyridine) copolymer nanoparticles of a size between 160 and 250 nm could be made with 2-hydroxyethyl methacrylate, acrylamide, or methacrylamide at a total monomer concentration of 2%. 10% *N,N'*-bis-methyleneacrylamide always served as a cross-linking agent (47).

Instead of 4-vinylpyridine, other vinylpyridines such as 2-vinylpyridine or 2-methyl-5-vinylpyridine also may be used for the production of these types of nanoparticles (57). The use of stabilizers such as poly(ethylene oxide), polysorbate 20, or polysorbate 85 enabled the polymerization in organic solvent-free aqueous systems.

Poly(vinyl pyridine) nanoparticles have the advantage of being able to incorporate or bind metals through complexing with their aromatic nitrogen (57). The ring itself can undergo chemical modification to introduce other functional groups.

6. Polyacrolein Nanoparticles

Polyacrolein nanoparticles can be produced by aqueous polymerization of acrolein using γ-irradiation (0.5 Mrad, ^{60}Co source) (58) or by polymerization under alkaline conditions with NaOH (2 hours, pH 10.5) (59). The particle size may be controlled by surfactants. In the case of polymerization by γ-irradiation, sodium lauryl sulfate was used; under alkaline conditions sodium sulfite-polyglutaraldehyde conjugate was the surfactant of choice. In the latter case, a number of surfactants including sodium lauryl sulfate, poly(ethylene oxide), poly(vinyl alcohol), and polysorbate 20 did not stabilize the nanoparticles (59). The size of these nanoparticles ranged from 40 to 8000 nm.

Ligands such as protein drugs, enzymes, and antibodies can be bound covalently through their amino group via formation of Schiff's bases to the polyacrolein nanoparticles (58,59).

7. Polyglutaraldehyde Nanoparticles

Polyglutaraldehyde particles ranging in size from 50 nm to 1500 nm could be obtained by aldol polycondensation of monomeric glutaraldehyde at alkaline pH (60). The rate of the reaction increased significantly at $pH > 7$. The mechanism of the aldol polycondensation is shown in Fig. 2.

In the presence of surfactants such as Aerosol® 604 or a mixture of 1% Guar C-13 and 5% poly(ethylene oxide) with a molecular weight of 10,000 and in basic aqueous solutions, polyglutaraldehyde precipitated out in the form of spherical colloidal particles. The diameter of the particles increased with increasing amounts of monomer or with decreasing surfactant concentrations. Increasing pH also caused a decrease in particle size, which is most

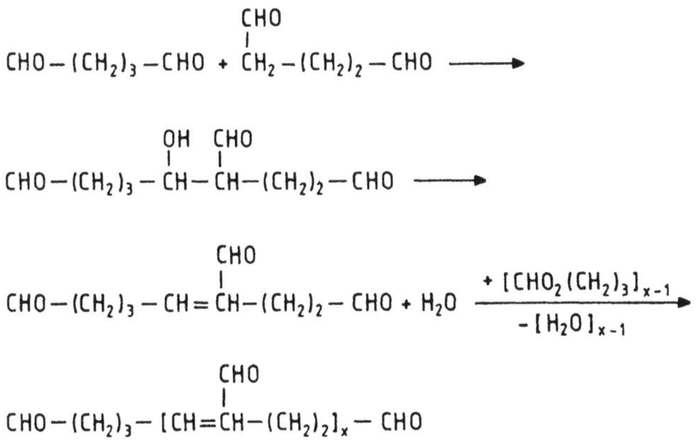

$$CHO-(CH_2)_3-CHO + \underset{}{\overset{CHO}{\overset{|}{C}}}H_2-(CH_2)_2-CHO \longrightarrow$$

$$CHO-(CH_2)_3-\overset{OH}{\overset{|}{C}}H-\overset{CHO}{\overset{|}{C}}H-(CH_2)_2-CHO \longrightarrow$$

$$CHO-(CH_2)_3-CH=\overset{CHO}{\overset{|}{C}}H-(CH_2)_2-CHO + H_2O \xrightarrow[-[H_2O]_{x-1}]{+[CHO_2(CH_2)_3]_{x-1}}$$

$$CHO-(CH_2)_3-[CH=\overset{CHO}{\overset{|}{C}}H-(CH_2)_2]_x-CHO$$

Figure 2 Mechanism of the aldol polycondensation.

probably due to the occurrence of a Cannizzaro reaction (60). Highly fluorescent particles were obtained if the polymerization was carried out in the presence of 0.05% 9-aminoacridine, 0.05% propidium bromide, 0.05% aminofluorescein, or 0.01% fluorescein isothiocyanate. Addition of 1–5% Fe_3O_4 to the initial reaction mixture resulted in magnetic polyglutaraldehyde nanoparticles of varying iron content (60).

5-Fluorouracil containing polyglutaraldehyde nanoparticles of a size of about 270 nm were produced by polymerization at a pH of 8.2 for 24 hours (61). 5% polysorbate 80 and 0.1% sodium carboxymethyl cellulose were used as surfactants/stabilizers.

A thorough characterization of these types of particles was carried out by Margel (62). An optimization of the process of polyglutaraldehyde nanoparticle preparation was performed by McLeod et al. (63).

8. Poly(alkyl Methylidenemaloneate) Nanoparticles

Dialkyl methylidenemalonic acid esters can also be polymerized by OH^--induced cationic polymerization similar to the poly(alkyl cyanoacrylates) (64). The advantage of the methylidenemalonic acid esters is that this polymerization can be carried out at neutral pH. De Keyser et al. (64) prepared poly(diethyl methylidenemalonate) nanoparticles of sizes of 140 to 250 nm depending on the pH (6.7–8.7) employed for polymerization. However, in contrast to the polycyanoacrylates, these nanoparticles are not rapidly biodegradable: at least 90% of the administered dose persisted in the body for 90 days after intravenous injection to mice (64).

C. Emulsion Polymerization in a Continuous Organic Phase

Emulsion polymerization in a continuous organic phase was one of the first processes for the production of nanoparticles (65–69). In this process, the phases are reversed and very water soluble monomers are employed. Initially acrylamide and the cross-linker *N,N'*-bisacrylamide were used as monomers. The monomers were solubilized by surfactants. The initiation of this polymerization can be carried out chemically using *N,N,N',N'*-tetramethylethylene-diamine and potassium peroxodisulfate as starters (66–69) or by γ-, UV-, or light-irradiation (65). In the case of light initiation, riboflavin-5'-sodium phosphate and potassium peroxodisulfate have to be added as catalysts of the reaction.

The process of polyacrylamide nanoparticle manufacturing was previously reviewed in detail (9). Especially because of the high toxicity of the monomers, but also because of the relatively high amounts of organic solvents and surfactants required for this process, it is of lesser importance today.

The cyanoacrylate monomers are of much lower toxicity than acrylamide, and their polymers are rapidly biodegradable. For this reason, more recently, the process of emulsion polymerization in a continuous organic phase was adapted for the production of poly(alkyl cyanoacrylate) nanoparticles. However, in this case the monomer was added to the continuous (organic) phase due to high solubility in organic solvents. As a consequence, nanoparticles with a shell-like wall (nanocapsules) as well as solid, monolithic particles were obtained (70). These solid particles were observed more frequently than the nanocapsules. The reason for the possibility of production of nanocapsules by this process is the following: The drug dissolved in a small amount of water is solubilized by surfactants within the organic phase. As a result, a microemulsion with water-swollen micelles containing the drug is formed. Added alkyl cyanoacrylate monomer diffuses to these micelles, and the OH^- ions and/or the basic drugs initiate the polymerization. This polymerization process in some case is so rapid that only a rather impermeable polymer wall may be formed at the organic/water interface, preventing the diffusion of further monomer molecules into the interior of these particles. However, as mentioned above, in most cases the interior of the particles is also polymerized and solid monolithic particles are formed.

For this process, isoctane (70), cyclohexane-chloroform 4:1 (71), isopropyl myristate-butanol 10:1 (72), and hexane (73,74) were used as the organic phase, sorbitane trioleate (70), Arlacel® A (71), or dioctyl sulfosuccinate (72–74) as surfactants. The drugs to be encapsulated included triamcinolol (70), doxorubicin (71,73,74), fluorescein (71), and methylene blue (72).

Also, as mentioned above, the toxicity of the employed cyanoacrylates is much lower than that of acrylamide. However, the high amounts of solvents and surfactants required limit the usefulness of this process.

D. Interfacial Polymerization

Polymerization of alkylcyanoacrylates in an organic solvent containing water-swollen micelles may lead to the formation of a polymer wall at the solvent-micellar water interface (see previous chapter). However, the majority of particles formed by this process are continuous monolithic particles. Hence alternative methods are required for the adequate formation of nanocapsules. Two of these processes are described below; another possibility, solvent disposition, in Sec. E below.

1. Poly(N^{α},N^{ϵ}-L-Lysinediylterephthaloyl) Nanoparticles

Electrocapillary emulsification was used to prepare poly(N^{α},N^{ϵ}-L-lysinediylterephthaloyl) nanoparticles (75). In this process, an electrical potential is applied between an oil and a water phase. When this potential exceeds a certain value, the interfacial tension is reduced to almost zero and spontaneous emulsification occurs. Thus stable monodisperse emulsions of a particle size below 100 nm can be formed (76).

Arakawa and Kondo entrapped sheep erythrocyte hemolysate in poly(N^{α},N^{ϵ}-L-lysinediylterephthaloyl) nanoparticles of a size of 380 nm (75). The water phase containing the hemolysate, L-lysine, and sodium carbonate was injected slowly (0.042 mL/min) by a motor-driven syringe into the oil phase consisting of cyclohexane-chloroform 3:1. In the oil phase terephthaloyldichloride, 1.5×10^{-4} M tetraethylammonium chloride, and 5% sorbitan trioleate were dissolved. A potential of 850 V was applied between the needle and a platinum wire immersed in the oil phase. The polycondensation reaction between the amine and the terephthaloyl chloride should be expected to occur exclusively at the interface of the formed nanometer-sized droplets, leading to the formation of wall-type nanocapsules. However, until now we were unable to prove the formation of nanocapsules by electron microscopy.

2. Poly(alkyl Cyanoacrylate) Nanoparticles

Poly(alkyl cyanoacrylate) nanoparticles may be formed by interfacial polymerization in an aqueous surrounding phase. This process was introduced by Al Khouri Fallouh et al. (77,78). In this process the cyanoacrylate monomer as well as the oil-soluble drug are dissolved in a mixture of an oil and ethanol. The ratio of oil to ethanol is about 1:10 to 1:200. The oils employed can be Miglyol®, benzylic acid, or another oil. The organic solution containing the drug and the monomer is then added slowly (about 0.5 mL/min) through a tube or a needle into water or a buffer solution (pH 3–9) containing surfactants such as poloxamer 188 or 407 (79) or phospholipids. Nanocapsules (Fig. 3) consisting of an internal oil droplet surrounded by a polymeric wall (80) are thus formed spontaneously by anionic polymerization of the cyano-

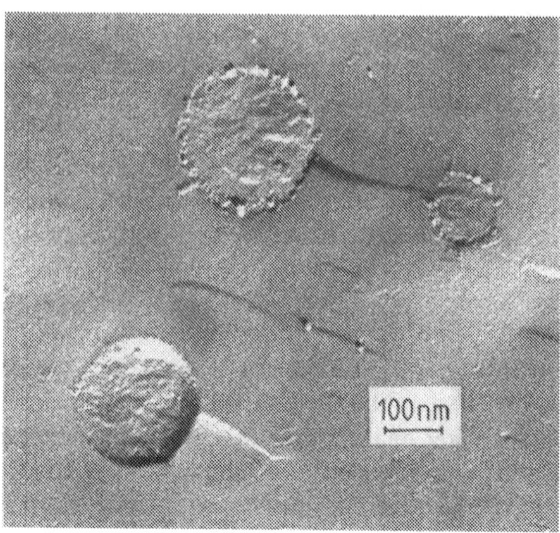

Figure 3 Transmission electron micrograph of poly(isobutyl cyanoacrylate) nanoparticles produced by the method of Al Khouri Fallouh et al. (78) after freeze fracture. The structures visible on the micrograph indicate that a shell-like wall is formed. (Reproduced from Ref. (1) with permission of the copyright holder.)

acrylate in the oil after contact with the initiating OH^- ions of the water. The diameter of these capsules is mainly controlled by the Miglyol® concentration in the ethanol, leading to smaller nanocapsules of about 180 nm at very low (0.5%) Miglyol® concentrations and of about 450 nm at high (8%) concentrations (79). The pH of the water phase as well as the saturation of the monomer with sulphur dioxide, a polymerization decelerator, had no significant effect on the particle size. Increasing monomer concentrations increased the density but not the size of the resulting nanocapsules, indicating that nanocapsules with a thicker wall were formed (79).

E. Solvent Deposition

Poly-(D,L-lactide) nanocapsules may be produced by a process called solvent deposition. In this process poly(D,L-lactide) polymer as well as phospho-phospholipids are dissolved in acetone (81). A solution of the drug (indomethacine) in benzyl benzoate is then added to the organic phase, and this mixture (about 25 mL) is subsequently poured into water (50 mL) containing 0.5% poloxamer 188 under moderate stirring. Nanocapsules with a poly(lactic acid) wall surrounding an oily core are formed instantaneously (81,82). This suspension then has to be concentrated to about 10 mL final volume by evaporation of the acetone and partial removal of water under reduced pressure.

Poly-ϵ-coprolactone as well as poly(lactic-co-glycolic acid) nanocapsules may be produced using the same method (83). Betaxolol was encapsulated in poly-ϵ-caprolactone (MW 42,000) and in the poly(D,L-lactic-co-glycolic acid-50-50) (MW 40,000) polymers by this process using Miglyol® 812 instead of benzoyl benzoate.

F. Solvent Evaporation

Solvent evaporation is a well-established method for the preparation of microspheres with sizes above 1 μm (84). However, with this process also particles of less than 1000 nm can be produced. The preformed polymer is dissolved together with the drug in an organic solvent, which is then emulsified in water and subsequently evaporated by heating and/or reduced pressure. For this reason, the particle size of the resulting polymer particles depends on the size of the emulsion droplets prior to solvent evaporation. This size can be controlled by a number of factors including the stir rate, the type and amount of dispersing agent, the viscosity of the organic and aqueous phases, the configuration of the vessel and stirrer, the quantity of the organic and aqueous phases, and the temperature (84).

The first polylactic acid nanoparticles containing a drug, testosterone, were produced by solvent evaporation by Gurny et al. (85). Poloxamer 188 was used as an emulsifier. The polymer-drug-solvent mixture was emulsified using a conventional laboratory homogenizer. The particle size of the dried polylactic acid nanoparticles was 450 nm (85). Later Krause et al. (86) prepared triamcinolone acetonide polylactic acid nanoparticles using a modified version of the evaporation process. The drug and the polymer were dissolved in chloroform and this solution was emulsified by sonication for 45 min at 15°C in an aqueous 0.5% w/w gelatin solution. The solvent evaporated during 45 min at 40°C under continuing sonication. After centrifugation, washing, and lyophilization, particles of a size around 500 nm resulted. Other emulsifiers that have been employed for this process include poly(vinyl acetate) (87), polysorbates (88), cetyltrimethyl ammonium bromide (89), and sodium lauryl sulfate (85). Besides standard laboratory homogenizers that require relatively high quantities of emulsifiers, ultrasonication (86,87, 90,91), microfluidization (88,92), or a French press can be used in order to achieve a submicron particle size. Jeffery et al. (89) presented an investigation of the influence of nature and concentration of emulsion stabilizers, polymer concentration, ratio organic/aqueous phase volume, and rate and duration of agitation during emulsification on the resulting particle size.

The solvent evaporation process has also been used for the preparation of polyacrylic and ethyl cellulose nanoparticles incorporating indomethacin as the model drug (92). Some of the polymers, including Eudragit® RS, Eudragit® RL, and ethyl cellulose, did not require any surfactants or polymeric stabilizers. In other cases, polysorbate 80, poloxamer 188, sodium lauryl sulfate, Brij® 35, Brij® 78, Myrj®52, or poly(vinyl acetate) had to be employed. The particle size decreased with increasing homogenization pressure and number of cycles, reaching a minimum plateau after 5 cycles, and with increasing surfactant concentrations (92).

Poly(β-hydroxybutyrate) nanoparticles were prepared by Koosha et al. (90,91) by solvent evaporation. This rather novel polymer is biodegradable and seems to possess a very good biocompatibility. The solvent for the polymer that was used by the above authors was chloroform. However, it should be substituted by other better tolerated solvents. The size of the nanoparticles after evaporation of the chloroform depended on the surfactant and the emulsification method. A number of polyhydroxyethylene-fatty-alcohol-ethers (Brij®) yielded particle sizes between 500 and 8400 nm after sonication. High pressure emulsification reduced the particle sizes to 170 to 220 nm (91). If, however, sodium dodecyl sulfate was used as the emulsifier, sonication for 30 min already yielded nanoparticles of a size of 170 nm (91).

The process of solvent evaporation recently has been adapted for the production of chitosan nanoparticles. Chitosan, a deacetylated chitin derivative, is a naturally occurring polysaccharide (93). Using this process, chitosan (5%) was dissolved in acetic acid (10%). The drug and, after formation of a clear solution, colloidal magnetite particles (ferrufluid) may be added at this stage (94). The chitosan then was cross-linked by addition of glutaraldehyde and ethanol. After that, this mixture was emulsified in mineral oil by sonication using Arlacel® 83 as an emulsifier. The aqueous phase was then evaporated at 70°C under reduced pressure in a nitrogen atmosphere. The resulting nanoparticles then can be stored as dry particles. Hassan et al. (94) optimized the preparation of chitosan nanoparticles containing an anticancer drug, oxantrazole, using a central composite experimental design. The particle size varied between 300 and 1400 nm depending on the formation variables. Despite this optimization that led to the inclusion of the chitosan cross-linking step with glutaraldehyde prior to the emulsification step, the drug loading with oxantrazole never exceeded 3–5%.

G. Polyacrylic Nanoparticles Preparation by Desolvation from an Organic Polymer Solution

Polyacrylic nanoparticles can be produced after dissolution of relatively hydrophilic copolymers (Eudragit® RS or Eudragit® RL) in water-miscible solvents such as acetone and ethanol (95). The solutions of the polymer and of the drug to be entrapped, i.e., ibuprofen, indomethacin, propanolol HCl, in the above solvents were then poured into water, resulting in the spontaneous formation of nanoparticles of sizes between 90 and 205 nm. The drug entrapment efficacy was rather high, ranging between 73% (Eudragit® RL) and 80% (Eudragit® RS) for propanolol and up to 94% and 97% respectively for indomethacin and ibuprofen. The nanoparticles produced by this method, however, were not redispersible in water after spray-drying or freeze-drying (95).

H. Production of Albumin Nanoparticles in an Oil Emulsion

Nanoparticles consisting of albumin or other macromolecules may be produced by emulsification of aqueous solutions of these macromolecules and of the drug to be incorporated into the particles in an oil. The resulting droplets can then be hardened by cross-linking with aldehydes or other cross-linkers or by denaturation of the molecules at high temperatures. This process was previously used to manufacture microspheres (96). The use of high efficacy homogenization or ultrasonication as in the case of solvent evaporation enabled the production of nanometer-sized emulsion droplets and, after hardening, the preparation of nanoparticles. This process was first introduced by Zolle et al. (97) and Scheffel et al. (98). These authors used the nanoparticles as carriers for radionucleotides for diagnostic purposes in nuclear medicine. Kramer (11), a couple of years later, was the first to use this process for the production of albumin nanoparticles for drug delivery purposes. These particles additionally can be made magnetic by incorporation of magnetite particles (12).

In order to prepare nanoparticles by this process, albumin or similar macromolecules in concentrations between about 100 and 500 mg/mL are dissolved in water. The water-soluble drug and—if the manufacture of magnetic particles is desired—the magnetite particles are then added to the aqueous phase. This phase is then emulsified in an oil or another lipophilic medium by high efficacy homogenization as mentioned above. After this, hardening to nanoparticles is performed by addition of glutaraldehyde or another cross-linker. Alternatively, the hardening can be carried out by pouring the emulsion into an equal volume of hot oil, and then this mixture has to be kept at elevated temperatures for 10 to 15 minutes. The high temperatures in the hot oil lead to the irreversible denaturation of the protein and the formation of the particles. Subsequently, the mixture is cooled to room temperature. In both cases the resulting nanoparticles then can be separated and washed with (volatile) organic solvents. The degradation velocity of the particles as well as their release rate decrease with increasing denaturation temperature and time.

Gallo et al. (14) investigated and optimized the above process of albumin nanoparticle preparation. The mean particle size in their experiments varied between 400 and 800 nm with high standard deviations varying between 60 and 100% relative to the mean. Although, as stated above, high-energy emulsification is required for the production of nanoparticles, the sonication energy that in the mentioned study ranged between 78 and 125 W for 2 min had relatively little influence in that the particle size decreased by only 33% from 720 to 540 nm with the higher power input. The emulsification time was of even lower importance in that a prolongation from 1 to 10 min reduced the particle size by only 17%. These findings correspond with our experience, showing that it is very difficult to obtain particle sizes below 500 nm. Other production variables, such as the albumin concentration, temperature of the emulsion before addition to the hot oil, aqueous-to-nonaqueous phase volume ratio, stirring rate during denaturation, and heat stabilization time also had very little influence on the mean particle size. The biggest effect was observed when cottonseed oil was substituted with other oils: the particle size increased from 560 nm to 710 nm with maize oil and to 820 nm with paraffin oil (14). Our own experience shows that it is necessary to add the W/O emulsion very rapidly drop by drop to the hot oil, because the W/O emulsion droplets are very instable prior to denaturation and coalesce. As a result, larger particles with a wide particle size distribution would be formed (1). The addition of surfactants such as polysorbate 20 or 80 or other surfactants not only stabilizes the emulsion at this critical stage but also helps resuspension of the final dry nanoparticles in water.

As already mentioned, the release rate of drugs decreases with increasing denaturation temperature. This effect was studied in detail by Widder et al. (99) and by Gupta et al. (100).

I. Production of Gelatin Nanoparticles in an Oil Emulsion

A method for the production of gelatin nanoparticles that is very similar to the method for the preparation described in the previous chapter was developed by Yoshioka et al. (101). These authors emulsified 0.3 mL of a 30% gelatin solution containing about 1.8 mg of drug (mitomycin C or mitomycin C-dextran conjugate) or 0.54 mg of ^{131}I-human serum albumin in 3 mL sesame oil using 6.6% sorbitan sesquioleate and 1.5% polyoxyethylene derivative of hydrogenated castor oil as emulsifiers. Instead of heating, the resulting emulsions were then cooled in an ice bath, resulting in complete gelation of the gelatin droplets. After dilution with acetone, the emulsion was filtered through a membrane filter with a pore size of 50 nm. In order to remove the oil phase, the resulting particles were washed with acetone and then hardened for 10 min with 30 mL of a 10% solution of formaldehyde in acetone, followed by washing with acetone and air-drying. The resulting particles had a size distribution between 100 nm and 600 nm and an average diameter of 280 nm.

In this procedure, the stabilization of the non-cross-linked gelatin particles was achieved by cooling below the gelation point. The hardening and fixation was then carried out with formaldehyde. It is, however, surprising that the gelated, although non-cross-linked, particles pass through a 50 nm filter, while the particles obtained after hardening have a much greater diameter (100–600 nm). Since the resulting particles are spheric in shape without any sign of agglomeration of smaller particles, the particles obviously change their shape and possibly start to coalesce upon removal of the oil phase and possibly of the emulsifier during the washing procedure. The velocity of the coalescence process, however, seems to be so much reduced in that the final particle size after hardening does not exceed 1 μm. In this context, it would be of interest to study the influence of different drugs and drug concentrations on this process.

A similar procedure was employed by Tabata and Ikada to incorporate muramyl dipeptide (102) and interferon (103) into nanoparticles. In this procedure a 1:1 mixture of chloroform and toluene represented the organic phase and sorbitan monooleate was used as emulsifier. Again gelation was performed by ice-bath cooling. In this case, glutaraldehyde in the form of a saturated solution in toluene was used as a cross-linker. The particles were then washed in succession with chloroform 25%: toluene 75%-mixture, isopropanol, and phosphate-buffered saline.

When the microspheres were subjected to degradation in phosphate-buffered saline solution containing collagenase, the digestion of microspheres was found to decrease with increasing cross-linking. Interferon was incorporated in the microspheres with a high trapping efficiency. The rate of interferon release from the microspheres was regulated by the extent of cross-linking with glutaraldehyde (103).

J. Nanoparticles Produced by Desolvation of Macromolecules

Macromolecules can be desolvated by charge changes, pH changes, or by the addition of a desolvating agent causing the so-called salting-out phenomenon. This desolvation results in precipitation of the macromolecules or in the formation of a coacervate (104). Consequently, desolvation leads to the formation of a new phase. Both effects, precipitation and coacervation, can be considered part of the general area of solubility and phase equilibria (105). Before phase separation occurs, a conformation change of the macromolecules takes place: in a dilute solution, the macromolecule is subject to the osmotic action of the surrounding solvent, which tends to swell it to a larger average size than it would otherwise assume (106). The better the solvent the greater is the swelling of the molecule. Addition of desolvating agent reverses this process, and the diameter of the macromolecule coil becomes smaller and smaller. After a certain degree of desolvation is obtained, the molecules begin to aggregate. When sufficient desolvation has occurred, phase separation will take place.

The formation of these coiled macromolecules can be monitored by turbidity measurements (107). Macromolecule solutions always induce a slight turbidity known as the Tyndall effect. Addition of a desolvating agent at first decreases the turbidity due to the decrease in the size of the macromolecules and also due to the dilution effect of the solvent. The phase separation, occurring after addition of sufficient amounts of desolvating agent, is coupled with a very significant, rapid increase in turbidity. Consequently, the area of tight coiling shortly before this increase in turbidity can easily be found. If too much desolvating agent has been added, addition of a resolvating agent will bring the system to the desired composition (107).

Marty et al. (107) used the process of reversible swelling of macromolecules to produce nanoparticles. Drug molecules can be bound to the swollen macromolecules by protein binding. The tightly coiled macromolecules with thus bound drugs can then be fixed and hardened by the addition of an aldehyde.

Later authors have studied extensively the optimal conditions for nanoparticle formation (107,108). They used gelatin, human serum albumin, bovine serum albumin, casein, and ethyl cellulose as the macromolecular materials. Marty et al. placed special emphasis on their work with gelatin and human serum albumin, whereas with bovine serum albumin, casein, and ethyl cellulose only preliminary experiments were carried out.

The desolvation of gelatin can be achieved with ethanol or sodium sulfate (107,108). The choice of the desolvating agent depends mainly on the drug to be attached to the nanoparticle. Ethanol has the advantage that it is easily removable during lyophilization. In some cases a surfactant, polysorbate 20 or polysorbate 80, must be present in order to solubilize certain drugs. The surfactants also facilitate the redispersion of the final freeze-dried product. After desolvation, monitored by turbidity measurements as described above, the addition of a small amount of a resolvating agent yields optimal products: at the endpoint of the turbidity monitored titration, i.e., the point before the nephelometer readings increase significantly, optimal conditions for nanoparticle formation are already exceeded. Hardening of this system would cause aggregation and flocculation of the particles because interparticulate cross-linking by the aldehyde takes place. This problem can be avoided by the above-mentioned addition of a small amount of resolvating agent before hardening. Glutaraldehyde proved to be the hardening agent of choice because it is a bifunctional aldehyde (107). Excessive aldehyde has to be removed with sodium sulfite or sodium metabisulfite so that further hardening or aggregation of gelatin nanoparticles does not occur. The crude product can then be freeze-dried.

Purification from low molecular weight materials may be achieved by passing over a Sephadex® 50 column using either a 0.04% w/v chlorbutanol solution or doubly distilled water as the eluent. A column packed with Bio-Gel®A-5 m proved to be moderately successful in separating polysorbate 20 micelles and extraneous cross-linked gelatin from the nanoparticles.

The production of human serum albumin nanoparticles requires slight modifications of the above-described procedure for gelatin nanoparticles (108). Desolvation may be achieved by addition of ammonium sulfate. Desolvation with ethanol seems to be less favorable, because hardening of thus obtained systems caused the formation of big aggregates (108). Sodium sulfate is not soluble enough to cause suitable desolvation of human serum albumin. Sodium sulfate, however, can be used for desolvation of human serum albumin if 2% polysorbate 20 is present. Polysorbate 20 prevents or delays the precipitation of the human serum albumin macromolecules. Marty (108) therefore suggested that a configuration change takes places in the presence of the surfactant. However, it seems more likely that the surfactant again acts as a stabilizer as in the emulsion polymerization systems described previously: surfactants can increase the surface charge (109), thus enhancing the electrical repulsion between particles. Hardening, removal of excess hardening agent, and purification can be performed using the same agents and methods as with gelatin nanoparticles, although the suitable amount of hardening agent is lower with human serum albumin nanoparticles than with gelatin nanoparticles (108).

Bovine serum albumin nanoparticles also may be produced using the above techniques. It was found to be necessary to acidify the bovine serum albumin solution slightly in order to desolvate it (108).

The desolvation of an alkaline casein solution can be achieved with hydrochloric acid or with sodium sulfate: nanoparticles of an average particle size of 500 nm (using hydrochloric acid) or 120 nm (using the sodium sulfate at a pH of 12.3) were obtained after lyophilization of the desolvated system without addition of a hardening agent (108).

Ethyl cellulose nanoparticles may be obtained by desolvation of a 1% solution of this macromolecule in carbon tetrachloride with cyclohexane: Marty (108) generated a faintly turbid system with 0.5% ethyl cellulose, 54% carbon tetrachloride, and 45.5% cyclohexane. This system flocculated over several hours. It was also possible to freeze-dry this type of system. Scanning electron microscopy of newly formed and freeze-dried systems showed some 200 nm particles mixed with many large sheets of material.

The method of desolvation of gelatin was adapted by El-Samaligy and Rohdewald to prepare triamcinolone (110,111), dactinomycin, and doxorubicin (111) nanoparticles. The drugs were solubilized in water together with the gelatin (1 g/100 mL) using 2% (w/v) polysorbate 80. Desolvation was then carried out with a 20% (w/v) sodium sulfate solution, and cross-linking was performed with a 25% glutaraldehyde solution (110). The size of the particles was about 300 ± 25 nm. Alternatively, empty nanoparticles may be prepared by adjusting the pH of the gelatin solution to 5.5–6.5. In this case, the drugs were adsorbed after cross-linking and Sephadex® G 25 m and subsequently G 10 f column purification (111). Albumin nanoparticles (size 170 ± 30 nm) were produced by the same authors using human serum albumin instead of gelatin. In this case, 5 g albumin/100 mL solution was required, and the desolvation was performed with ammonium sulfate (110).

Drug uptake was similar in both cases, when the drug was present during cross-linking or when the drug was adsorbed to previously prepared empty nanoparticles. The drug release was mainly governed by drug desorption. Incorporation into the nanoparticles by cross-linking in the presence of drugs resulted in about 20% slower release rates than preparation by adsorption to empty particles (110).

Mukherji et al. (112) prepared ethyl cellulose nanoparticles containing 5-fluorouracil with a size ranging between 190 and 1100 nm. In this case, the drug 5-fluorouracil (1.2 mg/mL), ethyl cellulose (1% w/v), and polysorbate 80 (1% w/v) were dissolved in ethanol, and dissolution was performed by the addition of water. The authors report that it was also possible to prepare methyl cellulose nanoparticles, but from the electron micrograph and the sizing experiments it is not clear if the resulting particulates were methylcellulose nanoparticles or crystallites of other materials. No cross-linking was performed during the reported process.

K. Carbohydrate Nanoparticles

Carbohydrate nanoparticles consisting of acryloylated dextran, maltodextran, mannan, or other starch derivatives were produced by polymerization of the acryloyl side chain after emulsification of the aqueous starch derivative solution in a toluene:chloroform (4:1) solution (113–128). As a first step, the polysaccharides have to be derivatized with acrylic acid glycidyl ester (Fig. 4): the polysaccharide is dissolved in 0.2 M phosphate buffer, acrylic acid glycidyl ester is added, and the system is stirred at room temperature for 10 d. After phase separation by centrifugation (3000 g, 15 min),

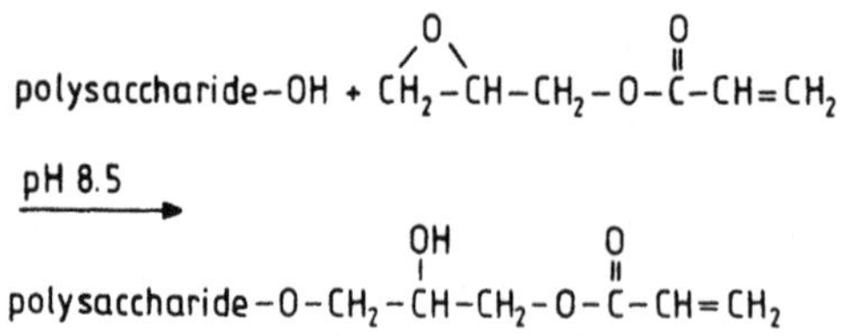

$$\text{polysaccharide–OH} + \overset{\displaystyle\overbrace{\quad O \quad}}{CH_2\text{–}CH}\text{–}CH_2\text{–}O\text{–}\overset{\displaystyle O}{\overset{\|}{C}}\text{–}CH{=}CH_2$$

$$\xrightarrow{\text{pH 8.5}}$$

$$\text{polysaccharide–}O\text{–}CH_2\text{–}\overset{\displaystyle OH}{\overset{|}{C}H}\text{–}CH_2\text{–}O\text{–}\overset{\displaystyle O}{\overset{\|}{C}}\text{–}CH{=}CH_2$$

Figure 4 Acryloylation of polysaccharides. (Reproduced from Ref. (1) with permission of the copyright holder.)

the unreacted acrylic acid is discarded and the aqueous solution containing the derivatized polysaccharide extracted with toluene to remove remaining unreacted glycidyl ester.

The derivatized polysaccharides can be characterized by the *D-T-C* expression (69,115). In this expression the *D* value denotes the concentration of the derivatized macromolecule (g/100 mL solvent), the *T* value denotes the total concentration of acrylic monomer (g/100 mL solvent), and the *C* value the amount of cross-linker (mostly *N,N'*-methylenebisacrylamide) expressed as the percentage (w/w) of the total amount of acrylic monomer.

The drug, and in some cases the cross-linker bisacrylamide, is dissolved in the solution of the acryloylated starch derivative at a pH of 8.5. After the addition of auxiliary substances such as EDTA, ammonium peroxodisulfate, and poloxamer 188, this aqueous phase (5 to 10 mL) is deoxygenated and emulsified in 600 mL of the above-mentioned toluene:chloroform 4:1 mixture (600 mL), resulting in a W/O emulsion. Polymerization is initiated by addition of *N,N,N',N'*-tetramethylethylenediamide, resulting in nanoparticles of about 500 to 2000 nm. Starch produced larger particles than lichenan, mannan, or dextran [Table 4 (117)]. The particles can be collected by centrifugation. They are purified from the organic solvents by repeated washing with phosphate buffer. Macromolecules such as enzymes or other proteins are incorporated by their addition to the unpolymerized aqueous acryloylated starch solution, whereas smaller drug molecules such as primaquine or trimetoprime are bound to the polymerized empty particles via tri-, tetra-, and pentapeptide spacer arms. The coupling of the spacer arm to the particles was achieved using the carbonyldiimidazole method again using *N,N,N',N'*-tetramethylethylenediamine (118,120).

An alternative manufacturing method for carbohydrate nanoparticles was employed by Schröder et al. (129). These authors used corn, rape seed, or cottonseed oil instead of the toluene/chloroform mixture. The aqueous carbohydrate solution was emulsified in the above oils by stirring, and the droplet size was then reduced by sonication for 40 s in an ice bath. The resulting microdroplet emulsion was poured slowly into acetone containing 0.1% polysorbate 80 while stirring at 1000 rpm. This led to the crystallization and particle formation of the carbohydrate. The particles precipitated and were then washed with an aqueous 0.1% polysorbate 80 solution. Finally, they were suspended in a 1% polysorbate 80 solution in acetone and were dried at room temperature. A continuous repetition of the above-described process led to the formation of nanoparticles (129).

IV. DRUGS BOUND TO NANOPARTICLES

Drugs may be bound to nanoparticles either by production in the presence of the drugs or by adsorption after the preparation of empty nanoparticles.

Table 4 Characteristics of the Nano- and Microparticles

Particle	*D-T-C*[a]	Degradation %[b]	Size distribution, %[c]		
			<0.5 μm	0.5–2.2 μm	2.2 μm
Starch(1,6-α-branched 1,4-α-glucan)	15–0.8–0	100 ± 4	0	89	11
Dextran(1,3-α-branched 1,6-α-glucan)	15–0.5–0	15 ± 6	12	88	0
Mannan(1,2-α- and 1,3-α-branched 1,6-α-mannan)	15–0.8–0	17 ± 4	0	93	7
Lichenan(1,3-β-glucan)	15–0.7–0	13 ± 2	0	91	9

[a]The *D-T-C* values denote the amounts of the components in the monomer solution used for particle formation. *D* denotes the concentration of derivatized macromolecules (i.e., polysaccharides; g per 100 mL); *T* denotes the total concentration of acryloyl groups (g per 100 mL); and *C* denotes the relative amount of the cross-linker, expressed as the percentage (w/w) of the total amount of acrylic monomers.
[b]Microspheres were degraded with amyloglucosidase as described in Ref. (113).
[c]The size of the autoclaved spheres was determined from photographs taken by scanning electron microscopy.
Source: Reproduced from Ref. (117) with permission of the copyright holder.

Production in the presence of the drug can lead to covalent coupling to the polymer (130,131), or it may yield the formation of a solid solution or a solid dispersion of the drug in the polymer network. After addition of the drug to previously prepared empty nanoparticles, again the drug may be coupled covalently (118,120), or alternatively the drug may be bound by sorption. The sorption can lead either to the diffusion of the drug into the polymer network and to the formation of a solid solution (132–135) or to surface adsorption of the drug (29,136). The type of binding may result in different release mechanisms and different release rates (110,134,136). In general, covalent coupling yields much slower release rates or no release at all (130,131). However, even if the drug is contained in the form of a solid solution or a solid dispersion, its release characteristics may mainly depend on the degradation rate of the polymer (137).

Table 5 gives a complete overview of drugs and other biologically active materials that have been bound to nanoparticles.

V. CHARACTERIZATION OF NANOPARTICLES

A. Physicochemical Characterization

A number of physicochemical methods exist for the characterization of nanoparticles. These methods are listed in Table 6.

Size is the most prominent feature of nanoparticles. However, other parameters such as density, molecular weight, and crystallinity will largely influence their release and degradation properties, whereas surface characteristics such as surface charge and hydrophilicity and hydrophobicity will significantly influence the interaction with the biological environment after administration to humans and animals, and thus will influence the resulting body distribution.

As said above, particle size is the most prominent feature of nanoparticles. Sizing in the nanometer size range, however, is still a problem. The fastest and the only routine method presently applicable is photon correlation spectroscopy or dynamic light scattering (233). Laser diffractometry using Fraunhofer diffraction is more useful for larger, micrometer-sized particles (234). Photon correlation spectrometry determines the hydrodynamic diameter of the nanoparticles via Brownian motion. For this reason, the measured size is influenced by the interaction of the particles with the surrounding liquid medium, and consequently the exact viscosity of the medium must be known. While some years ago only the mean and a somewhat dubious dispersity index could be calculated, modern correlators and computer equipment allow the determination of the size distribution and of multimodal distributions. The precision of the size distribution measurements further can be improved using scattering factors derived from Mie theory

Table 5 Drugs and Biologically Active Materials Bound to Nanoparticles (Incorporation or Adsorption)

Drug	Polymer	Reference
Adenine	Poly(isohexyl cyanoacrylate)	138
Amikacin	Poly(butyl cyanoacrylate)	139–141
Amphotericin B	Poly(isobutyl cyanoacrylate)	142
Ampicillin	Poly(isobutyl cyanoacrylate)	41,143,144
	Poly(isohexyl cyanoacrylate)	41,144–147
Antibodies	Poly(isobutyl cyanoacrylate)	148
	Poly(isohexyl cyanoacrylate)	148
Antisense oligonucleotides	Poly(isobutyl cyanoacrylate)	149
	Poly(isohexyl cyanoacrylate)	149
Betaxolol	Poly(isobutyl cyanoacrylate)	150,151
	Polylactic-co-glycolic acid	151
	Poly-epsilon-caprolactam	151
Blood hemolysate	Poly(N_α,N_ϵ-L-lysine-diylterephthalamide)	75
Cyclosporin	Poly(isohexyl cyanoacrylate)	152
Dactinomycin	Poly(methyl cyanoacrylate)	137,153,154
	Poly(ethyl cyanoacrylate)	137,153,154
	Poly(butyl cyanoacrylate)	155
	Poly(isobutyl cyanoacrylate)	38
	Gelatin	111
Darodipin	Poly(isobutyl cyanoacrylate)	156
Dehydroemetine	Poly(isohexyl cyanoacrylate)	157
Dexamethasone	Poly(isobutyl cyanoacrylate)	41
	Poly(isohexyl cyanoacrylate)	41
DNA	Poly(butyl cyanoacrylate)	158
Doxorubicin	Polyacrylic copolymer	135,159–161
	Poly(butyl cyanoacrylate)	71,136
	Poly(isobutyl cyanoacrylate)	38,162–164
	Poly(isohexyl cyanoacrylate)	165–167
	Albumin	99,100,168–187
	Gelatin	111
	Fibrinogen	188
Enzymes	Polyacrylamide	67,189,190
Fluorescein	Polyacrylamide	191
	Poly(alkyl cyanoacrylates)	71
5-Fluorocytosine	Poly(isobutyl cyanoacrylate)	142
5-Fluorouracil	Albumin	192–195
	Ethyl cellulose	112
	Poly(butyl cyanoacrylate)	196
	Poly(methyl methacrylate)	133
	Polyglutaraldehyde	61
Gentamycin	Poly(isobutyl cyanoacrylate)	144
Growth hormone releasing factor	Poly(isobutyl cyanoacrylate)	197
	Poly(isohexyl cyanoacrylate)	197,198
Ibuprofen	Polyacrylic copolymers	95
Indomethacin	Ethylcellulose	92
	Poly(methyl methacrylate)	199,200
	Polyacrylic copolymers	92,95
	Poly(isobutyl cyanoacrylate)	201–203
	Poly(D,L-lactide)	81,82,201,202
Insulin	Poly(isobutyl cyanoacrylate)	204–208
Interferon	Gelatin	103
Lipiodol	Poly(isobutyl cyanoacrylate)	77,78,209,210

Table 5 (Continued)

Drug	Polymer	Reference
Mercaptopurine	Serum albumin	11
Methotrexate	Albumin	211
	Poly(methyl cyanoacrylate)	137
	Poly(ethyl cyanoacrylate)	137
Methylene blue	Poly(methyl cyanoacrylate)	72
β-Methylumbelliferone	Polyacrylamide	212
Metronidazole	Albumin	213
	Gelatin	213
	Polyglutaraldehyde	213
Miconazole	Poly(isobutyl cyanoacrylate)	142
Mitomycin C	Gelatin	101
Muramyl dipeptide	Gelatin	102
	Polylactic-polyglycolic acid	87
Norephedrin	Polyacrylamide	130,131
Oxantrazole	Chitosan	94
Pilocarpine	Poly(ethyl cyanoacrylate)	214
	Poly(butyl cyanoacrylate)	134,215–217
Primaquine	Albumin	213
	Gelatin	213
	Polyglutaraldehyde	213
	Poly(isohexyl cyanoacrylate)	218
	Poly(diethylmethylidene malonate)	219
	Polyacryl starch	118,120
Progesterone	Poly(butyl cyanoacrylate)	220
Propanolol	Polyacrylic copolymers	95
Proteins	Polyacryldextran	69
	Polyacrylamide	190
	Polyacryl starch	113–128
Pyrene butyric acid	Polyacrylamide	212
Rose bengal	Polyacrylic copolymers	53
	Poly(butyl cyanoacrylate)	136
Testosterone	Polylactic acid	85
Timolol	Poly(ethyl cyanoacrylate)	214
Triamcinolon	Albumin	110
	Gelatin	110,111,221
	Polylactic acid	86
	Poly(methyl cyanoacrylate)	70,110
Trimetoprim	Polyacryl starch	118
Verapamil	Poly(butyl cyanoacrylate)	222
Vidarabin	Poly(isohexyl cyanoacrylate)	138
Vinblastine	Poly(methyl cyanoacrylate)	154
	Poly(ethyl cyanoacrylate)	154
Vincamin	Poly(hexyl cyanoacrylate)	223,224
Antigens		
Bovine serum albumin	Poly(methyl methacrylate)	225,226
HIV-2 split antigen	Poly(methyl methacrylate)	227
Immunoglobulin G	Polyacrylamide	10
Influenza whole virus	Poly(methyl methacrylate)	33,34,228–231
Influenza split antigen	Poly(methyl methacrylate)	226,232
Tetanus toxoid	Polyacrylamide	10

Table 6 Physicochemical Characterization Methods for Nanoparticles

Parameter	Method	Reference
Particle size	Photon correlation spectrometry	243,260
	Transmission electron microscopy (TEM)	36,80,81,233
	Scanning electron microscopy (SEM)	36,70,130,186, 233,236
	SEM combined with energy-dispersive x-ray spectrometry	210
	Scanned-probe microscopes	250–258
	Fraunhofer diffraction	234
Molecular weight	Gel chromatography	30–32,37,38,157
Density	Helium compression pycnometry	233
Crystallinity	X-ray diffraction	133,215,253
	Differential scanning calorimetry	254,266
Surface charge	Electrophoresis	233,255
	Laser Doppler anemometry	255,267
	Amplitude weighted phase structuration	255
Hydrophobicity	Hydrophobic interaction chromatography	259–262
	Contact angle measurement	233,258
Surface properties	Static secondary ion mass spectrometry (SSIMS)	268
Surface element analysis	ESCA (x-ray photoelectron spectroscopy for chemical analysis)	269,270

or a suitable approximation. Calculation of Mie factors requires the knowledge of the refractive index of the particles, which may differ from the bulk refractive index of the polymer. Shankland and Whateley (235) developed a method for the determination of refractive indexes of polyalkylcyanoacrylate particles. However, the presence of small amounts of different particle species can still obscure the measurements. Especially larger particles such as dust, accidental microbial contamination, crystallization of some ingredients, or secondary particle agglomerates can lead to erroneous results.

For this reason, it is important to verify the results obtained with photon correlation spectroscopy by the other important method, electron microscopy. Two electron microscopy methods have mainly been used, scanning electron microscopy and transmission electron microscopy. Scanning microscopy requires the coating of the dryed sample with a conductive material, usually gold. The thickness of the gold layer, however, is difficult to determine and normally can vary between 25 and 60 nm. Since this gold layer is coated on all sides onto the particles, and since the particle size is determined by this technique as the surface diameter, the uncertainty in particle size with this method amounts to 70 nm, an amount that is intolerable with actual particle sizes below 50 nm (233). Newer methods (low voltage scanning electron microscopy plus platinum coating) are being developed that enable a thinner coating and hence a more precise scanning electron microscopic particle size determination (236). However, auxiliary materials such as surfactants that may be present in the nanoparticle suspension can deposit on the surface of the nanoparticles upon drying during preparation for electron microscopy (130,233). This too can lead to errors in sizing.

Transmission electron microscopy without or with staining is a relatively easy method of particle size determination. Some nanoparticle materials, however, are not electron-dense enough, cannot be stained, or melt and sinter when irradiated by the electron beam of the microscope and therefore cannot be visualized by this method.

These nanoparticles can be visualized by transmission electron microscopy after freeze-fracturing or freeze-substitution (233). This method is optimal because it also allows fracturing of the particles and consequently an observation of their interior. For this reason, it is the most versatile sizing and morphology characterization method. Unfortunately, it is also the most laborious and time-consuming method and therefore not useful for routine measurements.

Recently, new types of high-resolution microscopes, so-called scanning probe microscopes, have been developed, the atomic force microscope, the laser force microscope, and the scanning tunneling microscope (237–250). These microscopes may be useful in investigations of nanoparticle surfaces.

Other methods such as mercury porosimetry are not useful for nanoparticle sizing (233).

The molecular weights of nanoparticles are mainly determined after the dissolution of the particles in an appropriate solvent and the subsequent gel permeation chromatography. The validation of the obtained results is a problem, however. So far, mainly polystyrene and some poly(methyl methacrylate) standards are available that are only of limited use for the molecular weight determination of other polymers. Accurate results can be obtained only if the polymer standards have a very similar structure and similar properties as the test material. With the technique of gel chromatography it is necessary that the polymer be totally dissolved. For this reason, the molecular weight determination of cross-linked polymers and of nanoparticles from natural macromolecules is not possible.

Density measurements can be performed by helium compression pycnometry (233) and by density gradient centrifugation. A comparison of these two methods may offer information about the internal structure of nanoparticles. Unfortunately, the possibilities of these methods so far have not much been exploited.

Information about nanoparticle structure also may be obtained by x-ray diffraction (233) and thermoanalytical methods such as differential scanning calorimetry (DSC), differential thermal analysis (DTA), thermal gravimetric analysis (TGA), thermal mechanical analysis (TMA), and thermal optical analysis (TOA) (251,252). In most cases, small drug molecules are entrapped in the polymer network in the form of an amorphous solid solution (133, 253). In some cases, however, small drug crystallites exist within the polymer network (254). It is very interesting to note that crystallization of pilocarpine in poly(butyl cyanoacrylate) nanoparticles led to another polymorphic modification that was not observable outside of the polymer network (215).

The surface charge of nanoparticles is mainly determined by electrophoretic mobility. A number of methods exist for the determination of this parameter, including ultramicroscopic examination of the movement of nanoparticle diffraction patterns under laser light (233), laser Doppler anemometry, and amplitude weighted phase structuration (255). The surface charge seems to have a major influence on the phagocytic uptake of nanoparticles by cells in vitro. The influence of the charge on the in vivo distribution, however, seems to be much less pronounced (256,257).

Hydrophobicity of the nanoparticles' surfaces seems to have a much larger influence on their body distribution after i.v. injection than surface charge (see Sec. VI.A). Two major methods for the determination of the hydrophobicity exist: water contact angle measurements (258) and hydrophobic interaction chromatography (259). Contact angle measurements, however, can only be performed on flat surfaces, not directly on hydrated nanoparticles in their dispersion media. For this reason, hydrophobic interaction chromatography seems to represent the better although more laborious method: a differentiation between nanoparticles with different surface properties can be obtained by loading the particles on columns with alkyl-sepharose and eluting them with a Triton® X-100 gradient (260–262).

B. Drug Loading Analysis

As mentioned in Sec. IV, drugs may be loaded onto the nanoparticles either by nanoparticle manufacture in the presence of the drugs or by addition of drugs to previously prepared empty particles. In summary, as discussed there, both methods can lead to

A solid solution of the drug in the polymer (133,253)
A solid dispersion of the drug in the polymer (136,215,254)
Surface adsorption of the drug (29,136)
Chemical binding of the drug to the polymer (130)
Practically no binding or incorporation at all, as in the case of inorganic salts with polyacrylic nanoparticles

The resulting payload, the type of interaction with the nanoparticle polymer, the binding of the drug to the nanoparticles, and the rate of this interaction depend on the drug, on the polymer, and on the preparation conditions.

The adsorption isotherm of the drug on the nanoparticles is an important criterion for the determination of the type of binding and of the binding capacity of the carrier, i.e., the payload of the nanoparticles (136). Linear sorption isotherms are characteristic for solid solutions (253), Langmuir-type isotherms, or S-type isotherms indicative for surface adsorption (222). The latter isotherms, however, also may be observed in cases of drug incorporation into the polymer (unpublished observation).

The precise drug content determination of the nanoparticles can be a problem due to the colloidal nature of the drug carrier. These problems are addressed in more detail in a review article by Magenheim and Benita (237). The method of choice for the drug content determination is separation of the particles by ultracentrifugation and following quantitative analysis of the drug after dissolution of the pelleted polymer. Other useful separation methods for nanoparticles are ultrafiltration and gel filtration (263). The major disadvantage of ultracentrifugation as well as ultrafiltration is that undissolved or secondarily precipitated drug particles may be removed from the liquid phase together with the carrier particles. This problem can in most cases be avoided by using gel filtration as the separation process (263). The disadvantage of the latter method is that some drug release may occur during passage over the gel column. The relatively short separation time of 10–35 min necessary for this method, however, reduces the possible error caused by this drug release.

Alternatively, the drug content can be determined in the supernatant or in the filtrate. The amount of drug bound to the particles can then be calculated by subtraction of this amount from the total amount of drug present in the suspension.

Generally, the construction of drug loading sorption isotherms is relatively time consuming. A more rapid determination of the plateau of the isotherm is the measurement of the dielectric parameters of nanoparticle suspensions at high frequencies (264). So far, however, the applicability of this method has only been demonstrated for a limited number of drugs and polymers.

Illum et al. (265) used very interesting alternative drug content analysis methods for nanoparticles that allowed the determination of bound drug without prior separation. These methods utilized the bathochromic shift or the quenching of the fluorescence caused by that amount of drug that is bound to the particles. These methods, however, are drug specific and therefore not applicable for every drug.

C. Nanoparticle Degradation

Nanoparticles due to their small size degrade faster than larger microspheres. Nevertheless, the degradation mechanism for both types of particulates very likely is similar.

The degradation of poly(alkyl cyanoacrylates) is studied best. Two degradation pathways exist (Fig. 5). One pathway is degradation by erosion of the polymer backbone under formation of formaldehyde. The second pathway is cleavage of the ester under formation of a soluble polymer acid. Initially it was believed that only the first degradation pathway existed (271–274). Later, however, Lenaerts et al. (275) found that in vivo the major degradation pathway of poly(alkyl cyanoacrylate) nanoparticles is the cleavage of the side chain ester bonds. This pathway was later verified by Stein and Hamacher (276), who observed a much larger degradation rate in dog serum than in corresponding buffer systems. By the second degradation mechanism, the polymer chain remains intact, but it becomes more and more hydrophilic until it is water soluble. Due to the low molecular weights (31,37) the resulting water soluble polymer acid is then rapidly excreted from the body (35). This degradation pathway is consistent with the production of the corresponding alcohol during the bioerosion of the nanoparticles in vitro in the presence of esterases (275–277). The involvement of rat liver microsomes and tritosomes in this degradation pathway of poly(cyanoacrylate) nanoparticles was clearly demonstrated earlier (275).

The degradation of these nanoparticles was found to be a surface erosion process, because the particle size decreases immediately after incubation and does not show any lag period (278,279). This observation is in agreement

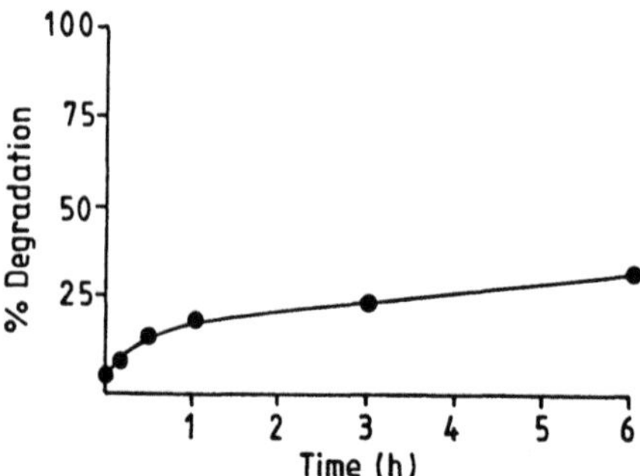

Figure 5 In vitro degradation of poly(hexyl cyanoacrylate) nanoparticles in tears of the albino rabbit. (Reproduced from Ref. (371) with permission of the copyright holder.)

with an earlier finding that the release rate of drugs such as dactinomycin depends on the cyanoacrylate side chain ester length (137). The release as well as the degradation rate decreases with increasing ester side chain length (137,271). It has to be mentioned that this decrease in degradation rate with increasing ester chain was observable with both pathways, backbone degradation as well as via side chain ester hydrolysis. The release of drugs, therefore, can thus be monitored by mixing the two different cyanoacrylate monomers in certain ratios (137).

Other types of nanoparticles may degrade somewhat differently. Polylactic acid microspheres, for instance, degrade from their center, again by hydrolysis (84,280). In the case of polylactic acid, of course, hydrolysis results in the erosion of the polyester backbone. It is, however, presently not known if polylactic acid nanoparticles also will degrade similarly to microspheres. Their smaller size increases their specific surface area, which in turn might significantly facilitate surface erosion and as a result might favor this degradation mechanism.

Human serum albumin nanoparticles as well as human serum albumin microspheres of a mean size of 1.5 μm and a maximal size of 5 μm definitely degrade from the center after their phagocytosis by human macrophages (281). Their rapid phagocytosis by these macrophages cultivated in tissue cultures and their degradation was easily followed by transmission microscopy due to the high electron density of these types of particles (Fig. 6). The degradation was mostly terminated after about 3–4 days. Only small numbers of intact microspheres or microsphere or nanoparticle fragments were found in the cytoplasma of the cells after 7 days.

D. Drug Release

Nanoparticles exhibit their special drug delivery effects (217,282) in most cases by direct interaction with their environment, i.e., their biological environment. Drug release may occur by

Desorption of surface-bound drug
Diffusion through the nanoparticle matrix (Fig. 7)
In case of nanocapsules, diffusion through the polymer wall (Fig. 8)
Nanoparticle matrix erosion
A combined erosion diffusion process (Fig. 9)

The release mechanism, the diffusion coefficient, and the biodegradation rate are the main factors governing the drug release rate (137). The release rate of drugs from nanoparticles is also strongly influenced by the biological environment. This environmental influence is far more intensive than with larger dose forms; nanoparticles may be coated by plasma proteins (283); this can establish a significant additional diffusion barrier and lead to a

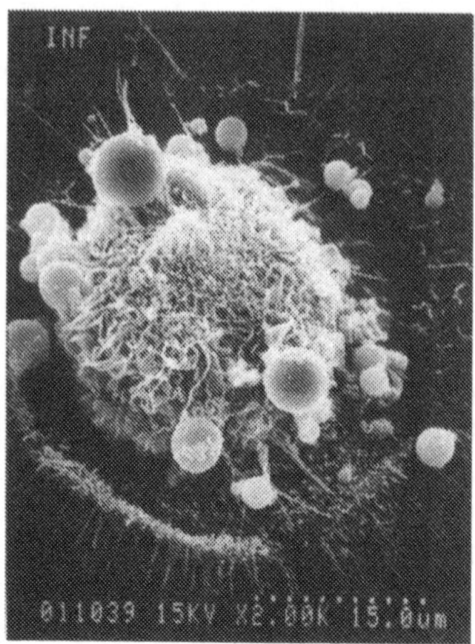

(a)

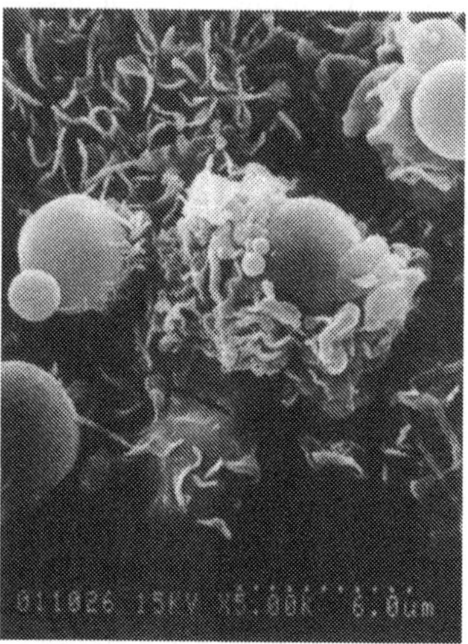

(b)

Figure 6 Phagocytic uptake and macrophage degradation of human serum album nanoparticles by human macrophages in tissue cultures 14 days after start of the cultures. (a and b) Uptake of the particles determined by scanning electron microscopy. Phagocytic uptake (c), total engulfment (d), and progressive degradation (e,f,g) starting from the particle interior determined by transmission electron microscopy. (e) 1 day, (f) 2 days, and (g) 3 days after addition of the particles.

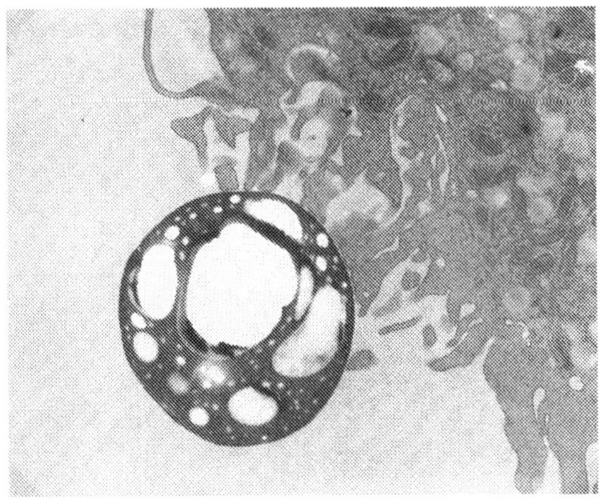

(c)

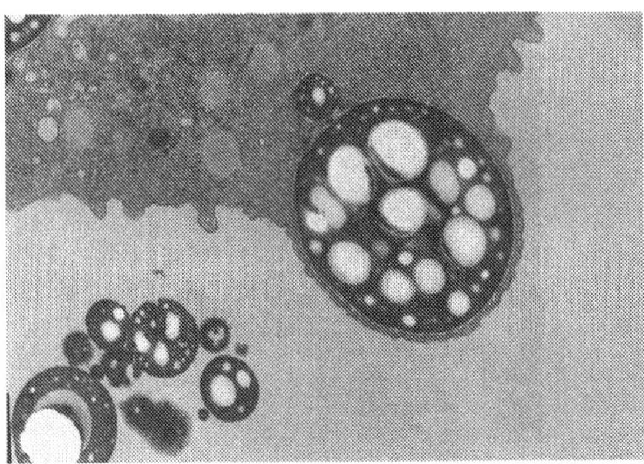

(d)

Figure 6 (Continued)

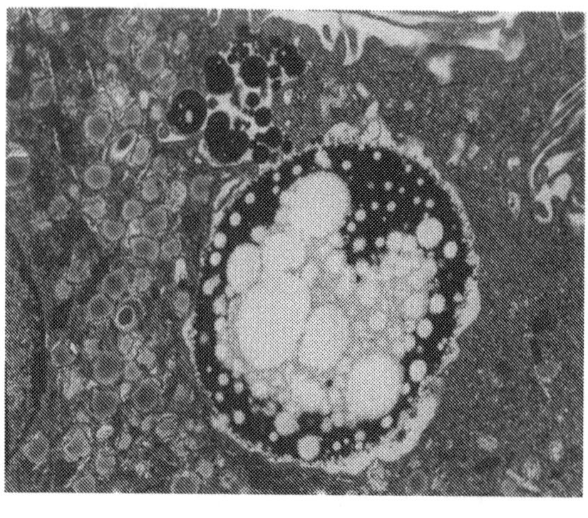

1,4 μm

(e)

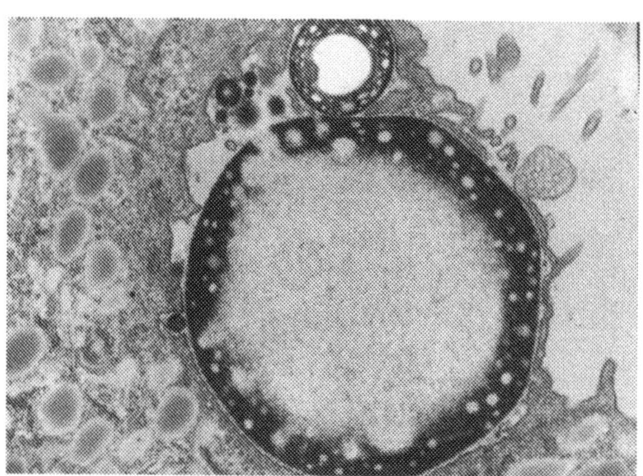

1,1 μm

(f)

Figure 6 (Continued)

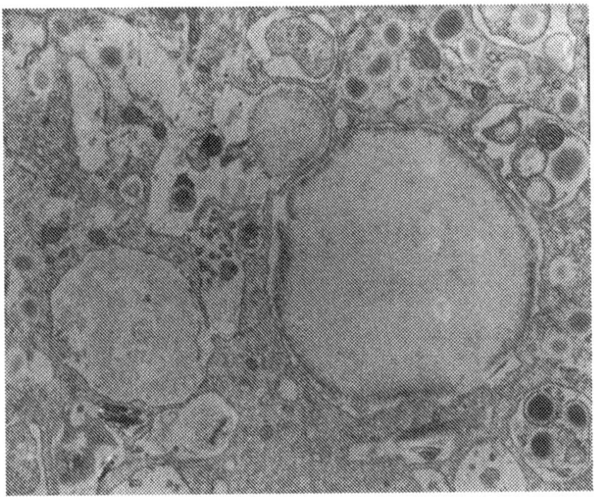

Figure 6 (Continued)

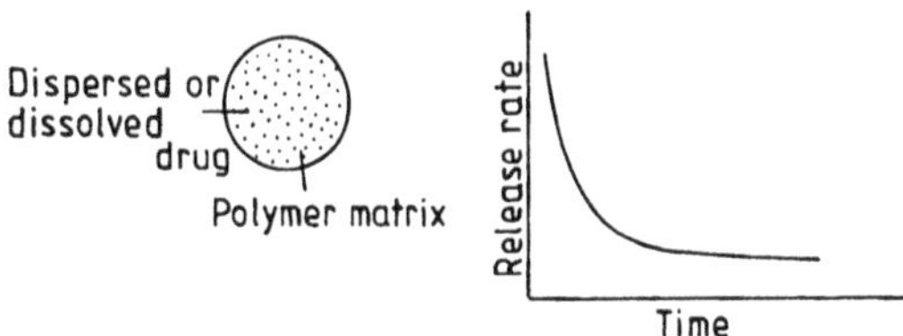

Figure 7 Monolithic device and a typical plot of release rate of a drug versus time.

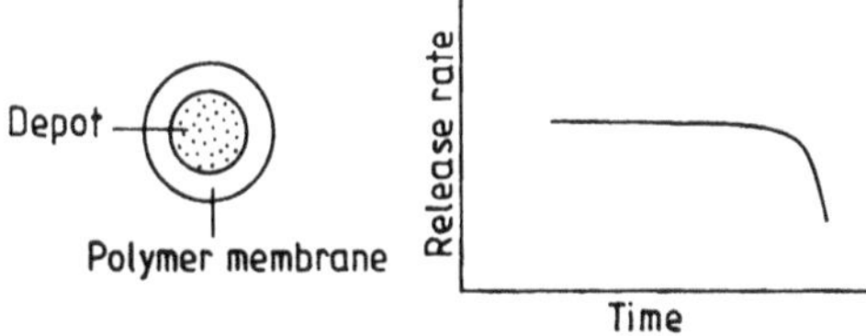

Figure 8 Reservoir device and a typical plot of release rate of a drug versus time.

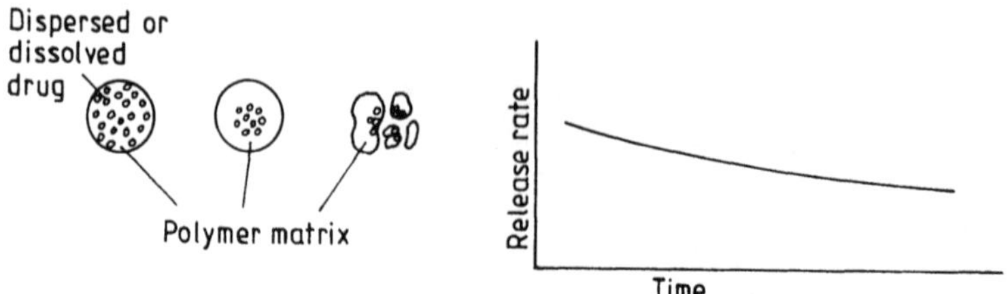

Figure 9 Eroding monolithic device and a typical plot of release rate of a drug versus time (combination of diffusion and degradation).

retardation of the release. Nanoparticles also may interact by direct contact with biological or artificial membranes, in this case leading to an enhanced delivery of drugs through these membranes in comparison to a simple solution (217,282). As a consequence, the in vitro drug release may have very little in common with the delivery and release situation in vivo, as was frequently observed (216,217). Nevertheless, for characterization purposes and for quality control reasons, the determination of the in vitro release of nanoparticles is important.

The characterization of the in vitro drug release from a colloidal carrier is technically difficult to achieve. This can be attributed to the inability of effective and rapid separation of the particles from the dissolved or released drug in the surrounding solution owing to the very small size of this dose form (237).

Two reviews have appeared recently that analyze the in vitro release situation from nanoparticles in detail (237,284).

The following methods for the determination of the in vitro release have been used:

1. Side by side diffusion cells with artificial or biological membranes (216, 217,282,285)
2. Dialysis bag diffusion technique (70,71,76,199)
3. Reverse dialysis sac technique (285)
4. Ultracentrifugation (41,110,144,186)
5. Ultrafiltration (70)
6. Centrifugal ultrafiltration technique (237)

Analysis of nanoparticle release profiles frequently shows a biphasic release pattern that can be described using a biexponential function:

$$C_t = Ae^{-\alpha t} + Be^{-\beta t} \tag{1}$$

where C_t is the concentration of compound remaining in the nanoparticles at time t, A and B are system-characteristic constants, and α and β are rate constants that can be obtained from semilogarithmic plots (136). The mathematical problems of drug release from nanoparticles were further analyzed in detail in another study using doxorubicin as the model drug (175).

E. Nanoparticle Bioacceptability and Toxicity

Nanoparticles distribute very rapidly into the phagocytic cells of the reticuloendothelial system, especially the liver (35,281,287–289). Here the radioactivity mainly appears in the Kupffer cells (290). The distribution ratio in the liver 30 min after intravenous injection between Kupffer, endothelial, and parenchymal cells was approximately 100:30:1 with 80 nm and 215 nm sized poly(isobutyl cyanoacrylate) nanoparticles, the ratio expressed in amount of particles per mg of tissue. Although a certain percentage of this amount just may have adhered loosely to the cells (290), and while no significant uptake by rat hepatocytes was observed in vitro (291), later in vivo studies definitely demonstrated uptake into these cells in vivo after intravenous injection (292).

The poly(cyanoacrylate) nanoparticles possess a slight but definite toxicity toward a number of cells including fibrioblasts (293–295), endothelial cells (290), hepatocytes (296), macrophages (281,296), osteogenic sarcomas (293), round cell (Ewing's) sarcomas (293), and malignant fibrous histiocytoma (293). The toxicity decreases with increasing side chain ester length with the exception of the methyl ester, which is less toxic than the ethyl ester (293, 295). Degradation products of the nanoparticles seem to be involved in the toxicity, because partly degraded nanoparticles were more toxic than undegraded particles (291). The LD_{50} of poly(butyl cyanoacrylate) nanoparticles for hepatocytes was calculated to be 0.4 mg nanoparticles/mL cell suspension containing 2×10^6 cells/mL (290). This result was in agreement with the data reported earlier (296) where hepatocyte toxicity was observed with 0.15 mg poly(butyl cyanoacrylate) nanoparticles/mL of cells starting after more than 2 hours of incubation, and no toxicity with 0.075 mg nanoparticles/mL of cells for 4 hours.

Nevertheless, the above results have to be taken with some caution. All the above experiments were carried out by so-called dye exclusion tests with vital dyes that in solution do not have access to living cells, whereas they easily penetrate into dead cells. However, nanoparticles with adsorbed dyes seem to enable the uptake of these dyes into living cells (236). Zimmer et al. (236) observed uptake of rhodamine 6G or of propidium iodide bound to poly(butyl cyanoacrylate) nanoparticles as fluorescent markers into freshly excised conjunctival and corneal rabbit tissue without any damage of these tissues. Free, non-nanoparticle-associated rhodamine 6G and propidium iodide had no access to these cells. In addition, Gipps et al. (293) in the case of the fibroblasts after 1 day observed a regeneration of cells that seemed to have been dead. It is possible that the vital dyes in the tissue culture experiments may get adsorbed to the nanoparticles present in the tissue culture and thus may gain access to living cells simulating false toxic results.

On the other hand, a certain toxicity of the poly(alkyl cyanoacrylate) nanoparticles still was observable: after incubation of cells with higher concentrations of cyanoacrylate nanoparticles with a short side chain ester, the cells started to become rounder, resulting in dilatation of the extracellular space (293). The extent of rounding and of cell detachment varied between different cell lines. Within the next stages of toxic damage, the cells began to swell. The nucleus was characterized by disorganization of the chromatin structure. In the cytoplasma the rough endoplasmic reticulum profiles appeared dilated. This may be due to direct damage to the endoplasmic reticulum, or possibly the cells attempted to produce enough enzymes to cope with the environment produced by the nanoparticles. These cells would finally exhaust their capacity to synthesize the enzymes. These enzymes, particularly those for endocytosis, are produced by the rough endoplasmic reticulum (297). The mitochondria showed an electron-dense matrix, and the glycogen fields were loosened. However, the cell membranes remained intact (293).

Other nanoparticle materials such as poly(hexyl cyanoacrylate), poly(methyl methacrylate), and human serum albumin were by far less cell-damaging than the polybutyl or other short side chain ester cyanoacrylates (281). Actually, no visible toxicity was observed with these materials. It also has to be kept in mind that the amount of nanoparticles administered to the cell cultures was very high: the dose at the highest nanoparticle concentration added to one well was equivalent to what is normally given to one mouse. In addition, under in vivo conditions the degradation products are eliminated from the nanoparticle location, thus decreasing the contact time with individual cells. Consequently, in an in vivo study in which several organs were investigated after i.v. injection into mice using autoradiography, no signs of cell damage were observed, even at locations where a significant accumulation of the nanoparticles occurred (292).

In addition to cell toxicity, the muagenity of poly(alkyl cyanoacrylate) nanoparticles was tested using the Ames test. Neither the nanoparticles nor their degradation products showed any toxocity (163,296).

In animal experiments, the acute toxicity (LD_{50}) of poly(butyl, poly(isobutyl, and poly(hexyl cyanoacrylate) nanoparticles was tested after intravenous injection to mice and was found to be 230 mg/kg, 196 mg/kg, and 285 mg/kg. In addition, these nanoparticles were tested for histiotoxicity by subcutaneous administration of 0.2 mL of nanoparticle suspension (10 mg of polymer/mL) into the backs of mice. Neither necrosis nor signs of tissue irritation were observable after 24 hours (163).

The subacute toxicity of 185 nm-sized poly(isobutyl cyanoacrylate) and of 145 nm-sized poly(hexyl cyanoacrylate) nanoparticles was tested in Wistor rats (163). The rats obtained intravenously 20 mg of polymer per kilogram of body weight in either a two injections per week schedule over 4 weeks or a one injection per week schedule over 16 weeks. No significant difference in weight increase was observed between the untreated controls, the nanoparticle preparations, and another control, the polymerization medium. With the exception of the platelet counts, no significant differences were observed between all the experimental groups and controls for the other hematologic and serum parameters including serum creatinum, serum γ-GT, serum-GPT, sodium and chloride serum concentration, red blood cell counts, hemoglobin, and hematocrit. The platelet counts increased significantly in the long-term group. An increase in platelet count, however, was also observed in a third control group that did not receive any injection but was anaesthetized in parallel to the nanoparticle and the polymerization medium group. After sacrifice of the rats also no significant morphological changes between the groups were observable by microscopic investigation of organs and tissues including the cerebrum and cerebellum, stomach, small and large intestine, pancreas, testis, epididymis, adrenal glands, thymus, mesenteric lymph nodes, and femoral bone marrow (163). Binding of doxorubicin to poly(isobutyl cyanoacrylate) nanoparticles decreased its acute and subacute toxicity very significantly (162,163). Furthermore, cardiotoxicity was decreased due to poor uptake by the myocardium.

The capture and rapid uptake of doxorubicin-loaded poly(isohexyl cyanoacrylate) nanoparticles in vivo after injection to reticulosarcoma M 5076 metastasis-bearing mice by the Kupffer cells created a large gradient of drug concentration in these cells that allowed a prolonged diffusion of the free drug toward the neoplastic tissue (167). The high concentration of the doxorubicin-loaded nanoparticles in the Kupffer cells had no adverse effects on these cells. Rolland et al. (160) followed the uptake of doxorubicin-loaded poly(methacrylate) nanoparticles by cultured rat hepatocytes using electron microscopy. Nanoparticle binding enhanced the drug concentration in the hepatocytes by 25–64% in comparison to free drug. No cytotoxicity as evidenced by electron and light microscopy or by leakage of lactate dehydrogenase was detectable.

VI. BODY DISTRIBUTION OF NANOPARTICLES

A. Body Distribution After Intravenous Injection

After intravenous injection, nanoparticles, like other colloidal carriers, are taken up by the reticuloendothelial system (RES), also called the mononuclear phagocytic system (MPS). (This latter expression neglects the involvement of the endothelial tissue in the body distribution process of nanoparticles. For this reason, the older expression RES is preferable here and will be used throughout this chapter.) The intravenously injected colloids mainly distribute into the RES organs liver (60–90% of the injected dose), spleen (2–10%), lungs (3–20 and more percent), and a low amount (>1%) into the bone marrow (288,289). It has been acknowledged for over a decade (283, 298) that coating of the injected colloid particulates with serum components—opsonins—precedes the uptake by the RES and that this opsonization process strongly facilitates or even is necessary to induce phagocytosis of the particles (299–302). As shown by two-dimensional polyacrylamide gel electrophoresis (PAGE), the surface properties of the particles seem to influence strongly the spectrum of the adsorbed serum components (303). Serum complement is one of the most important components of this opsonic system that interacts with the injected particulates. In vitro experiments with liposomes (304,305) have demonstrated that complement indeed strongly activates phagocytosis by macrophages. In particular, C3 seems to promote phagocytosis, although other complements such as C5 also seem to be involved in the phagocytic process (306). Recently, it was shown that overnight incubation of poly(methyl methacrylate) nanoparticles in rat serum and especially additional heating of the overnight incubated particles to 56°C for 30 min prior to injection significantly reduced the liver uptake from 70% (plain particles in saline) down to 32% (serum incubation) or even 22% (inactivated serum) of the injected dose after intravenous injection to rats (302). The spleen uptake decreased from 2.8% to 0.68% and 0.62%, the bone marrow uptake from 1.15% to 0.51% and 0.58%, and the total RES uptake from 76% to 54% and 38% respectively. At the same time, the serum concentrations increased from 1% to 2.6% (serum incubation) and 10% (inactivated serum) and the muscle concentration from 4.5% to 7.7% and 30%. Coating with rat serum albumin (RSA), on the other hand, had no influence at all on the body distribution of the nanoparticles in comparison to the uncoated particles (302). These results strongly support the above-presented assumption that coating by serum opsonization largely determines the body distribution.

The adsorption of the serum components, mainly proteins, is competitive; fibrinogen, for instance, adsorbs faster to many artificial surfaces than albumin, and the latter protein in most cases faster than γ-globulin (307). One of the most puzzling aspects of protein adsorption is that, despite oft-reported Langmuir-like adsorption, adsorption is generally found to be at best only partially reversible and is frequently described as irreversible (308). However, more recent reports show that desorption of proteins occurs (309, 310); this results in replacement by another blood component at the adsorption site. Chuang et al. (311), for instance, observed replacement of the rapidly adsorbing thrombin on polyvinyl chloride surfaces by albumin. It therefore has to be concluded that the adsorption of blood components is a dynamic process and that the interfacial region "is a very busy place" as Coleman (312) calls it.

Soderquist and Walton (308) extensively studied the adsorption/desorption events at the polymer interface using albumin, γ-globulin, and fibrinogen. They found that this process is characterized by three stages:

Step 1. Fairly rapid and reversible uptake of the proteins occurs in a minute or so, reaching a pseudoequilibrium. Up to 50–60% surface coverage there is a random arrangement of adsorbed molecules, but at about this level some form of surface transition occurs that is probably in the direction of surface ordering, thus allowing further protein uptake.

Step 2. Each molecule on the surface undergoes a structural transition as a function of time that occurs in the direction of optimizing protein/surface interaction. As proteins increase their interaction with the surface (and their entropy), desorption becomes less likely.

Step 3. Although the probability of desorption decreases with increase in the period of surface residence, since the material that does desorb is denatured and is less able to undergo entropic processes favoring adhesion, protein slowly desorbs more or less irreversibly.

Soderquist and Walton (308) concluded that it is probably hydrophobic bonding that is the predominant force in the adhesion of the proteins to the foreign surface. Furthermore, it seems likely that the unfolding of the proteins reveals more of the hydrophobic interior, thus increasing the surface interaction. This process is probably driven predominantly by an increase in entropy of the adsorbed material, but it also optimizes hydrophobic interactions.

The opsonin coat around the injected particulates then seems to trigger their uptake by the cells of the RES (299–301). In the liver, the main target organ, the main target cells are the Kupffer cells (290,292,294). However, a small portion also distributes into hepatocytes (292).

Based on the above assumption that opsonin adsorption determines the body distribution of nanoparticles and other colloidals after intravenous injection, it can be speculated that different surface properties of the particles will lead to a different spectrum as well as to different quantities of adsorbed serum components (313). It can be further speculated that different types of adsorbed serum components will prevent or trigger uptake of different cells of the RES. Following along these lines it was early suggested to design or artificially alter the surface properties of the particles in order to alter, and later be able to monitor, their body distribution (256,288,313, 314).

Wilkins and Myers (256,313,314) were the first to attempt to modify the body distribution of polystyrene particles by altering the surface charge of the particles. They used polystyrene particles coated with different portions of polylysyl gelatin and gum arabic. Coating with polylysyl gelatin alone resulted in a (positive) charge of 1.30 μm cm sec^{-1} V^{-1}, and coating with gum arabic alone in a (negative) charge of -1.16 μm cm sec^{-1} V^{-1}, both in saline (256). By mixing both substances at different ratios it was possible to obtain intermediate charges (Table 7).

However, in serum the charges approached each other at a level of about -0.7 μm cm sec^{-1} V^{-1} (Table 7). This levelling of the charges after incubation in serum is another indication that serum components adsorb on the particle surface and, as a result, cause the disappearance of the charge differences. Nevertheless, despite the charge equality of particles in the serum, the organ distribution of the initially differently charged particles was clearly related to the initial surface charge: whereas 15 min after injection, negatively charged particles mainly were found in the liver (92%) and very little in the lungs (0.8%) and in the spleen (2.7%), a high portion of the positively charged particles were found in the lungs (15%) and a somewhat lower portion in the spleen (8%), while the liver concentration only reached 58% (256, Table 7). These findings were later confirmed with other gelatin derivatives (312,313). However, despite these differences in organ distribution, the organs involved all belonged to the RES. The total RES distribution was not altered to the same extent, and it can be assumed that non-RES organs received only relatively little of the injected dose. The process that occurred in the experiments of Wilkins and Myers (256,313,314) as a result of the surface charge alteration, therefore, was mainly a redistribution within the RES.

Later, Illum et al. (315–320), Leu et al. (321), and Tröster et al. (258,262, 322,323) used surfactants and polymers to coat nanoparticles in order to modify their body distribution. All authors could show that surfactants coated onto the surface of the particles reduce the nanoparticle liver uptake and simultaneously increase their blood serum concentrations as well as their concentrations in the other organs, including non-RES organs, for several hours. The methodology used for the determination of the body distribution of the particles differed between the group of Illum et al. (315–320) on one hand and the Leu et al. (321) and Tröster et al. (262,322,323) groups on the other. This difference in methodology is probably responsible for the sometimes different quantitative distribution data in the various organs and tissues between these groups. Illum et al. either grafted ^{131}I-iodine on to the nanoparticles using gamma-irradiation (315–317,319,320) or they adsorbed ^{99m}Tc-labeled dextran 10 onto nanoparticles (318) in order to label the nanoparticles. The use of these gamma-ray emitting isotopes enables the determination of the body distribution by a gamma camera. This technique has

Table 7 Organ Distribution of Charged Polystyrene Particles (Mean Diameter 1.3 μm) in % of the Injected Dose 15 min and 72 h After i.v. Injection to Rats

Ratio of PLG[a]/GA[b] in coating	Electrophoretic mobility[c]		Organ					
			Liver		Spleen		Lung	
	In saline	In rat serum	15 min	72 h	15 min	72 h	15 min	72 h
GA alone	−1.16	−0.70	92.52	95.14	2.70	1.90	0.77	0.12
10/90	−1.15	−0.71	77.83	85.17	10.96	8.36	7.31	0.68
20/80	−1.00	−0.74	72.54	79.18	6.84	9.98	10.97	0.67
60/40	+0.20	−0.78	68.60	82.79	5.03	8.19	11.31	0.82
80/20	+0.63	−0.66	68.99	87.73	6.21	8.82	14.01	0.79
PLG alone	+1.30	−0.73	58.00	73.90	8.22	14.05	15.19	1.07

[a]Polylysyl gelatin.
[b]Gum arabic.
[c]Units are in μm cm sec^{-1} V^{-1}.
Source: Adapted from Wilkins and Myers (256).

the advantage that results are obtained and analyzed rapidly and that the animals can be used over the whole observation period. However, considerable uncertainty exists about label stability, especially after injection into the living body. In addition, it is not possible to determine the exact amount of radioactivity in organs that are in close vincinity. The radioactivity in organs or tissues containing only small amounts of particles cannot be determined by this method. The method of Leu et al. (321) and Tröster et al. (262, 322,323) has the disadvantage that it is very laborious. However, because the poly(methyl methacrylate) particles used by the latter two groups are labelled within the polymer chain with ^{14}C, and because the particles biodegrade extremely slowly (324), a very high label stability of this material exists. Therefore radioactivity found anywhere in the body within 7 days with a very high probability ($p < 0.01$) reflects intact nanoparticles. In addition, the radioactivity in tiny amounts can be determined in all organs and tissues without any interference from any tissue in the vincinity.

In their very extensive study, Tröster et al. (262,322,323) investigated 13 surfactants and polymers as coating material for nanoparticles and examined the radioactivity in 13 organs and tissues after 30 min, 2, 6, and 24 hours, and 7 days after intravenous injection to rats. The organs and tissues examined included blood serum, liver, spleen, lungs, bone marrow, lymph nodes, heart, kidneys, brain, ovaries, testicles, muscles, and intestine. An excerpt of the results is given in Table 8. As mentioned above, all surfactants reduced liver uptake and increased blood and all other organ concentrations of the nanoparticles. Among these surfactants two lead substances may be identified by the work of Tröster et al., poloxamine 1508 and polysorbate 80. Poloxamine 1508 is the optimal substance in reducing liver uptake and keeping the particles in the blood circulation. Since this material presently seems not to be produced any more, poloxamine 908 may be used. The latter material is as good in reducing liver uptake (or even better; see Table 8, although differences between these substances were not statistically significant) and almost as good in maintaining the particles in the blood circulation, confirming earlier data by Illum et al. (319). Polysorbate 80 was the overall optimal material for increasing the nanoparticle concentration in non-RES organs. [This finding is not obvious from the limited data in Table 8, but it can be deduced from the whole body of data in the work of Tröster et al. (322).]

Although the surfactant coating induced reduction in the liver uptake of the nanoparticles was combined with a simultaneous increase in the other RES organs spleen, lungs, bone marrow, and lymph nodes (262,322,323), the total RES uptake was significantly reduced by most surfactants after 30 min. Poloxamine 1508, for instance, reduced the total RES uptake from 91% with the uncoated particles down to 58%, poloxamine 908 down to 55%,

Table 8 Body Distribution of Surfactant- or Polymer-Coated Nanoparticles 30 min and 7 days After i.v. Injection of Poly(methyl-2-^{14}C-Metharylate) Nanoparticles in % of Injected Dose

Organ	Uncoated controls	Coating material: Polysorbate 80	Poloxamine 908	Poloxamine 1508	Brij® 35	Klucel® EF
			30 min			
Blood serum	0.28	0.55	22.8	37.5	0.36	0.22
Liver	83.2	26.9	24.4	28.7	90.0	41.5
Spleen	1.8	4.6	16.0	16.4	3.4	2.9
Lungs	2.1	43.2	12.6	8.9	2.6	54.0
Heart	0.06	0.61	0.56	0.66	0.06	0.27
Brain	0.05	0.72	0.47	0.54	0.06	0.30
Muscles	0.40	2.2	2.0	2.8	0.62	1.59
			7 days			
Blood serum	0.01	0.01	0.02	0.02	0.02	0.02
Liver	73.1	53.3	46.9	51.3	86.5	49.7
Spleen	2.9	4.2	19.0	20.5	3.7	4.3
Lungs	0.26	30.9	13.1	14.2	5.0	40.8
Heart	0.00	0.04	0.04	0.03	0.01	0.03
Brain	0.01	0.01	0.03	0.01	0.01	0.01
Muscles	0.12	0.35	0.41	0.25	0.22	0.28

Source: Adapted from Tröster et al. (262,322,323).

and polysorbate 80 down to 77%. With increasing time, however, the total RES uptake increased, and no significant differences between uncoated and coated particles existed after 7 days. This redistribution of the particles into the RES may be due to total or partial desorption of the surfactants. This desorption could lead to a different composition of the body serum coat of the particles, which in turn may induce an uptake by different cells in the body. Another possibility is that the time-dependent redistribution of the particles may be the result of a natural relocation process. Previous work by Adlersberg et al. (325) with uncoated polystyrene particles is an indication of such natural relocation processes. The problem of redistribution of the particles with increasing time, however, may be solved only after attaining a better knowledge of the sorption/desorption processes of the surfactants at the particle surface that occur not only in vitro but more importantly in the life situation. Alternatively, covalent coupling of the surfactants should enable an answer to the question whether the particles are redistributed because of surfactant desorption or because of natural redistribution processes.

An interesting but also presently unsolved problem is the enhanced lung accumulation of the nanoparticles occurring after coating with some surfactants or other coating materials. Wilkins and Myers observed an increased lung uptake with positively charged particles. Adlersberg et al. (325) and Singer et al. (298) as well as Kreuter et al. (287) observed with uncoated polystyrene or poly(methyl methacrylate) nanoparticles a comparatively high lung accumulation shortly after injection that decreased with increasing times: immediately after the injection, in the first minutes of its passage through the blood, a considerable portion was lodged in the lung capillaries in the form of aggregates (298,325). There these embolus-like formations were partially broken up within 5 min, and after 15 min the emboli started to disintegrate. These emboli were not locally formed thrombi, as evidenced by the absence of any inflammatory lesions until 15 min after injection. Four hours after injection, the lung appeared to be particle free; the residual radioactivity probably resulted from the radioactive content of the lung's circulating blood.

Leu et al. (321) later, with the same particles as used by Kreuter et al. (287), were able to decrease the lung concentration after 30 min from 22% down to 1.4% simply by injecting the same amount of particles as Kreuter et al. in a four times larger saline buffer volume and by simultaneously increasing the injection time from 10 s to 2 min. Tröster et al. (262,322,323) achieved similar low lung concentrations with the same poly(methyl methacrylate) nanoparticles simply by slow injection without increasing the volume. These results sults are indicative of a formation of aggregates of these hydrophobic particles after injection in the earlier studies (287,298,325). These aggregates were then filtered by the lung capillaries, since these capillaries represent the first capillary bed to be reached after i.v. injection and possess a diameter with a lower level of about 4 μm. These aggregates then were dispersed after prolonged times and hence left the lungs.

However, the formation of aggregates and subsequent lung filtration does not explain the lung accumulation frequently observed after coating of the nanoparticles with a number of surfactants. As mentioned above, poly(methyl methacrylate) nanoparticles as well as polystyrene nanoparticles are very hydrophobic and therefore possess a strong tendency to aggregate in hydrophilic media. Coating with surfactants decreases their hydrophobicity very significantly (258). Consequently, these nanoparticles possess very little or in most cases no agglomeration tendencies at all as observed by light microscopy and laser light scattering (322,326, Fig. 10). Figure 10 shows that there was no correlation between poly(methyl methacrylate) nanoparticle aggregate formation and lung uptake. For this reason, it has to be concluded that other, presently unknown factors must be responsible for the lung accumulation.

Adlersberg et al. (325) in their experiments with polystyrene particles observed that after reaching very low lung levels of below 2%, the radioactivity in the lung started to increase gradually, reaching 5.4% of the injected dose after 6 weeks. Adlersberg et al. attributed this radioactivity to the redistribu-

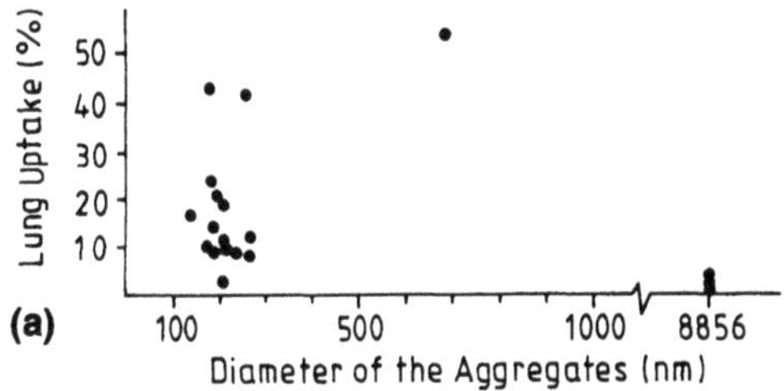

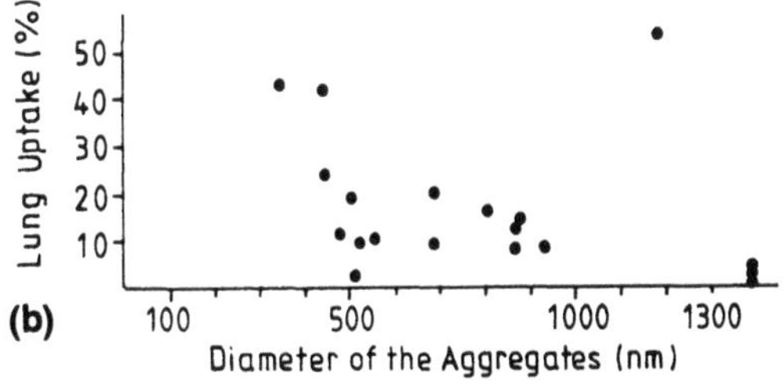

Figure 10 Particle aggregate diameter in different media in vitro and lung uptake in rats after i.v. injection of corresponding poly(methyl methacrylate) nanoparticles (primary particle diameter = 130 nm) after coating with different surfactants. (a) Aggregate formation in phosphate buffer; (b) aggregate formation in plasma.

tion and recirculation of phagocytic cells and their particle content. They suggested that the particles also reentered the circulation by reverse phagocytosis and were then picked up by locally proliferated phagocytic cells in the lungs. The alveolar macrophages with their ingested particulate matter then are extruded into the alveolar lumen. Subsequently, they could move up the bronchial tree and mix with the saliva and were at least partially excreted in the feces (325).

Illum and Davis (320) observed that the coating of 60 nm sized ^{131}I-labeled polystyrene nanoparticles with poloxamer 407 leads to a massive deposition in the bone marrow. Results by Müller (327) showed that this was not the case with larger 142 nm sized ^{131}I-polystyrene particles. He attributed this lack of bone marrow accumulation with the larger particles to the size of the bone marrow sinusoidal fenestration that would prevent the passage of these larger particles. However, this explanation does not seem to be satisfactory. Tröster et al. (262,322,323) observed with only very slightly smaller, 131 nm sized poly(methyl methacrylate) nanoparticles also a significant increase in bone marrow uptake after coating with poloxamer 407, poloxamer 184, poloxamer 904, and Klucel® EF. Especially poloxamer 184, which reached peak bone marrow concentrations of 17% in the injected dose after 24 hours, but also poloxamine 904 were even superior to poloxamer 407 in increasing the bone marrow accumulation. It seems likely that transport through or exclusion by the sinusoidal fenestration is not the reason for the observed differences in bone marrow concentrations between the different nanoparticle preparations. It has to be kept in mind that other mechanisms such as adhesion to or uptake by the endothelial cells lining the blood vessels of the bone marrow could be responsible for the different nanoparticle concentrations observed in this tissue. In these cases, size would not play such a dominant and exclusive role. Experiments by Müller have already suggested that the surface coating with some of his surfactants including poloxamer 407 may not be as thick or as effective with larger particles as with smaller ones (327). Consequently, the interaction of the larger particles with the endothelial cells might be less pronounced than with the smaller ones.

The mechanism of the reduction of the RES uptake and especially the liver uptake and the simultaneous increase in blood circulation time and increased distribution into other organs and tissues is presently not totally explained. Hydrophilicity of the particles definitely plays an important role. However, hydrophilicity does not explain all the in vitro and all body distribution data. Tröster and Kreuter (258) determined the contact angles of surfactant solutions on poly(methyl methacrylate) nanoparticles. By comparing with body distribution data taken from the literature they observed that low advancing, but especially low receding contact angles were combined with long blood circulation times and a low liver uptake of particles coated with these surfactants. They challenged their observations experimentally and found, however, that polyoxyethylene(23)laurylether, which fulfills these contact angle requirements and therefore should have induced a long blood circulation time and a low liver uptake, in reality had almost no influence on the body distribution of the nanoparticles in comparison to the uncoated particles (322,323). Other attempts to explain and predict the efficacy of surfactants in preventing the liver and RES uptake of particulates were made by Illum et al. (317). Illum et al. suggested that an increasing length of the hydrophilic polyoxyethylene chains protruding into the aqueous serum phase, and consequently the increasing thickness of the hydrophilic coat, would reduce the opsonization and subsequently the RES uptake by steric stabilization. They hypothesized that dysopsonic substances like the previously mentioned surfactants must satisfy two requirements: a sufficiently large hydrophobic anchoring moiety to prevent displacement by plasma proteins, and a sufficiently large hydrophilic moiety to prevent adhesion by steric stabilization. However, Carstensen et al. (328), Wesemeyer et al. (329), and Tröster et al. (262) later observed that in the case of poloxamers an increased polyoxyethylene chain length and coating layer thickness indeed reduced the hydrophobicity as well as the RES uptake; in the case of the Antarox® surfactants (ethoxylated nonylphenols), on the other hand, an increase in chain length increased the coating layer thickness but decreased hydrophobicity and also the efficacy in preventing the RES uptake in vivo (262).

Hydrophobic interaction chromatography, which was used for the determination of the hydrophobicity of the surface of coated and uncoated particles, presently is the best method for predicting the potential of substances to prevent the liver uptake of the particles by coating. This method is suitable to provide a negative selection of coating materials, i.e., with this method it is possible to identify and exclude from animal experiments substances with no potential for preventing liver uptake. As shown with hydroxypropylcellulose (Klucel® EF) (262,330) this method is better than contact angle measurements. Hydrophobic interaction chromatography, however, is not able to predict the accumulation potential of coated nanoparticles in other RES organs such as spleen, lungs, and bone marrow (261,262). Factors such as complement activating groups and the effect of conformational changes of the coating surfactants cannot be predicted by this method.

As already postulated by Illum et al. (317), the efficacy of the anchoring moiety, i.e., the hydrophobic part of the coating substance, also plays an important role in enabling a modification of the body distribution. Of course, insufficient anchoring will result in a rapid desorption and displacement of the coat. Indeed, Douglas et al. (318) found that poloxamer 338 or poloxamine 908 did not significantly influence the biodistribution pattern of more hydrophilic poly(butyl 2-cyanoacrylate) nanoparticles. However, their results have to be taken with some caution. It is possible that unsuitable labeling—labeling with ^{99m}Tc-dextran interacting with the nanoparticle surface—may be responsible for their negative findings. Beck et al. (132) instead later observed that coating with poloxamine 1508 indeed could alter the efficacy as well as the body distribution of unlabeled and of ^{14}C-labeled poly(butyl-3-^{14}C-cyanoacrylate) nanoparticles containing mitoxantrone in comparison to uncoated particles.

Müller and Wallis (331) tried to coat a number of surfactants including poloxamine 908 and poloxamer 338 and 407 on polyhydroxybutyrate, polylactic acid, and polylactic acid-polyglycolic acid-copolymer nanoparticles produced by solvent evaporation using poly(vinyl alcohol) as a stabilizer. None of the surfactants adsorbed to these particles to a sufficient degree. The surfactants, however, could be adsorbed to the nanoparticles when other stabilizers such as sodium dodecyl sulfate, sodium tridecyl sulfate, sodium tetradecyl sulfate, or sodium hexadecyl sulfate were used during the solvent evaporation process. The coating layer thickness, however, was thinner as with the polystyrene and with poly(methyl methacrylate) nanoparticles. In addition, the polyester particles in vitro interacted much more extensively with blood serum than polystyrene particles coated with the same surfactants, as indicated by a significant increase in thickness after serum incubation (331).

B. Magnetic and Antibody Directed Targeting of Intravenously Injected Nanoparticles

The incorporation of magnetic Fe_3O_4 particles of a size of 10–20 nm in diameter into albumin (12,99,168–171,177,180–184), poly(isobutyl cyanoacrylate) (332), and chitosan (94) nanoparticles enabled the preparation of magnetically responsive particles. If the particles were injected into the tail vein of rats and a magnet was placed 6.5 cm distal to the point of injection, the particles were retained at the location of the magnet depending on the field strength of the magnet (Table 9). By the same method, placing a magnet in close vincinity of some organs or extremities, it was possible to increase the concentration of the magnetic particles and hence of particle bound drugs in the kidneys (332), lungs (174), or hind limbs (333). Drugs that were bound to these particles were then released at the target site, leading to higher concentrations at this site and to lower concentrations in other parts of the body in comparison to a solution of the same drug after the intravenous injection of these preparations (181–184).

Widder et al. (334) used a rat tail Yoshida tumor as the target site and performed an ultrastructural investigation of the tumor site after injection of magnetically responsive albumin nanoparticles and placing a magnet with a

Table 9 Distribution of Magnetic ^{125}I-Labeled Human Serum Albumin Nanoparticles in Rats in % of the Injected Dose After 30 min

Organ or tissue		Magnetic field strength (T)			
		0 (Control)	0.4	0.6	0.8
Liver		76–85	72–85	57–70	30–48
Spleen		3–9	3–7	2–8	2–6
Kidneys		<1	<1	<1	<1
Lungs		6–19	5–18	6–18	5–17
Heart		<1	<1	<1	<1
Tail	1	0	0	0	0
segment	2	0	0	0–4	0–3
number	3[a]	0	0–3	10–25	37–65
	4	0	0	0	0–2

[a]Location of the magnet.
Source: Adapted from Widder et al. (168).

field strength of 0.55 T at this site (334). Transmission microscopy revealed that at this site nanoparticles were endocytosed by endothelial cells as early as 10 min after infusion. By 30 min, microspheres were seen in the extravascular compartment, sitting adjacent to tumor cells and occasionally in tumor cells. By 24 hours, the majority of the particles had been endocytosed by tumor cells. Microspheres were still observed within tumor cells as late as 72 hours after administration (334). Incorporation of doxorubicin into the nanoparticles achieved a very impressive remission of the Yoshida tumors (171): of the 22 animals receiving magnetically localized doxorubicin nanoparticles 17 had total histological remission of the tumor. The remaining animals demonstrated marked tumor regression representing as much as 500–600 mm^2 decrease in tumor size. While no deaths or metastases occurred in the groups receiving localized drug, animals treated with free doxorubicin, placebo nanoparticles, or nonlocalized doxorubicin nanoparticles exhibited a significant increase in tumor size with metastases and subsequent death in 90–100% of the animals (171).

A similar but somewhat less impressive increase in efficacy was observed with slightly larger 1–7 μm sized doxorubicin albumin microspheres in an AH 7974 lung metastasis tumor (174). By placing a magnet at each side of the rat close to the lungs, the doxorubicin concentration of magnetic microspheres was increased about 10-fold after application of a magnetic field of 0.6 T in comparison to free drug and about 1.3-fold in comparison to doxorubicin microspheres without application of a magnetic field. However, the mean survival time of animals bearing an AH 7974 tumor was increased by 44% with doxorubicin incorporated into the magnetic particles with placing the magnet but only by 7% without the magnet and by 29% with the free drug.

It should be noted, however, that the described animal experiments dealt with a somewhat therapeutically unrealistically favorable situation. The localization of the target site in the tail and the use of small animals such as mice and rats enable the use of a relatively small magnetic field. In a larger three-dimensional patient body it will not be possible to focus the magnetic field exclusively on a restricted target area in the interior of the body, because the magnetic field is strongest at the magnetic poles and cannot be focussed and concentrated somewhere in its interior. In addition, unless the particles are injected directly into tissue or organs lying within the magnetic field or into a location where blood flow reaches the magnetic field area immediately after injection, the particles will come into contact with the reticuloendothelial system, which would probably remove them from the circulation before they could pass through the magnetic field (283,335,336). Moreover, the use of a magnetic field requires the knowledge of the existence and of the exact position of the tumor or target site. Undetected metastases cannot be treated with this method (289). Nevertheless, the use of magnetic particles is an interesting approach that is worth studying, especially when in addition rapid removal of the particles by the reticuloendothelial system can be prevented by other methods.

An alternative approach that has the advantage that it possibly may solve the problem of targeting undetected metastases is the coupling of monoclonal antibodies to the nanoparticles. Of course it is necessary that such antibodies be specific for the target site. In the case of cancer this would mean that the tumor and the metastases carry tumor-specific antigens.

So far, monoclonal antibodies or normal IgG have been bound to nanoparticles by adsorption via the hydrophobic Fc portion (55,148,337–339), by using staphylococcal protein A incorporated into (340), or surface-bound to, the nanoparticles (341) as a linker, or by involving chemical methods (55, 342,343). Short reviews about chemical coupling of antibodies to nanoparticles can be found in the literature; see (55) and (342). In vitro, the monoclonal antibody particle conjugates were binding very efficiently to a variety of cells (55,148,337–343). By using magnetically responsive spheres it was possible to remove selectively target cells such as erythrocytes (340), neuroblastoma cells (55), or Burkitt lymphoma cells (55) from cell suspensions. The method of selective in vitro removal of neuroblastoma cells from bone marrow of patients that was briefly taken out and then retransfused after removal of these cells is today used clinically (55). This method enables the removal of malignant cells that survived the preceding treatment of the bone marrow by other cancer treatment methods.

Especially adsorptive binding of the antibodies to the nanoparticles may not be sufficient for these particles to reach their target site after injection in vivo. Serum proteins may easily displace the antibodies from the nanoparticle surface (337): no significant uptake of nanoparticles in 788 T tumors xenografted to mice was observable after i.v. injection of ^{14}C-labeled poly(hexyl cyanoacrylate) nanoparticles to which ^{125}I-labeled monoclonal 791 T/36 antibodies that recognize 788 T osteogenic sarcoma cells were bound (338). Instead the particles accumulated in the liver and the spleen. The liver and spleen accumulation was even increased two- to threefold by adsorption of the monoclonal antibodies to the nanoparticles or to a slightly lesser extent by adsorption of normal mouse IgG in comparison to nanoparticles without attached antibodies. These findings would suggest that coating the particles with antibodies enhances the uptake in the liver and spleen (338). It has to be mentioned, however, that Kubiak et al. (148) observed that radiolabeling of antibodies with iodine significantly reduced the stability of the binding of the antibodies to the nanoparticles. These authors note that the instability of the iodine labeling could be responsible for the above negative results of Illum et al. and suggest that nanoparticles with unlabeled antibodies might have been more successful in reaching their target site. Nevertheless, the result that antibody binding to nanoparticles increased liver and spleen uptake is an indication that the immunogenicity of these nanoparticles may be enhanced. The antibodies that may be used for drug targeting in many or in most cases will be heterologous and therefore have a potential to act as antigens. Nanoparticles, on the other hand, act as adjuvants (225,232) although their adjuvancy decreases with increasing hydrophilicity (226). For this reason, these antibodies may induce an immuno reaction against themselves that would significantly increase after multiple dosing and that would inactivate their targeting properties.

In the context of drug targeting it has to be mentioned that nanoparticles without attached antibodies or magnetic guidance have a tendency to accumulate in certain tumors (33,132,196,344,345) and may reduce tumor growth (132,153,163,165–167). This tumor accumulation occurred only in living tumor tissue, not in necrotic tumor areas (344). Since a large number of tumors contain large amounts (>90%) of necrotic tissue, a tumor accumulation of such particles and of bound drug may easily be overlooked in experiments, because often the whole tumor or necrotic parts only are analyzed for drug or particle content, leading to the false conclusion that little or no tumor accumulation had occurred. It is also very likely that nanoparticles or other colloidal carriers such as liposomes do not accumulate in all sorts of tumors (289).

The mechanism of tumor accumulation has so far not been totally explained. It has been observed that nanoparticles accumulate in areas of inflammation in the body in significantly enhanced amounts (346–348). It is possible that accumulation in tumors occurs by similar processes. The particles then (1) may be attached to the blood vessel linings within the tumor due to an enhanced bioadhesiveness at this site, (2) may be endocytosed by the endothelial cells lining the blood vessels, or (3) may escape the vasculature by frequently found leaky or open blood vessels in the tumor. As mentioned above, even after primary magnetic fixation of such particles within tumor blood vessels (334) or even in the interior of normal blood vessels (182), the nanoparticles traverse the vascular endothelium within minutes up to a few hours after dosing. The enhanced tumor concentration of the particles and the bound drug then can improve the antitumoral therapy (see Sec. VII). A limited overview of drug targeting in cancer chemotherapy is given in the literature (349).

C. Body Distribution After Intramuscular or Subcutaneous Injection

After intramuscular (287) or subcutaneous injection (350) of ^{14}C-labeled poly(methyl methacrylate) nanoparticles, over 99% of the injected radioactivity stayed at the injection site. Apart from this site, no radioactivity was detectable in any other part of the body for up to 70 days by autoradiography except for some isolated black radioactivity dots that very likely represented local lymphatic areas and lymph nodes (287). Very high lymph node concentrations were sporadically observed by radioactivity determination of single nodes (350). During an initial phase that lasted for a few days, about 1% of the administered dose per day was excreted in the urine and feces. This excretion rate decreased very rapidly and plateaued off after about 80 days at a very low level of about 0.005% per day via the feces and about 0.0005% per day via the urine. After 200 to 300 days, the fecal excretion rate started to increase exponentially, whereas the urinary excretion stayed at the low level. The increase in fecal excretion was combined with a 100- to 300-fold increase in the radioactivity levels in all organs. This increase occurred in the observation interval between 157 and 287 days after injection (termination of experiment). The organ-to-blood ratio, however, did not change. The initial excretion during the first days after injection can be attributed to methacrylate oligomer excretion, because residual methacrylate monomers were not detectable by gas chromatography. The increase in excretion and in residual body distribution after about 200 days is a strong indication that a biodegradation occurred at the injection site: the slope of the elimination pattern—of course not the time frame—was very similar to the elimination pattern observed with polylactic acid microspheres of a much larger particle size (351). In addition, a lag phase of over 54 weeks was already seen previously with ^{14}C-poly(methyl methacrylate) implanted into rats in form of films (352). After this time, the films were left in place for an additional 40 weeks and were then removed. The excretion of radioactivity in the latter study started after 54 weeks and continued for the following 40 weeks and disappeared after removal of the film. The mentioned experiments with polylactic acid that also showed a lag phase and that also involved molecular weight measurements were interpreted by the following scenario (351): Biodegradation started rather early after implantation and as a result gradually led to a 50 to 85% decrease in molecular weight of the polymer, whereas only a very tiny amount of about 1% of the total polymer was eliminated from the injection site. The elimination of the residual material started only after a certain low polymer weight was reached. It can be assumed that a similar process may have happened with the poly(methyl methacrylate) nanoparticles and films. The degradation may have started rather early after the injection or implantation, but the excretion occurred only after a lower molecular weight was reached. This suggests that the degradation of biodegradable polymers follows along similar lines in a number of different polymers, although the mechanism of the chain degradation itself may be different: polylactic acid is degraded by hydrolytic cleavage, while such a mechanism is not possible for the degradation of the poly(methyl methacrylate) main chain (1).

One year after i.m. injection of an influenza vaccine containing poly(methyl methacrylate) nanoparticles as an adjuvant, histological examination of the tissue at the injection site showed no abnormalities (232). Absorption tissue with eosinophiles and giant cells were found. However, the number of these cells and the appearance of the tissue was the same as with the control fluid vaccine (353).

D. Distribution of Nanoparticles After Peroral Application

After peroral administration of ^{14}C-poly(hexyl cyanoacrylate) nanoparticles, the particles were observed after 30 min by autoradiography in the stomach, and after 4 hours a large quantity of radioactivity was found to be distributed in clusters in the intestine without macroautoradiographic evidence of accumulation at specific intestinal sites (163). At this time, a persistent film of nanoparticles on the stomach walls still was also observable. The distribution in the intestine of individual mice was very irregular and could vary up to a factor of 1000 between neighboring sections (354). However, these distribution variations appeared to be totally random and were different from animal to animal. The overall distribution was more homogenous with a distinct accumulation in the areas before the ileocecal junction after 90 min and in the cecum after longer time periods. The amount of radioactivity dropped to 30–40% of the 90 min value within 4 to 8 hours and to 5% 24 hours after dosing. Radioactivity (0.04%) was still detectable after 6 days. Histological investigations revealed radioactivity adjacent to the brush border, inside goblet cells, and in intestine muscle cells up to 6 hours after administration. The histological analysis also gave evidence that labeled material was translocated to the circulation (354).

Evidence for a peroral uptake of intact nanoparticles was provided by extensive histological investigations by Damgé et al. (209) and Aprahamian et al. (210). Nanoparticles appeared in the intestinal lumen close to the mucus, then in intracellular spaces and defects of the mucosa, and finally in the lamina propria and blood capillaries.

There are three possibilities for the mechanism of gastro-intestinal uptake of nanoparticles: (1) intracellular uptake; (2) intracellular/paracellular uptake; and (3) uptake via the M-cells and Peyer's patches in the gut (355). The uptake of nanoparticles by all of these pathways was observed. In a number of studies (207,210,356–358), the simultaneous occurrence of more than one pathway was reported. Damgé et al. (207) observed a preferential uptake of Lipiodol-loaded nanoparticles via intercellular spaces between the enterocytes in the jejunum 10–15 min after administration to dogs and rats, whereas in the ileum after the same time the particles passed through M-cells in large quantities and were found in the intercellular spaces around lymph nodes. It appears, therefore, that the major uptake pathway may be different in different parts of the intestine. Indeed, Michel et al. (208) observed with insulin-containing nanocapsules that the particles were taken up to a different extent or interacted differently with the intestinal wall after administration at different locations of the gastrointestinal tract. These authors measured the effect of these nanocapsules on fasted glycemia in diabetic rats. The intensity and duration of the glycemia depended on the site of administration (65% ileum, 59% stomach = normal experimental peroral administration site, 52% duodenum and jejunum, 34% colon; the percentages are relative to a theoretical total ≅ 100% glycemia = blood glucose levels of zero (208)).

Although all three above-mentioned uptake pathways are involved in the gastrointestinal uptake of intact nanoparticles, the major pathway seems to be uptake via the M-cells and Peyer's patches in the gut. Scherer et al. (359) observed by using laser confocal microscopy that fluorescence appeared in localized patches when fluorescein isothiocyanide (FITC) nanoparticles were applied in vitro onto porcine or rabbit small intestine, whereas the fluorescence was distributed homogeneously over this tissue when FITC or FITC-dextran solution was applied. If porcine small intestinal tissue was mounted into two-chamber side-by-side diffusion cells and ^{14}C-labelled poly(butyl cyanoacrylate) nanoparticles were charged into the chamber facing the brush border side of the intestinal tissue, no radioactivity was transported into the other (acceptor) chamber within 4 hours if the tissue came from the upper part of the small intestine. When the lower part of the small intestine was used, that possessed a considerable number of M-cells and Peyer's patches, a very significant amount of radioactivity appeared in the acceptor chamber during this time (359). Further evidence for the importance of the Peyer's patches for the intestinal uptake of particles was provided by Eldridge et al. (360) and Jani et al. (356,361,362). Eldridge et al. showed that Peyer's patch uptake in mice increased with increasing hydrophobicity of the particles. The uptake was restricted to particles below 10 μm in size. Particles above 5 μm remained fixed in the Peyer's patches, whereas those below this size were transported within macrophages through the efferent lymphatics (360). Jani et al. (356,361,362) used 50 nm, 100 nm, 300 nm, 500 nm, 1 μm, and 3 μm polystyrene particles. They observed that uptake in rats into Peyer's patches and passage via the mesentery lymph supply and lymph nodes to the liver and spleen, and to the general circulation, increased with decreasing particle size. Their findings indicate that the size limit for the transport into the lymphatics in rats may even be lower, with a cutoff at about 1 μm. Particles of this size, but mainly of smaller sizes, were found in stomach, small intestine with Peyer's patches, appendix and mesentery, liver, spleen, blood, bone marrow, and kidneys (Table 10). Peroral administration of fluorescent particles to rats with experimental inflammatory air pouches in the neck lead to the recovery of 0.2% of the particles in these pouches (363).

The first quantitative study about the gastrointestinal uptake of nanoparticles was performed by Nefzger et al. (364). After peroral administration of 130 nm sized ^{14}C-labeled poly(methyl methacrylate) nanoparticles to bile-canulated rats, between 10 and 15% of the radioactivity was found to be gastrointestinally absorbed and appeared in the bile (~58% of the absorbed amount) and in the urine (~42%). The absorbed radioactivity was excreted rather rapidly, namely about 80% within 24 hours and 95% after 48 hours. It has to be taken into consideration that this absorbed and excreted amount of the nanoparticle radioactivity was quantified by determination of the bile and urine radioactivity. Nanoparticles may also be excreted via exocytosis directly through the intestinal wall into the intestinal lumen (365). In addition, elimination can occur via the lungs where the particles traverse into the alveoli after being phagocytized by macrophages (325). After that they travel up the bronchial tree and are then swallowed with the saliva and excreted with the feces or may be reabsorbed from the gut. The significance and quantitative contribution of these pathways to the total excretion pathways taken up gastrointestinally is so far not known (355).

Similar quantitative results were found by Jani et al. (361). They calculated a total gastrointestinal uptake of 50 nm sized polystyrene nanoparticles amounting to 33% of the administered dose after peroral administration to rats within 10 days. The uptake decreased to 26% with 100 nm sized

Table 10 Body Distribution of Polystyrene Particles after Peroral Administration to Rats by Gavage

	Particle size					
Organs	50 nm	100 nm	300 nm	500 nm	1 μm	3.0 μm
Stomach	1.1 ± 0.189	0.65 ± 0.15	0.45 ± 0.07	1.013 ± 0.06	0.27 ± 0.12	1.376 ± 0.08
Small intestine with Peyer's patches and mesentery	12 ± 0.47	3.4 ± 0.21	2.027 ± 0.14	4.28 ± 0.52	1.082 ± 0.12	3.627 ± 0.05
Colon with Peyer's patches, appendix and mesentery	14 ± 1.46	16 ± 1.61	4.32 ± 0.47	6.53 ± 0.75	2.43 ± 0.46	7.53 ± 0.5
Liver	3.3 ± 0.93	3.8 ± 0.73	1.38 ± 0.35	1.38 ± 0.34	0.54 ± 0.03	n.d.
Spleen	0.92 ± 0.22	0.69 ± 0.07	0.21 ± 0.03	0.507 ± 0.03	0.24 ± 0.01	n.d.
Blood 2 mL per animal	2.2 ± 0.39	1.255 ± 0.44	1.1 ± 0.19	n.d.	n.d.	n.d.
Bone marrow 2 mL per animal	NE	0.1 ± 0.011	n.d.	n.d.	n.d.	n.d.
Kidney	0.2 ± 0.05	n.d.	n.d.	n.d.	n.d.	n.d.
Lungs	n.d.	n.d.	n.d.	n.d.	n.d.	n.d.
Heart	n.d.	n.d.	n.d.	n.d.	n.d.	n.d.
Total	33.72 ± 3.71	25.95 ± 3.21	9.487 ± 1.25	13.71 ± 1.2	4.562 ± 0.73	–

NE = not examined; n.d. = not detected.

Source: Reproduced from Ref. (361) with permission of the copyright holder.

particles, 9.5% with 300 nm particles, and 4.5% with 1 μm particles (Table 10). The amount in the liver decreased from 3.3% with the 50 nm particles to 0.5% with the 1 μm particles, that in the spleen from 0.9% to 0.24% respectively. In the blood, 1 μm sized particles were not detectable after peroral administration. The total amount appearing in the blood within 10 days was 2.2% with the 50 nm particles, 1.25% with the 100 nm particles, and 1.1% with the 300 nm particles (361).

The binding of the particles to the gastrointestinal mucosa and, possibly, the uptake may be enhanced by the use of mucoadhesive nanoparticles. Coating of poly(isohexyl cyanoacrylate) with poloxamers 407, 238, and 403 and with poloxamine 908 (366), and preparation of nanoparticles with hydroxypropyl methacrylate (367), were performed for this reason. The mucoadhesion was tested in vitro with rat ileal segments. Especially the coating agents poloxamer 238 and 407 and the poly(hydroxypropyl methacrylate) nanoparticles increased mucoadhesion substantially, leading to 50% stable adhesion of the particles to the ileal segments.

It has been mentioned already that not only peroral uptake of nanoparticles is occurring but that nanoparticles or nanoparticle fragments also may be excreted by exocytosis through the intestinal wall (365).

E. Distribution of Nanoparticles in the Eye After Ocular Application

The rapid tear turnover, lacrimal drainage, and dilution by tears leads to a rapid loss of drug solutions applied topically to the eye (368). The half-life of elimination of such solutions is about 1–3 min (369). As a consequence, only a small amount of the administered drug (1–3%) actually penetrates the cornea and reaches intraocular tissues (370).

Nanoparticles were eliminated much more slowly. The elimination half-life of poly(hexyl-2-cyano-[3-^{14}C]acrylate) nanoparticles from the tear film was about 20 min (371). A different labelling technique, sorption of ^{111}In-oxine to poly(butyl cyanoacrylate), yielded only a half-life of 10 min (372). This shorter half-life, however, seems to be attributable to release of label in the tears (373).

About 1% in total of the administered nanoparticle dose adhered to the cornea and to a larger extent to the conjunctiva for over 6 hours. Although this interaction with these ocular tissues can be mainly attributed to surface mucoadhesion, fluorescence-labeled nanoparticles were observed by laser confocal microscopy to penetrate into the first two cell layers of the conjunctival and to a lesser extent corneal tissue (236). No full tissue penetration or passage via penetration through tight junctions was detectable. The uptake of the particles by the conjunctiva and cornea cells occurred without any visible damage or alteration of the cells.

After the ocular application of the ^{14}C-labeled nanoparticles, a very small amount of radioactivity, probably degradation products, penetrated the cornea and could be found in the aqueous humor (371). The assumption that this amount was due to degradation products is supported by the finding that degradation of the poly(hexyl cyanoacrylate) nanoparticles in the tears is a relatively rapid process leading to 22% degradation within 1 hour.

The increased residence time of nanoparticles in the eye leads to a 30% increase in miosis time (216,217) and to about a two- to threefold increase

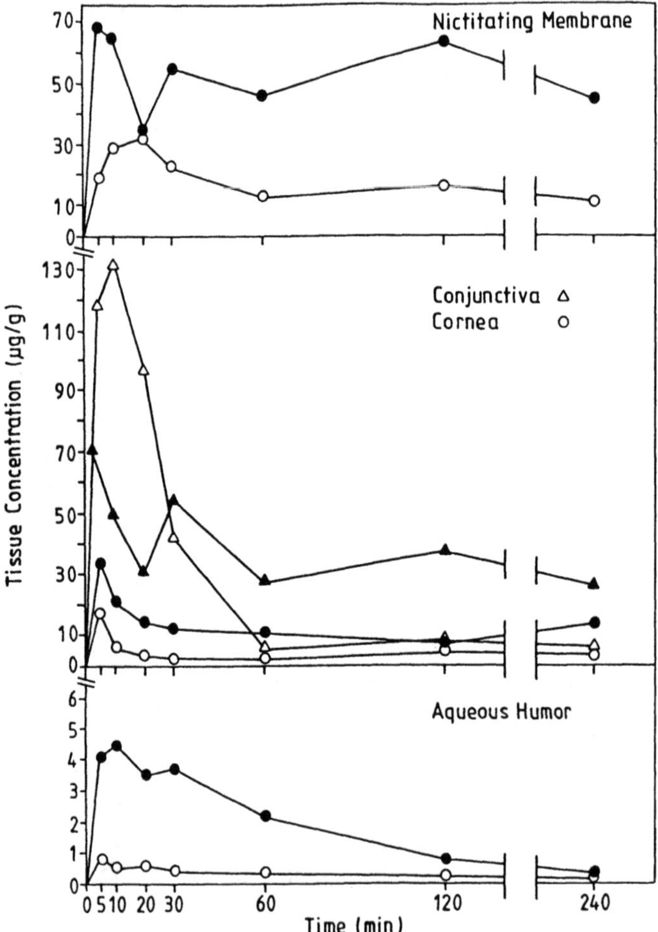

Figure 11 Concentration of ^{14}C-poly(hexyl cyanoacrylate) nanoparticles in healthy (open symbols) or inflamed rabbit eyes (closed symbols).

in intraocular pressure-reducing efficacy of pilocarpine bound to nanoparticles in comparison to a solution of this drug (217).

In chronically inflamed eye tissues, the adhesion of the poly(hexyl cyanoacrylate) nanoparticles was further increased and was about fourfold higher than in normal eye tissues (347) (Fig. 11). This demonstrates the enhanced bioadhesiveness of inflamed tissue to this type of polymer and this delivery system. Consequently, these nanoparticles hold promise as drug carriers for targeted ophthalmic delivery of antiinflammatory or antiallergic drugs (1, 373).

VII. APPLICATION AND USE OF NANOPARTICLES

A. Delivery of Cytostatic Drugs Bound to Nanoparticles

Nanoparticles exhibit a significant tendency to accumulate in a number of tumors after intravenous injection (35,132,163,344,345) (Sec. VI.B). This accumulation may be due to a variety of reasons. As already discussed in Sec. VI.B, one possibility for the accumulation is an enhanced bioadhesiveness of the vasculature in some or in many tumors similarly to the observed increased accumulation in areas of inflammation (346–348). This enhanced bioadhesiveness, that may be the result of inflammatory processes in the tumor, or may be an inherent feature of the tumor, would lead to an increased attachment of the particles to the inner surface of the blood vessels supplying the tumor. In addition to or as an alternative to this process the particles may be endocytosed by the endothelial cells lining these tumor blood vessels. The endothelial uptake of nanoparticles has already been reported (182,334). The enhanced endocytic activity exhibited by many tumors also probably is the explanation for the very pronounced increase in lipid clearance in the Lipofundin® Clearance Test in cancer patients (374). Moreover, the nanoparticles may escape the vasculature through leaky or open blood vessels in the tumor. All the above processes would lead to an enhanced concentration of the nanoparticle-bound cytostatic drugs in the tumor and to an enhanced efficacy of these drugs.

Accordingly, the results obtained with drugs bound to nanoparticles are very promising. The first cytostatic drug bound to nanoparticles was dactinomycin [actinomycin D (153)]. After adsorption of this drug to poly(methyl cyanoacrylate) nanoparticles and intravenous injection to rats bearing a soft tissue carcinoma (S 200), the growth of this tumor was significantly reduced in comparison to an intravenously administered solution of this drug at a concentration of 111 μg/kg body weight. At a concentration of 222 μg/kg the tumor size even decreased considerably after injection of nanoparticles but not with the solution. However, the toxicity was significantly increased with nanoparticles compared to the solution. Drug-free nanoparticles exhibited no effects (153). The increase in efficacy as well as in toxicity may be explained in part by a decreased elimination rate of dactomycin after binding to nanoparticles (154,155,163). Binding of the dactomycin to other cyanoacrylate polymer nanoparticles, i.e., poly(hexyl cyanoacrylate) (Fig. 12) or poly(isobutyl cyanoacrylate), not only increased the efficacy further but also decreased the toxicity down to levels below that of free drug (163).

Similar results were obtained with 5-fluorouracil (196). Binding to poly(butyl cyanoacrylate) nanoparticles significantly increased the efficacy against Crocker sarcoma S 180 but also the toxicity in mice in comparison to free drug after intraperitoneal injection. The efficacy was further enhanced by increasing the polymer-to-drug ratio. The nanoparticles yielded a prolonged persistence of the 5-fluorouracil in comparison to the free drug in all organs examined including the tumor. However, as mentioned, the increased efficacy was combined with a higher toxicity of the drug measured by induced leukopenia, body weight loss, and premature death (196).

Again, binding of the drug 5-fluorouracil to nanoparticles consisting of another polymer, albumin, increased the life span of tumor-bearing mice in comparison to free drug (194). In this case, 5-fluorouracil was incorporated into albumin nanoparticles (193) manufactured by the method developed by Zolle et al. (97) and Scheffel et al. (98) (Sec. III.H). The nanoparticles re-

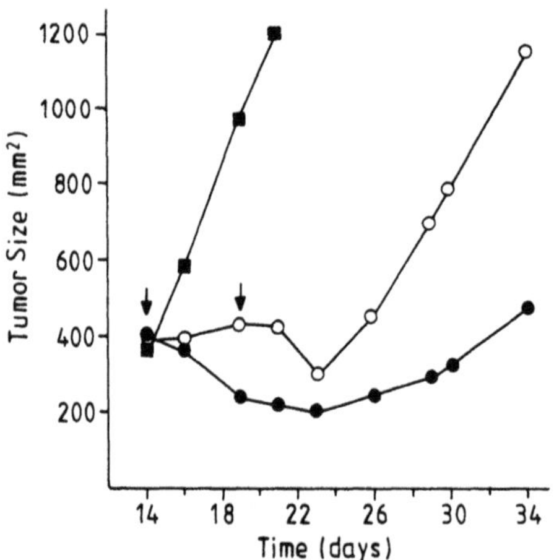

Figure 12 Tumor size of S250 treated with free actinomycin (○), or with actinomycin-D-poly(hexyl cyanoacrylate) nanoparticles (●). Dosage: two intravenous injections (↓) of 222 μg of drug (free or adsorbed) per kg body weight. (■) Represents the tumor size of control rats. (Reproduced from Ref. (163) with permission of the copyright holder.)

leased the drug in vitro for periods exceeding 1 week. The release rate decreased with increasing denaturation temperature during manufacture. After a single intraperitoneal injection, the life span of mice bearing an Ehrlich ascites carcinoma was increased by only 1.2% with free 5-fluorouracil but by 20.5% with nanoparticles containing this drug in comparison to control mice that had obtained 0.9% sodium chloride solution. After multiple-shot administration of nanoparticle-entrapped 5-fluorouracil to ascites-bearing mice, the increase in life span was over 50% compared with a free drug control (195). The therapeutic effect of nanoparticle-entrapped 5-fluorouracil on Ehrlich solid tumor after intratumoral injection was also studied. The nanoparticle-bound drug led to higher 5-fluorouracil levels in the tumor and caused a higher suppression of tumor growth than free drug. Only negligible side effects occurred in acute and chronic toxicity studies with 5-fluorouracil bound to the nanoparticles. No ulcerations and/or loss of hair, indicative of rejection, were observed at various dosage levels tested (194).

The toxicity situation was even more favorable in the case of doxorubicin (adriamycin). Here the toxicity was decreased with all nanoparticles tested. Adsorption of doxorubicin to poly(isobutyl cyanoacrylate) nanoparticles significantly reduced their acute toxicity in comparison to free drug after intravenous injection in mice (162,163). The acute toxicity in these experiments was determined in terms of survival time (Fig. 13) and of body weight as well as individual organ weight loss. Binding to nanoparticles also led to a decreased cardiotoxicity of doxorubicin due to the decreased heart accumu-

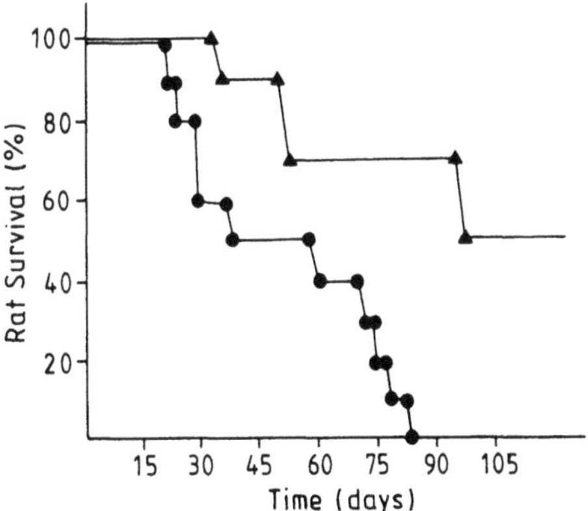

Figure 13 Survival rate of S130 myeloma-bearing rats after administration of free doxorubicin (●) or doxorubicin-poly(isobutyl cyanoacrylate) nanoparticles (▲). Dosage: one intravenous injection of 5 mg of drug (free or adsorbed) per kg body weight. (Reproduced from Ref. (163) with permission of the copyright holder.)

lation of this cardiotoxic drug. The decrease in doxorubicin toxicity due to nanoparticle binding also was observable after histological investigations of various organs (163).

Binding to nanoparticles did not significantly increase the efficacy of doxorubicin in S 130 myeloma compared to free drug (Fig. 14). However, due to the simultaneous decrease in toxicity, the therapeutic index nevertheless was significantly enhanced. In contrast to the S 130 myeloma, in other tumor models the decreased toxicity was combined with an increase in efficacy. In the L 1210 leukemia mouse model, the mean survival time was extended with nanoparticle-bound doxorubicin. Nanoparticles alone did not exhibit any effects (163). In hepatic metastasis-bearing mice, poly(isohexyl cyanoacrylate) nanoparticle-bound doxorubicin showed a much greater efficacy against liver metastases compared to free drug after tail vein injection (165). The metastases originated from the M 5076 reticular cell sarcoma. Empty nanoparticles had no effect; a mixture of empty nanoparticles and free doxorubicin prepared immediately before injection was as effective as free drug. The reduced number of liver metastases were combined with a significant increase in survival time. The observed increase in efficacy against liver metastases probably was due to the increased hepatic concentrations of the drug after binding to nanoparticles (165–167). Autoradiography showed that most of the radioactivity was found in the reticuloendothelial system as soon as a few minutes after intravenous administration of the doxorubicin-loaded nanoparticles. Quantitative determinations by liquid scintillation counting of fresh tissue (spleen, heart, kidneys, liver, lungs, bone marrow) and blood

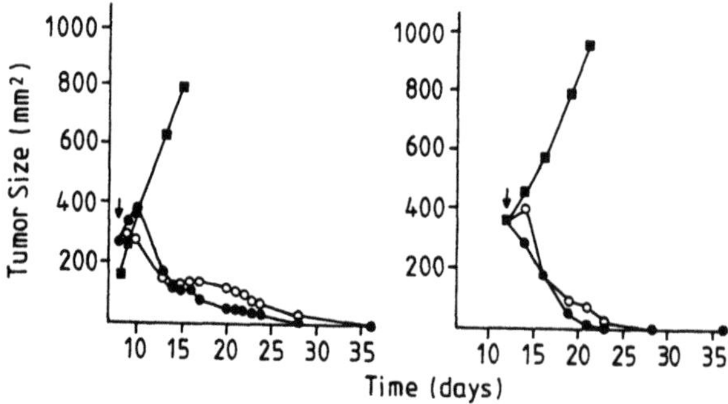

Figure 14 Tumor size of S130 treated with doxorubicin (○) or with doxorubicin-poly(isobutyl cyanoacrylate) nanoparticles (●). Dosage: one intravenous injection (↓) of 5 mg of drug (free or adsorbed) per kg body weight. (■) Represents the tumor size of control rats. The results of two separate experiments are presented. (Reproduced from Ref. (163) with permission of the copyright holder.)

samples confirmed these observations. When the drug was linked to nanoparticles, doxorubicin blood clearance was reduced during the first few minutes after injection, whereas heart and kidney concentrations were substantially decreased (166). Histological examinations using transmission electron microscopy revealed that the hepatic tissues represented a considerable reservoir of the drug when injected associated with nanoparticles. Accumulation of these particles with bound drug in the Kupffer cells enabled a prolonged doxorubicin delivery to the neoplastic tissue (167).

Binding of doxorubicin to poly(methacrylic) nanoparticles by adsorption decreased the clearance of this drug after intravenous injection to rabbits (135). Again the particle-bound drug accumulated in the liver. One hour after the injection of doxorubicin-bound nanoparticles, the concentration of doxorubicin in the liver was found to be 75% higher, and the levels of both doxorubicin metabolites, doxorubicinol and doxorubicinone, were twofold lower than those obtained with injected free drug. The liver accumulation of the drug was due to the accumulation of the nanoparticles as shown by ^{111}In- and ^{125}I-labeling (159) and by in vitro uptake studies with cultured hepatocytes (160). The cultured hepatocytes endocytosed the doxorubicin-loaded poly(methacrylic) nanoparticles as shown by electron microscopy. This led to an enhancement of the doxorubicin concentrations in these cells by 64% and 25% for initial concentrations of 50 and 500 ng doxorubicin/mL of medium respectively.

In preliminary clinical trials in four hepatoma patients, poly(methacrylic) nanoparticles loaded with doxorubicin achieved higher doxorubicin plasma levels and a prolonged elimination half-life in the β-phase of elimination in comparison to free drug. In addition, the administration of doxorubicin-loaded nanospheres led to a reduction of the distribution volume and of the total clearance of doxorubicin, probably owing to the passive drug targeting to the liver. The administration of nanoparticle-bound doxorubicin to patients with hepatomas also led to a reduction of the toxic side effects attributed to doxorubicin (nausea, vomiting, cardiomyopathy, medullar depression) and to an apparent improved therapeutic activity (161,375).

Slightly larger bovine serum albumin particles (1.44 μm) containing doxorubicin were prepared by heat denaturation of the albumin in a cottonseed oil emulsion and were compared to free drug in the AH 7974 liver metastasis model in rats. Intraportal vein injection of free drug did not lead to a significant increase in survival time (10 to 16 d) compared to the controls. Binding to nanoparticles, however, increased the median survival time by 150%. Two rats that had obtained particle-bound drug and were not included in this calculation of the survival time even survived for over 60 days after a single administration of these particles. Empty particles again had no effect.

Mitoxantrone was bound to poly(butyl cyanoacrylate) nanoparticles either by incorporation (polymerization in presence of the drug) or by adsorption (addition of the drug after polymerization) (132). The portion of mitoxantrone bound to the particles was about 15% of the initial drug concentration with the incorporation method and about 8% with the adsorption method. The nanoparticle formulations were tested in leukemia- or melanoma-bearing mice after intravenous injection. Efficacy and toxicity of mitoxantrone nanoparticles were compared with a drug solution and with a mitoxantrone-liposome formulation (small unilamellar vesicles with a negative surface charge). Furthermore, the influence of a coating surfactant, poloxamine 1508, which

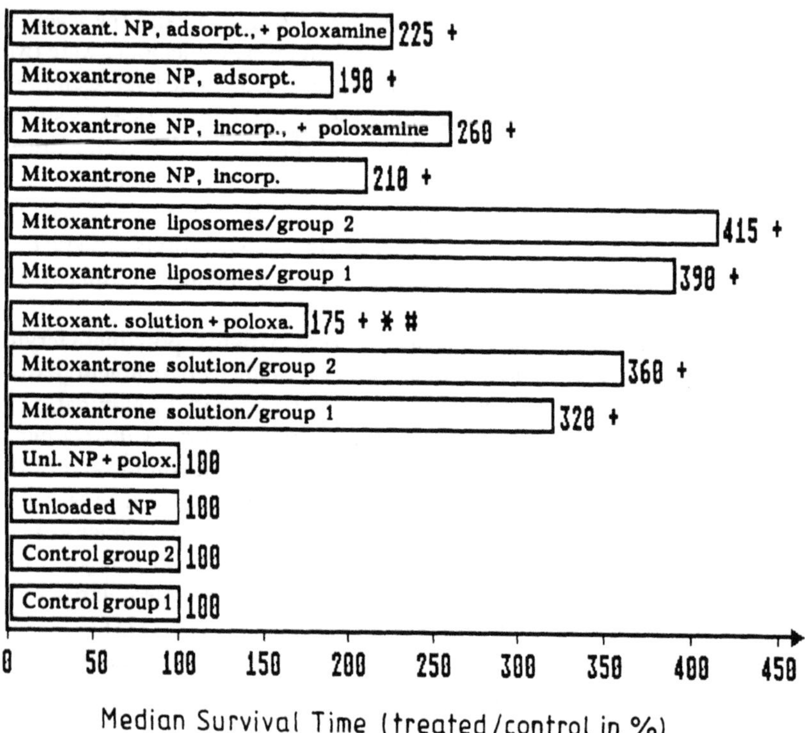

Figure 15 Effects of the different mitoxantrone formulations on the survival time in murine P388 leukemia. Values are median survival times of treated animals in percentages of median survival time of untreated control animals. (NP) nanoparticles; (+) significantly different from control groups (U-test); (*) significantly different from mitoxantrone solution (U-test); (#) significantly different from mitoxantrone liposomes (U-test). (Reproduced from Ref. (132) with permission of the copyright holder.)

has been shown to change the body distribution of other polymeric nanoparticles (323), was investigated. It was shown that poly(butyl cyanoacrylate) nanoparticles and liposomes influenced the efficacy of mitoxantrone in cancer therapy differently: liposomes prolonged the survival time in P 388 leukemia (Fig. 15), whereas nanoparticles led to a significant tumor volume reduction of the B 16 melanoma (Fig. 16). Neither nanoparticles nor liposomes were able to reduce the toxic side effects caused by mitoxantrone, namely leucocytopenia (132).

The use of magnetically targeted nanoparticles as carriers for cytostatic drugs was already discussed in chapter Sec. I.B. Widder et al. (171,334) were able to target magnetic albumin nanoparticles with bound doxorubicin to Yoshida sarcoma tumors located in the tails of rats. By placing a magnet at the tumor site during the infusion of the nanoparticles, total remission of

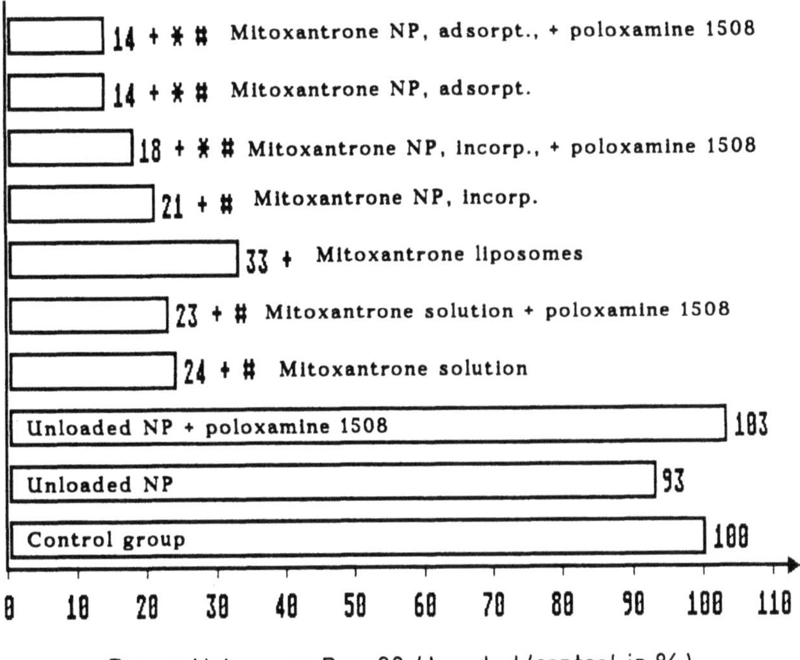

Figure 16 Tumor size reduction of the murine B16 melanoma achieved by the different mitoxantrone formulations. Values are the tumor volumes of the treated animals in percentages of the tumor volume of untreated control animals on day 22. For symbols see Fig. 15. (Reproduced from Ref. (132) with permission of the copyright holder.)

the tumors was obtained with a single dose. Endocytosis of these nanoparticles kept at the tumor site by the magnetic field was observable already after 10 min and led to an almost total uptake of the particles after 24 hours. Particles were still observable within tumor cells as late as 72 hours after administration (334).

These results clearly demonstrate that nanoparticles represent very promising carrier systems for the targeting of cytostatic drugs to tumors. The targeting properties of these carriers may even be enhanced by magnetic targeting or the binding of tumor-specific antibodies. However, even without the latter features, nanoparticles were shown to accumulate in a variety of primary tumors and metastases and to enhance efficacy and decrease toxicity of a number of cytostatics. In general, albumin as well as more slowly degrading polyacrylic derivatives such as poly(methacrylate) derivatives and poly(hexyl cyanoacrylate) nanoparticles were shown to be less toxic than the more rapidly degrading polymers such as poly(methyl cyanoacrylate), poly(butyl-), and poly(isobutyl cyanoacrylate). However, it has to be kept in mind that the slow biodegradability and the resulting slow body elimination of the poly(methacrylates) (287,288,350) and of the poly(hexyl cyanoacrylates) (376) may limit their applicability to single shot administration. Nevertheless, albumin on the one hand, and with some limitations the poly-(butyl cyanoacrylates) and probably the polylactic polymers on the other, will enable a considerable improvement of the therapeutic index in the treatment of an important number of tumors. In addition, the finding that nanoparticles are more effective in one type of tumor, whereas liposomes containing the same drug are more effective in another, is very important. It shows that for optimal therapy the treatment of different tumors may require different delivery systems.

B. Delivery of Antiinfective Drugs Bound to Nanoparticles

The treatment of a number of intracellular infections with chemotherapeutic agents is very often impeded or impossible because of the relative inability of these agents to penetrate infected cells (377–379). Barriers to the penetration of these agents into cells include strong protein binding, an unfavorable lipid-water distribution coefficient, an unfavorable pH-gradient between different cellular compartments, and the existence of active transport pump mechanisms that prevent the accumulation of sufficient antibiotic concentrations in the interior of the infected cells (380,381).

Cells that are frequently infected and are inaccessible to a number of relevant antibiotics include phagocytic cells (382). As mentioned in previous chapters, colloidal carriers are easily taken up by these cells and find access to the lysosomes. Therefore drug carriers such as liposomes and nanoparticles may be useful antibiotic delivery systems for the treatment of infections of these cells.

Target cells for this type of drug delivery include fixed macrophages in the liver (Kupffer cells) and in the spleen as well as circulating monocytes. In the lung, colloidal drug carrier loaded monocytes subsequently migrate to the alveoli to become alveolar macrophages. By manipulation of the surface properties of the carriers as well as their wall or matrix composition, not only the targeting properties but also their intracellular degradation and consequently the drug release and delivery mode and rate may be monitored.

A number of antiinfective drugs were bound to nanoparticles including amikacin (139–141), amphotericin B (142), ampicillin (41,143–147), dehydroemetine (157), gentamycin (144), metronidazole (213), miconazole (142), primaquine (118,120,213,218,219), and vidarabin (138). For most of these compounds only their method and degree of binding and their in vitro release was reported.

Very important and interesting in vitro as well as in vivo studies were performed with ampicillin (144–146). Ampicillin and also gentamycin were bound to poly(isobutyl cyanoacrylate) and poly(isohexyl cyanoacrylate) nanoparticles by polymerization of the cyanoacrylates in the presence of the drugs (144). The maximum carrier capacity was 184 μg of ampicillin/mg of polymer for poly(isobutyl cyanoacrylate) and 256 μg of ampicillin/mg polymer for poly(isohexyl cyanoacrylate) nanoparticles. In the case of gentamycin, 70% of a 30 μg/mL solution was bound to the nanoparticles.

The release rate of these drugs was studied both with and without esterases. For poly(isobutyl cyanoacrylate) nanoparticles, the drug release wase enhanced in esterase-containing medium. In contrast, poly(hexyl cyanoacrylate) nanoparticles showed no enhancement caused by enzymatic degradation. Both drugs retained their in vitro antimicrobial activity after binding to the particles.

The effectiveness of nanoparticle-bound ampicillin was demonstrated in mice against *Listeria monocytogenes* (145) as well as against salmonellosis (146). Nanoparticle-bound ampicillin at a concentration of 2.4 mg was more effective than 48 mg of free ampicillin against *L. monocytogenes*. The liver was free of bacteria 7 d after the beginning of the treatment with nanoparticle-bound drug. The bacterial count, however, increased again after this time (Fig. 17). Nevertheless, the therapeutic index of ampicillin in the liver increased by a factor of 20. In the spleen, only a slight decrease in bacterial counts was observed, but again the nanoparticle-bound ampicillin was more effective than free drug (Fig. 18). It is interesting to note that while nanoparticles were more effective than liposomes in reducing the bacterial counts of *L. monocytogenes* in the liver, liposomes decreased the bacterial counts more effectively in the spleen (383,384). This effect was very likely due to

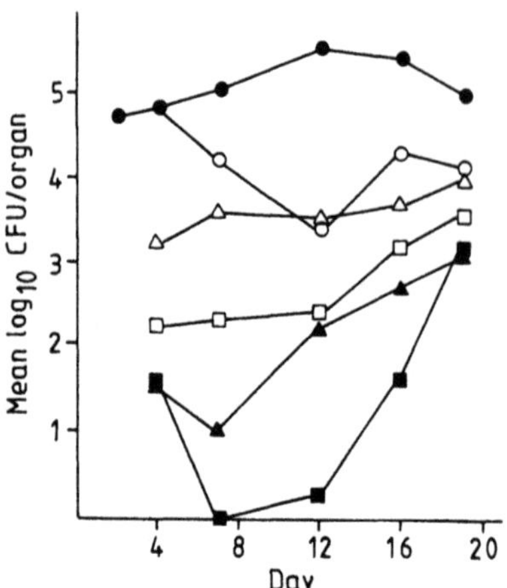

Figure 17 Effects of nanoparticle-bound ampicillin, free ampicillin, and empty nanoparticles on counts of *Listeria monocytogenes* in the liver of nude mice at various times after bacterial inoculation. (●) untreated controls, (○) empty nanoparticles, (△) free ampicillin 2.4 mg, (▲) free ampicillin 48 mg, (□) nanoparticles + free ampicillin 2.4 mg, (■) nanoparticle-bound ampicillin 2.4 mg. (Adapted from Ref. (145).)

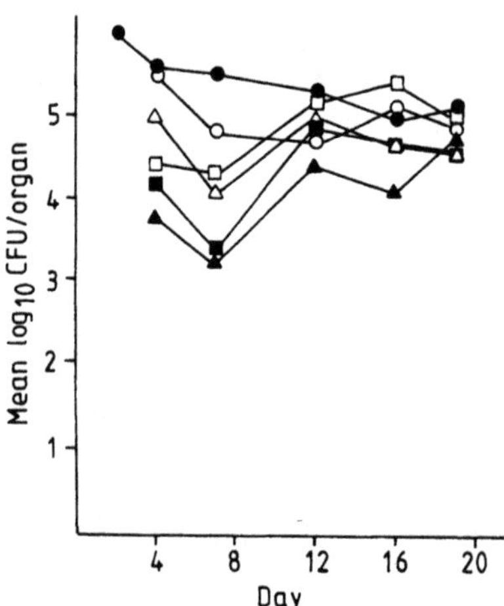

Figure 18 Effects of nanoparticle-bound ampicillin, free ampicillin, and empty nanoparticles on counts of *Literia monocytogenes* in the spleen of nude mice at various times after bacterial inoculation. (●) untreated controls, (○) empty nanoparticles, (△) free ampicillin 2.4 mg, (▲) free ampicillin 48 mg, (□) nanoparticles + free ampicillin 2.4 mg, (■) nanoparticle-bound ampicillin 2.4 mg. (Adapted from Ref. (145).)

an enhanced delivery of ampicillin by these carrier systems to the respective organs as a result of the different surface properties of these carriers: as noted earlier (322) (Sec. VI.A), the surface properties of colloidal carriers exhibit an important influence on their body distribution.

The nanoparticle-bound ampicillin showed a similar effectiveness against salmonellosis, i.e., 0.8 mg of the nanoparticle-bound drug was as effective as 32 mg of free drug (146,385). However, neither the liver nor the spleen was bacteria-free in any of the surviving mice even after 60 d from the start of the treatment, although drug concentration measurements revealed significantly enhanced drug levels in the liver and spleen in comparison to injection with free drug up to 24 hours. Nevertheless, after injection of C 57 black mice with *Salmonella thyphimurium* all of those mice survived that had obtained 0.8 mg or 2.4 mg ampicillin bound to nanoparticles or 96 mg free ampicillin (384) (Fig. 19). Incorporation of small doses of ampicillin into liposomes yielded only a survival of 50 or 60% of the mice after 20 days.

The improvement of ocular delivery of amikacin for the treatment of ocular infections by binding to poly(butyl cyanocrylate) nanoparticles was reported by Losa et al. (139). Increased corneal and aqueous humor concentrations of this drug in comparison to free drug and to other formulations were observed. More details about this study will be given in Sec. VII.E.

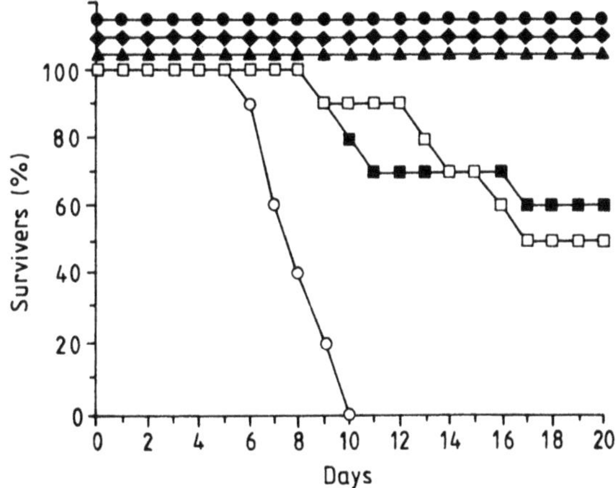

Figure 19 Survival of C57 black mice ($n = 10$) after infection with *Salmonella thyphymurium* and treatment with the following preparations: (▲) free ampicillin 96 mg; (◆) ampicillin 0.8 mg bound to nanoparticles; (●) ampicillin 2.4 mg bound to nanoparticles; (■) ampicillin 0.8 mg bound to liposomes; (□) ampicillin 2.4 mg bound to liposomes; (○) controls. (Adapted from Ref. (384).)

Visceral leishmaniasis (kala-azar), one of the major tropical diseases, is caused by *Leishmania donovani*, an intracellular parasite of the reticuloendothelial system. Since nanoparticles accumulate in the reticuloendothelial system, they hold promise as drug carriers for the treatment of leishmaniasis. Dehydroemetine is one of the drug candidates for this treatment but has some side effects involving the heart (386). Dehydroemetine poly(isohexyl cyanoacrylate) nanoparticles were produced by Fouarge et al. (157). The acute toxicity of dehydroemetine in mice was significantly reduced after intravenous administration. The nanoparticle-bound drug accumulated rapidly in the liver and in the spleen, while the cardiac concentration was reduced in comparison to free drug.

Another drug that may be used for the treatment of leishmaniasis is primaquine. This drug was bound to a number of different nanoparticles (118, 120,213,218,219). Primaquine when bound covalently via amino acid spacers to polyacryl starch nanoparticles was shown to kill *Leishmania donovani* in cultured mouse peritoneal macrophages (120). The drug-carrier complex did not kill free promastigotes in suspension but was effective against amastigotes in the macrophages. Stjärnkvist et al. (120) conclude from these results that lysosomal processing of the drug-carrier complex is necessary in order to liberate the pharmacologically active drug. Binding of primaquine to poly(isohexyl cyanocrylate) nanoparticles yielded a significant reduction of the acute toxicity in NMRI mice (218). The binding of this drug to poly(diethylmethylene maloate) nanoparticles achieved an increased life span of NMRI mice infected with *Plasmodium berghei* compared to free drug (219).

Empty isobutyl cyanoacrylate nanoparticles exhibit antiparasitic activity in vitro and in vivo in mice against *Trypanosoma brucei brucei* (387). However, the mechanism of action presently is not totally explained. It is possible that peroxide production occurring during phagocytosis of the nanoparticles is responsible for their antiparasitic activity: Gaspar et al. (388) assessed the effect of poly(isohexyl cyanoacrylate) nanoparticles on the induction of the respiratory burst in a macrophage-like cell line (J774G8) in noninfected macrophages and in macrophages infected with amastigotes of *Leishmania donovani infantum* by measuring nitroblue tetrazolium reduction and hydrogen peroxide production. Phagocytosis of the nanoparticles led to a respiratory burst, which was more pronounced in infected than in uninfected macrophages. The production of reactive oxygen intermediates associated with the respiratory burst was inhibited by addition of superoxide dismutase and catalase to the cell suspensions. The addition of catalase to the culture medium together with poly(isohexyl cyanoacrylate) nanoparticles significantly reduced the antileishmanial activity of the carriers. Moreover, poly(isohexyl cyanoacrylate) nanoparticles did not induce interleukin-1 release by macrophages. No significant difference in antileishmanial activity could be detected between nanoparticles of five poly(alkyl cyanoacrylates) with differing alkyl side chains, suggesting that the main degradation products of the cyanoacrylates are not involved in their antileishmanial action. Gaspar et al. (388) conclude that the antileishmanial action (and possibly other antiparasitic action) of the poly(alkyl cyanoacrylate) nanoparticles results from the activation of the respiratory burst in macrophages.

Cells of the reticuloendothelial system frequently are infected by both strains of the human immunodeficiency virus, HIV-1 and HIV-2 (389–391). For monocytes/macrophages the virus seems to be much less cytocidal than for T-cells. Therefore, cells of the monocyte/macrophage lineage most likely play an important role in the immunopathogenesis of the HIV infection, by serving as a reservoir for the virus and its dissemination throughout the body and brain (392,393).

Because nanoparticles are easily phagocytosed by macrophages they represent promising drug delivery systems for the treatment of human immunodeficiency virus persisting in these cells. Schäfer et al. (394,395) have shown in macrophage cultures that human macrophages easily phagocytose different types of polyacrylic and albumin nanoparticles. The best uptake was seen for the hydrophobic poly(methyl methacrylate) nanoparticles, followed by large (1.5 μm) human serum albumin particles produced by the method of Scheffel et al. (98), poly(butyl cyanoacrylate) nanoparticles, small (200 nm) human serum albumin nanoparticles produced by the method of Marty et al. (107,108), and poly(hexyl cyanoacrylate) nanoparticles. The uptake was significantly altered by coating with surfactants.

The most important finding of this study, however, was the observation that HIV-infected macrophages show a higher phagocytotic uptake of nanoparticles than noninfected macrophages (Fig. 20). This higher amount of phagocytosis is probably due to an activated state of these infected cells (396) and may allow a preferential phagocytosis of drug-loaded nanoparticles, resulting in a targeted delivery of drugs to these cells. In addition, the observation of good phagocytotic activity of macrophages from HIV-infected patients argues that successful drug targeting should be possible. However, in two of seven AIDS patients, lower phagocytotic activity was observed, while in three of seven a higher activity was found, which was compatible with the enhanced activity seen for macrophages infected with HIV in vitro (Table 11). Further studies are needed to show whether AIDS patients fall into two different groups with enhanced or reduced phagocytotic activity (395).

In summary, nanoparticles hold great promise for targeting antiviral substances to infected cells of the reticuloendothelial system.

C. Peptide Drugs

Over the past years peptide drugs have been gaining more and more importance. In many cases these peptides are quite efficiently bound to nanoparticles (148,337,339).

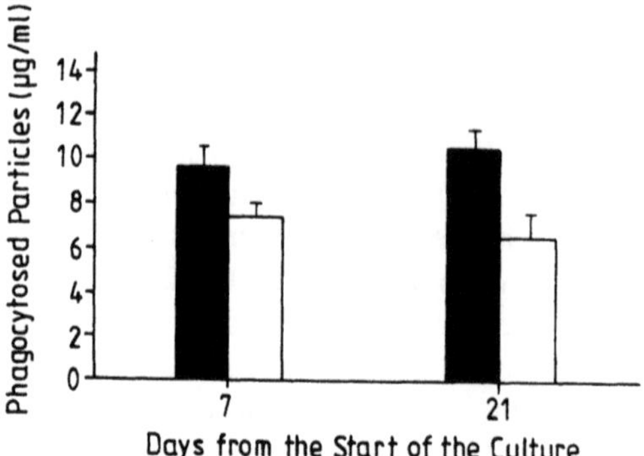

Figure 20 Influence of HIV infection on phagocytosis of poly(butyl cyanoacrylate) nanoparticles by human MAC. MO/MAC were infected with HIV-1 at day 1 after start of culture. At day 7 or day 21 the nanoparticles (200 nm diameter) were added at a final concentration of 0.5 mg/mL to the infected cultures and incubated for 6 h. Mean values of three parallel cultures are presented. Bars indicate standard deviation. Light columns = uninfected macrophages; dark columns = HIV-1-infected macrophages. (Adapted from Ref. (395).)

Table 11 Phagocytotic Activity of Macrophages Isolated from HIV-Infected Patients

Patient no.	State	CD4-positive cells/μL	CD8-positive cells/μL	Phagocytosis of PBCA particles by MAC (μg/mL ± SD)[a]
1	Uninfected control	ND[b]	ND[b]	8.42 ± 0.86
2		ND	ND	7.91 ± 0.70
3		ND	ND	7.84 ± 0.44
4		ND	ND	9.98 ± 0.51
5	CDC/II/III	505	603	7.02 ± 1.02
6		472	1935	6.25 ± 0.97
7		362	1729	5.87 ± 0.72
8		627	493	5.86 ± 1.88
9		669	698	4.29 ± 0.15
10		709	902	3.75 ± 0.15
11		672	797	5.62 ± 0.33
12	CDC IV	504	875	2.12 ± 0.34
13		36	1022	1.71 ± 0.20
14		44	453	11.22 ± 0.42
15		343	1090	12.81 ± 1.61
16		191	722	8.10 ± 0.75
17		74	699	13.08 ± 1.33
18		56	774	10.30 ± 0.83

[a]SD, standard deviation of three independent cultures from each patient.
[b]Not determined.
Source: Reprinted from Ref. 395 with permission of the copyright holder.

Insulin can be bound by surface adsorption to poly(alkyl cyanoacrylate) nanoparticles manufactured by emulsion polymerization (204), or it can be encapsulated into poly(isobutyl cyanoacrylate) nanoparticles (205–208) according to the method of Al Khouri Fallouh et al. (77,78). In the latter case, the peptide phase was admixed directly with the lipophilic phase, consisting of miglyol (1 mL), isobutyl cyanoacrylate (0.125 mL), and ethanol (25 mL). This lipophilic phase was then added with a syringe to 50 mL of an aqueous 0.25% poloxamer 188 solution under stirring, leading to the formation of nanocapsules. In the resulting nanocapsule preparation 55% of the insulin was encapsulated; the rest was present in the formulation in free form. After subcutaneous injection in rats both methods, adsorption as well as encapsulation, achieved a significant prolongation of the blood glucose level reduction compared to free insulin. The prolongation was especially pronounced with the nanocapsules in normally fed and diabetic rats (Fig. 21). The duration of the glucose level reduction increased with increasing doses and lasted for over 24 hours with 50 U insulin/kg body weight after encapsulation into nanocapsules (the 50 U represent total-encapsulation plus free insulin) but only for 8 hours with free insulin (206).

After peroral administration, the nanoparticles with merely surface adsorbed insulin did not reduce the blood glucose levels (204). However, the peroral administration of insulin (12.5, 25, and 50 U/kg) encapsulated in poly(isobutyl cyanoacrylate) nanoparticles decreased glycemia in diabetic rats fasted overnight before administration of the nanoparticles by 50–60% by day 2 (206, Fig. 22). Peroral administration of nonencapsulated insulin had no

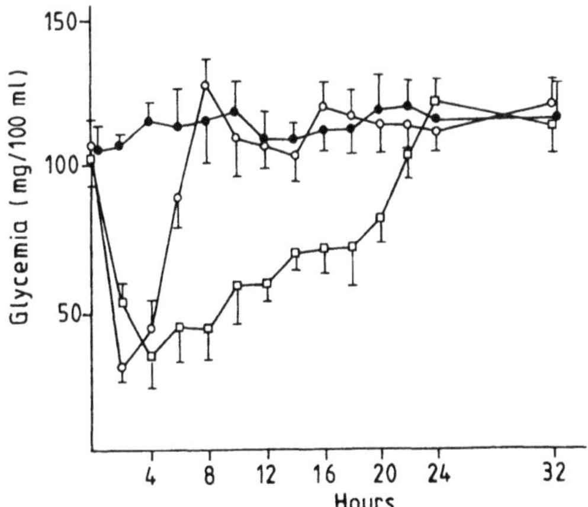

Figure 21 Effect of single subcutaneous administration of (○) nonencapsulated and (□) encapsulated insulin on glycemia in normally fed rats. Dose 25 U total insulin/kg body weight; (●) controls. (Adapted from Ref. (206).)

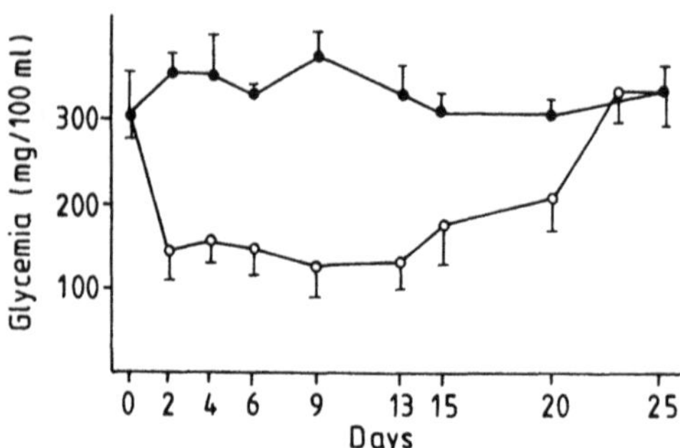

Figure 22 Effect of single intragastric administration of insulin nanocapsules on glycemia in diabetic rats fasted overnight. (○) 50 U total insulin/kg body weight; (●) controls. (Adapted from Ref. (206).)

effect. In fed diabetic rats insulin containing nanoparticles yielded a 25% reduction of the blood glucose levels only at a dose of 100 U/kg. The same dose reduced hyperglycemia induced by an oral glucose load of 2.5 g/kg by 50% in normal rats (206). These results show that encapsulation of insulin into nanoparticles in contrast to nanoparticle surface adsorption was able to protect insulin from degradation by gastric fluids. The nanoparticles then probably were taken up by the gut walls as discussed in detail in Sec. VI.D. The long-term hypoglycemic effect of oral insulin nanoparticles may be attributed in part to the formation of a depot in the gut wall, the Peyer's patches, and the draining lymphatic system and/or possibly other compartments of the body, as for instance the liver.

Grangier et al. (197) and Gautier et al. (198) studied the binding of growth hormone releasing factor (GRF) to poly(alkyl cyanoacrylate) nanoparticles. The moment during polymerization of the nanoparticles at which GRF was added to the polymerization medium was found to be a critical factor in avoiding the degradation of GRF as well as avoiding the covalent binding of the peptide with the polymer. Drug release experiments suggested that the liberation of GRF from nanoparticles was the result of bioerosion of the polymeric material rather than passive diffusion of the peptide through the polymer. Consequently, drug release appeared to depend on the esterase content of the incubation medium. In addition, the drug release was dependent on the alkyl chain length of the poly(alkyl cyanoacrylate) polymer (197).

Poly(isohexyl cyanoacrylate) nanoparticles with GRF were able to maintain rather constant plasma levels of this drug for over 24 hours after subcutaneous administration. In contrast, with the same dose of free GRF an about 7 times higher plasma peak appeared after 2 min that decreased very rapidly. After 100 min, no GRF was detectable any more. In macroautoradiographic studies with ^{14}C-labeled nanoparticles it was shown that after 24 hours the majority of the subcutaneously injected radioactivity was found at the injection site. At this time electron microscopy revealed that also undegraded nanoparticles still remained at this site (198). This finding demonstrates that poly(alkyl cyanoacrylate) or other types of nanoparticles may act as injectable sustained release systems for peptide drugs.

Cyclosporin was encapsulated into poly(isohexyl cyanoacrylate) nanocapsules (152) using a modification of the encapsulation method developed by Al Khouri Fallouh et al. (77–78). No cyclosporin was released from these capsules during storage in water. In plasma about 40% of the encapsulated cyclosporin diffused rapidly out of the capsules, the drug most likely interacting and associating with the fatty components of the plasma such as lipoproteins, etc. The initial rapid release was followed by a slower release of about 20%/3 hours (152).

Antisense oligonucleotides (pd$(T)_{16}$ and d$(T)_{16}$) were bound to poly(isobutyl cyanoacrylate) (149) and poly(isohexyl cyanoacrylate) (149,397) nanoparticles using tetraphenylphosphonium chloride (149) or cetyltrimethylalkylammonium bromide as anchoring moieties (397). The oligonucleotide adsorption in these cases was mediated by the formation of ion pairs between the negatively charged phosphate groups of the nucleic acid chain and the hydrophobic cations. These ion pairs then were bound to the nanoparticles probably by hydrophobic binding. In the case of using tetraphenylphosphonium chloride as the counterion for pd$(T)_{16}$, poly(isohexyl cyanoacrylate) was a more efficient sorptive than poly(isobutyl cyanoacrylate) (149).

The uptake of the oligonucleotides by the human histiocytic lymphoma cell line U 937 was increased eightfold by binding to poly(isohexyl cyanoacrylate) nanoparticles after 24 hours incubation. In addition, binding to nanoparticles greatly improved both the intracellular and the extracellular stability of the oligonucleotides. The appearance of previously intact nanoparticle-bound oligonucleotides in the cell nuclear fraction and cytosolic fractions after nanoparticle-mediated uptake suggests that part of the internalized oligonucleotides escape from the endosomal/lysosomal compartments (149,397) in which nanoparticles accumulate (191).

D. Peroral Administration of Drugs by Nanoparticles

So far only a few drugs bound to nanoparticles have been investigated for peroral use. The first experiments with peroral administration of a nanoparticle-bound drug, vincamine, were carried out by Maincent et al. (223–224). These authors administered poly(hexyl cyanoacrylate) nanoparticles of a size of about 230 nm with bound vincamine to rabbits and determined the bioavailability. About 82 mg of the drug was sorbed by 1 g of nanoparticles. The relative peroral bioavailability of the nanoparticle preparation in rabbits was significantly increased and amounted to 162% in comparison to a solution of the drug. The respective bioavailabilities in comparison to an intravenous injection were 36% for the nanoparticle formulation and 22% for the drug solution.

A much more impressive increase in bioavailability was achieved by Beck et al. (398) for avarol (Fig. 23). Avarol is a practically water-insoluble drug that can be administered perorally in form of a Solutol® solution. In rats peroral administration by gavage of an avarol solution in Solutol® led to a sharp maximum in blood levels after 1.5 hours. After that time blood levels decreased very rapidly. Binding to nanoparticles by polymerization in presence of the drug (100 mg avarol/100 mL) in a medium consisting of a 1:1 mixture of ethanol 96% and 0.2 N nitric acid and containing either 0.6% (formulation 1, Fig. 23) or 1.2% poloxamer (formulation 2, Fig. 23) and 1% sodium sulfate yielded an eight- to ninefold increase in peroral bioavailability. Both nanoparticle formulations behaved very similarly. A broad maximum that was about 1.5-fold higher than that of the solution occurred between 4 and 12 hours after administration. After this time the blood levels decreased slowly with a $t_{1/2}$ of about 36 hours. Investigations of the blood

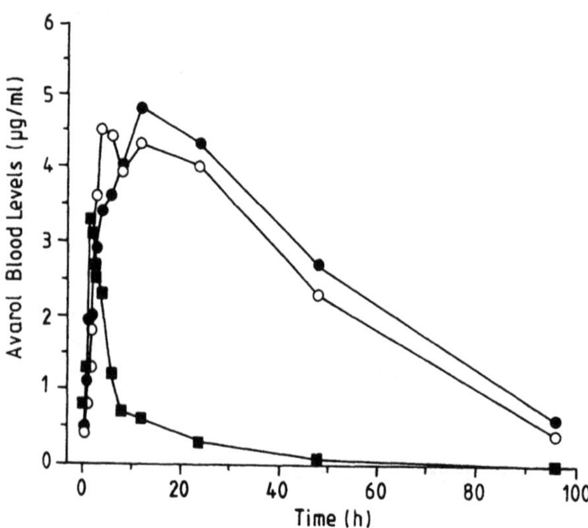

Figure 23 Avarol blood levels in rats after peroral administration of the drug in Solutol® solution (■), or bound to nanoparticles preparation 1 (●) or preparation 2 (○).

samples by differential centrifugation and ultracentrifugation demonstrated that the drug levels in the blood observed in this study were attributable to free avarol, not to nanoparticle-associated drug.

As mentioned and discussed in more detail in Sec. VII.D, insulin-containing nanocapsules induced a significant hypoglycemic effect for several days in fasted (up to 20 d with 50 U/kg) and fed diabetic rats (up to 6 d with 100 U/kg) (Fig. 22). This long duration of the hypoglycemic effect probably can be attributed to the prolonged interaction of the nanoparticles with the biological environment and the resulting release rate of the insulin.

Hydrocortisone was incorporated into albumin particles of a size of 3.2 ± 1.7 μm, slightly above the nanoparticle size range, by Alpar et al. (399), using the method of Gallo et al. (14). Particles in individual batches contained 12–15% hydrocortisone. These particles were administered orally to rats 1 hour before artificial inflammation of the paws with carrageenan. The particles were later found in inflammatory exudates and tissues. The hydrocortisone-loaded microspheres suppressed the carrageenan-induced paw inflammation at lower doses than free orally administered drug. At a dose level of 1.0 mg/kg, a 61% inhibition of inflammation was obtained with the nanoparticle formulation and only a 44% inhibition with the free drug. At lower doses, 0.1 mg/kg and 0.078 mg/kg, the inhibition was 50% and 49.5% respectively with the nanoparticles and 26% and 16% respectively with the free drug.

E. Ophthalmic Delivery with Nanoparticles

The first nanoparticulate system with pilocarpine that was developed was a cellulose acetate hydrogen phthalate (CAP) pseudolatex formulation introduced by Gurny et al. (400,401). In this formulation, the pH of the nanoparticulate pseudolatex was maintained at 4.5 during storage. After instillation into the eye, the tear fluids rapidly raised the pH of the system to 7.2, causing dissolution of the polymeric particles. The resulting polymer solution had a very high viscosity that prevented a rapid washout of the formulation. As a result, the action of pilocarpine was considerably prolonged. Consequently, the miosis time as well as the AUC of the miosis versus time curve were increased by 50% in comparison to a solution. The pilocarpine was present in the pseudolatex suspension in form of the hydrochloride salt. A binding of the drug to the CAP pseudolatex was not necessary because the prolongation of the action in this system was not caused by a prolonged release of the pilocarpine from the nanoparticles but by a decreased elimination rate from the eye due to the formation of a very viscous solution of the polymer after instillation.

The pseudolatex preparation was considerably more effective in rabbits than ophthalmic rods or a thermosetting gel (402). However, a solution of

0.125% hyaluronic acid, a strong bioadhesive substance, induced a significantly prolonged elimination of ^{99m}Tc-labeled sulfur colloids and of pilocarpine from rabbit eyes in comparison to the pseudolatex formulation (403). Ibrahim et al. (404) investigated and compared several pseudolatex systems including cellulose acetate phthalate, polyvinyl acetate phthalate, hydroxypropyl acetate phthalate, two copolymers of vinyl acetate and crotonic acid, a butyl monoester of poly(methylvinyl ether/maleic acid), a copolymer of methacrylic acid and methylmethacrylate, and carboxymethylethylcellulose. The best polymers for the formation of pH-sensitive latexes/nanoparticles were cellulose acetate phthalate and hydroxypropyl acetate phthalate (404).

Binding of pilocarpine to poly(butyl cyanoacrylate) nanoparticles by adsorption (216) or by partial incorporation (217) enhanced the miotic response by about 22% (216) or 33% (217). In the latter study by Diepold et al. (217) involving partially (15%) incorporated pilocarpine into poly(butyl cyanoacrylate) nanoparticles, the aqueous humor levels as well as intraocular pressure-lowering effects of the nanoparticles were measured in addition to the miotic response and compared to a solution. This study revealed very interesting effects obtained with the nanoparticles (Fig. 24). Although a considerable enhancement of the AUC, determined as the pilocarpine aqueous humor concentrations versus time, was observable with the nanoparticles (AUC $\cong$ 200 μg mL^{-1} min) in comparison to the solution (AUC $\cong$ 111 μg mL^{-1} min), the time period with higher pilocarpine concentrations with the nanoparticle preparation in the aqueous humor lasted for less than 60 min. The miosis time, on the other hand, was prolonged from 180 to 240 minutes in comparison to the solution. The intraocular pressure-lowering effect was investigated in three rabbit models, and its prolongation was even more pronounced. In these models the intraocular pressure was increased artificially because pilocarpine has little effect on the pressure in normostatic eyes. The three models used were the water-loading model, the alpha-chymotripsin model, and the betamethasone model. The results of this study concerning the duration of the intraocular pressure-lowering effect and the relative AUCs are shown in Table 12. Diepold et al. (217) concluded that the betamethasone model most closely resembles the course of the intraocular pressure curve in humans. As these data show, the duration of the intraocular pressure-lowering effects lasted over 9 hours with the nanoparticles in all models. This long duration and especially the differences in the curves between the nanoparticles and the solutions cannot be predicted by the pharmacokinetics determined by the measurement of the aqueous humor concentrations of pilocarpine. In contrast, these differences suggest that different absorption pathways for pilocarpine may exist if the drug is delivered in the form of a solution or if it is bound to nanoparticles.

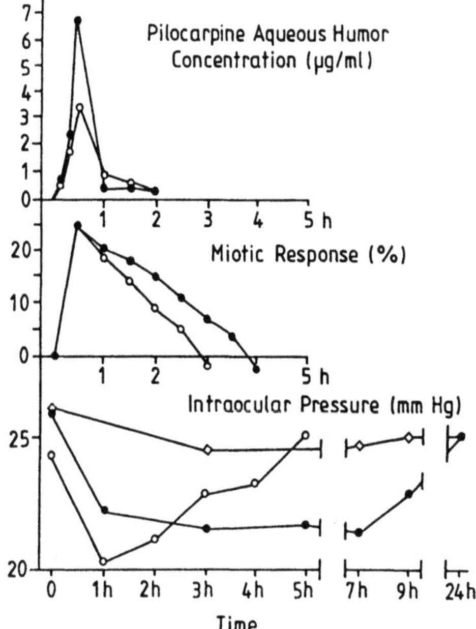

Figure 24 Pilocarpine aqueous humor concentration and miotic response in normal rabbits and intraocular pressure in betamethasone-treated rabbits' eyes after instillation of pilocarpine nitrate 1% (aqueous humor and miosis) or 2% (intraocular pressure) in the form of aqueous eyedrops (○) or bound to poly(butyl cyanoacrylate) nanoparticles (●); controls (□). (Reproduced from Ref. (217) with permission of the copyright holder.)

Table 12 Duration of Activity and Relative AUC (Area Under the Activity Versus Time Curve) of Pilocarpine (2%) in Three Different Rabbit Models for the Determination of the Intraocular Pressure. Pilocarpine Nitrate was Instilled into the Eyes in the Form of an Eyedrop Solution (S) or Incorporated into Nanoparticles (N)

	Duration (h)		Relative AUC	
Model	S	N	S	N
Betamethasone	5	>9	1.0	3.35
Alpha-chymotrypsine	>10	>10	1.0	1.28
Water-loading	>9	>9	1.0	1.42

Source: Data from Ref. 217.

One of the drawbacks of the Diepold et al. (217) study was that the aqueous humor levels as well as the miosis data were determined in normostatic rabbits, whereas the intraocular pressure was artificially increased in the three models. For this reason, this study was repeated and the aqueous humor concentrations as well as the miosis and the intraocular pressure were determined after instillation of the nanoparticles and of the solution in the betamethasone model. Preliminary results are shown in Figs. 25 and 26. Similar results were obtained as with the normostatic rabbits. The main difference in the two studies is that no significant differences were observed in the second study between the 30 min aqueous humor levels of both preparations. In addition, the miosis differences were less pronounced. The differences in the intraocular pressure remained, demonstrating the high reproduceability of the betamethasone model. These differences indicate that there is indeed a different pathway of absorption for pilocarpine delivered as a solution or bound to nanoparticles. The different absorption pathway seem to lead to different concentrations at the relevant receptor sites. The differences between the two pilocarpine formulations are especially significant, since it has to be taken into consideration that only 15% of the pilocarpine is bound to the nanoparticles whereas 85% is contained in the solution form. Nevertheless, the low amount of nanoparticle-bound pilocarpine combined with the slow elimination of the particles from the eye leads to

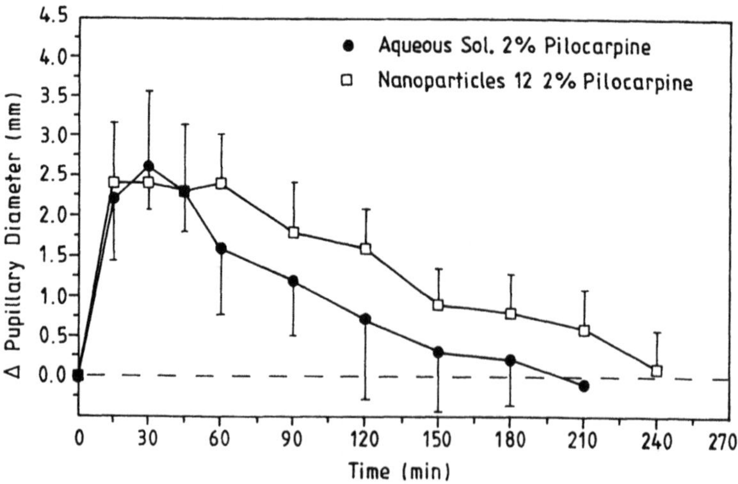

Figure 25 Miotic response of betamethasone-treated rabbits after instillation of pilocarpine nitrate 2% into the eyes. (Data generated by A. Zimmer and J. Kreuter.)

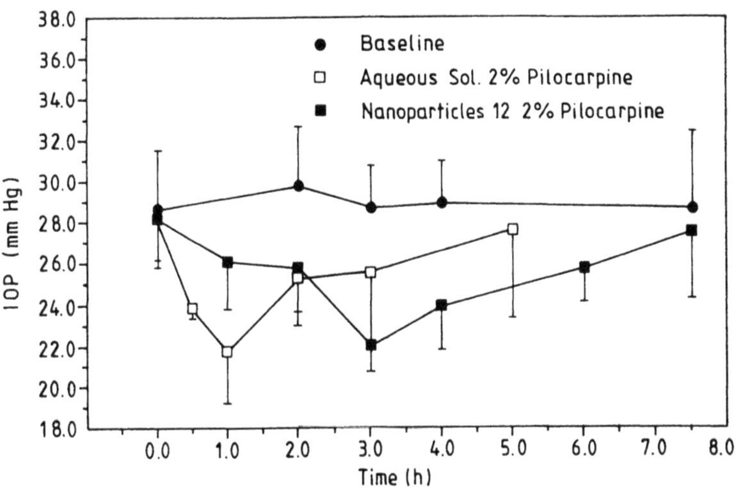

Figure 26 Intraocular pressure (IOP) of betamethasone-treated rabbits after instillation of pilocarpine nitrate 2% into the eyes. (Data generated by A. Zimmer and J. Kreuter.)

the observed significant effects. Improvement of the binding, therefore, may even enhance the effectivity of nanoparticles.

On the other hand, a study by Li et al. (220) demonstrates that a too efficient loading may reduce the drug delivery. Li et al. (220) sorbed progesterone to poly(butyl cyanoacrylate) nanoparticles yielding a solid solution with a polymer/water distribution coefficient of over 80,000. As a result, the drug was released too slowly and the nanoparticles were eliminated from the eye before effective progesterone delivery could take place.

Other drugs that were bound to nanoparticles included betaxolol (150, 151), timolol (214), amikacin (139), and metipranolol (405). Betaxolol chlorhydrated was adsorbed to poly(isobutyl cyanoacrylate) nanoparticles of a size of about 240 nm by Marchal-Heussler et al. (150), and the sorption isotherm, the drug release by dialysis, and the reduction of the intraocular pressure in alpha-chymotrypsin-pretreated rabbits were determined. An S-shaped sorption isotherm resulted for all preparations tested. Although an optimal loading of 0.45 g drug/g nanoparticles was obtained with a preparation containing 0.8% dextran 70,000 and 0.3% dextran sulfate, an optimal reduction in the release rate and a better intraocular pressure reduction was observed with a preparation containing only 1% dextran 70,000. The latter preparation yielded a maximal loading of 0.25 g drug/g polymer. The release rate determined by dialysis of the latter preparation in comparison to the solution of the drug was reduced by 45%. The intraocular pressure reducing effect also was most pronounced with the latter preparation. However, the difference from a commercial eyedrop solution (Betoptic®) was not statistically significant.

A significant improvement of betaxolol action compared to the Betoptic® eyedrops was achieved after adsorption of the drug to poly-ϵ-caprolactone nanocapsules manufactured by the method of Fessi et al. (81,82). This improvement was even more pronounced after encapsulation of the drug into these nanocapsules (151). While the 0.5% commercial betaxolol solution (Betoptic®) achieved a maximal reduction of the intraocular pressure in alpha-chymotrysin-pretreated rabbits of about 20% and approached baseline after about 8 hours, betaxolol at the same concentration encapsulated into the poly-ϵ-caprolactone nanoparticles yielded about a 30% intraocular pressure reduction and a much longer action: after 8 hours, the intraocular pressure reduction still amounted to 25%. A 0.1% betaxolol nanocapsule preparation with this polymer still was significantly better than the commercial 0.5% betaxolol solution. Moreover, systemic effects of betaxolol were reduced by encapsulation into the nanocapsules. Betoptic® inhibited the isoprenaline induced increase in heart frequency by about 50%, whereas no significant difference from the control was observed with same amount of encapsulated drug (151).

Encapsulation of another beta-blocker, metipranolol, into a number of nanocapsules consisting of different polymers and different nanocapsule oil core materials did not lead to a significant change in the reduction of the intraocular pressure in rabbits in comparison to commercial eyedrops containing the same amount of metipranolol (Betamann®) (405). However, again cardiovascular side effects determined by heart rate measurements were drastically reduced by encapsulation into the nanocapsules (405).

As already mentioned in Sec. VII. B, the binding of amikacin sulfate by adsorption to previously polymerized and resuspended freeze-dried empty poly(butyl cyanoacrylate) nanoparticles significantly improved the delivery of this antiinfective drug to the cornea and into the aqueous humor of rabbits (139). The efficacy of delivery to these ocular compartments depended on the polymerization stabilizer used. One percent dextran 70 in the polymerization mixture yielded nanoparticles that increased the concentrations of amikacin in the cornea about twofold and in the aqueous humor about threefold. On the other hand, the combination of dextran with sodium lauryl sulfate or the use of poloxamer 188 as the stabilizer did not significantly increase the corneal or aqueous humor amikacin levels compared to the solution of the drug (139).

F. Use of Nanoparticles for the Targeting of Drugs to Areas of Inflammation in the Body

As mentioned in Sec VI. E, the concentrations of poly(alky cyanoacrylate) nanoparticles in various tissues and compartments of the eye were about three to fivefold higher in chronically inflamed than in normal eyes of rabbits after ocular administration (347) (Fig. 11). This enhanced accumulation in the inflamed eyes was attributed to an enhanced bioadhesiveness of inflamed tissue. A similar attraction of nanoparticles and of other polymers was observed by Illum et al. (346), Alpar et al. (348), and Seale et al. (406). Illum et al. (346) observed that poloxamine 908-coated polystyrene nanoparticles accumulated at a very high level in the inflamed thigh muscles of rabbits after intravenous injection. The inflammation was induced by previous intramuscular injection of carageenan. In contrast, noncoated polystyrene nanoparticles behaved as in normal rabbits and were rapidly removed by the elements of the reticuloendothelial system, especially the liver and the spleen, and did not accumulate in the inflamed areas.

Fluorescent particles of a size of 1100 nm accumulated in artificially induced inflammatory air pouches in the necks of rats after peroral administration with a plastic stomach tube (348) (Table 13). Particles administered suspended in water were taken up to a larger degree than particles suspended

Table 13 Transfer of Particles from the Gastrointestinal Tract to the Circulation in the Rat and to Inflammatory Air Pouches. The Number of Particles in Tail Vein Blood Samples Detected for 100 Erythrocytes in the Counting Field Is Listed Below. The Time Represents the Interval Between Dosing and the Removal of a Blood Sample. Two Results are Quoted for the % of the Total Particles Found in the Lavage Fluids from the Air Pouch

	No. of particles/100 erythrocytes		% of particles found in air pouch	
	Particles administered		Particles administered	
Time	In water	In emulsion	In water	In emulsion
10 min	1.56	0.82	–	–
45 min	0.92	0.69	–	–
120 min	0.76	0.57	–	–
240 min	0.41	0.21	–	–
24 h	–	–	0.20	0.08

Source: Adapted from Alpar et al. (348).

in an emulsion consisting of acharis oil, water, Span® 83, and polysorbate 80 (8:8:1:1). This increased uptake in water may be due to a water overload-induced increased lymph flow and/or gastric motility (348).

The increased uptake of ^{125}I-labeled polyvinylpyrrolidone in the inflamed paws of adjuvant-induced arthritic rats after intravenous injection was reported by Seale et al. (406). The inflammation was induced by intradermal injection of *Mycobacterium butyricum* in heavy mineral oil. Especially high molecular weight polyvinylpyrrolidone (40 kDa and 360 kDa) accumulated to a high degree, i.e., 11 to 15 times more than in normal paws, in the inflamed paws of the rats (Table 14).

Similar findings of accumulation of particulates in areas of inflammation were observed with nanocolloids (407), liposomes (408,409), and lipid microspheres (410,411). Although the mechanism of this accumulation presently is not known, the above results clearly demonstrate that colloidal carriers such as nanoparticles hold promise for the targeted delivery of drugs to inflamed areas of the body after administration by a number of possible routes (1,346,347,399).

G. Nanoparticles as Adjuvants for Vaccines

Many antigens, especially smaller peptides, virus subunits, and antigens produced by genetic engineering procedures, are week antigens and produce little or no protection. For this reason, it may be necessary to add adjuvants to render these antigens potent enough to be useful for vaccines.

A large number of immunological adjuvants have been used including minerals, emulsions, peptides, lipids, and surfactants. Unfortunately, many of these adjuvants cause adverse toxicological effects or are difficult to manufacture in a reproducible manner (229). It has been shown that the particle size of the adjuvant can significantly influence the immune response (225,412). With emulsions, however, the size of the emulsion droplets may

Table 14 Accumulation of PVP within Inflamed Rat Paws. Tissue Uptake of i.v. [^{125}I]PVP (Mean % Accumulation of Total Dose ± S.D.)

PVP mol. wt (kDa)	Paws		Paw ratio arthritic/normal
	Normal	Arthritic	
10	0.13 ± 0.11	0.45 ± 0.31	3.5
40	0.07 ± 0.03	0.82 ± 0.22	11.7
360	0.23 ± 0.08	3.58 ± 1.43	15.6

Source: Adapted from Seale et al. (406).

change from preparation to preparation. Moreover, the particle size that will be relevant for the immune response may be further altered during and after injection. The consistency of the tissue in which the vaccine is injected may yield a smaller droplet size due to friction between the vaccine liquid and the tissue. On the other hand, for instance fatty tissue may induce a coalescence of the droplets resulting in a larger particle size.

The structure and the properties of aluminum compounds—the most commonly used adjuvants for commercial vaccines—may change significantly with slight alterations in production conditions and with aging (413–461). This can lead to significant differences in the immune response obtained (226,417–421). This is demonstrated by the fact that the adsorption properties of different qualities of aluminum hydroxides showed no correlation with the adjuvant effect obtained (418).

For this reason, adjuvants on the basis of nanoparticles were developed (33). The most promising polymer, poly(methyl methacrylate), has been used in surgery for over 50 years and was shown to be slowly biodegradable in the form of nanoparticles (Sec. VI. C) (350). The observed slow degradation rate of about 30% to 40% of the nanoparticulate polymer per year seems to be very promising for vaccines, because a prolongation of the contact of the antigen with the immunocompetent cells of the organism in most instances is required in order to maintain a long immunity.

Antigens like drugs may be either incorporated into the nanoparticles by polymerization of the antigen or they may be adsorbed by addition to previously polymerized empty particles.

1. Influence of Physicochemical Parameters on the Adjuvant Effect

a. Particle Size. The influence of the particle size of poly(methyl methacrylate) and of polystyrene nanoparticles on the adjuvant effect of influenza (229) and of bovine serum albumin (225) was investigated after intramuscular injection. The particle sizes of the nanoparticles in these experiments ranged from 62 nm to 10 μm. Smaller particles below about 350 nm exhibited a significant adjuvant effect (225). This adjuvant effect was better than aluminum hydroxide at sizes below 150 nm. The adjuvant effect of particles of a size of 62 nm was smaller than that of those with a size of 103 nm and 132 nm. However, the latter result was not statistically significant and may be caused by biological variations. For this reason, at present it is not clear if a particle size optimum exists around 100 nm or if particle sizes below this size yield equal or better adjuvant effects.

b. Hydrophobicity. The influence of hydrophobicity on the adjuvant effect using bovine serum albumin or influenza subunits as antigens was investigated by determination of the antibody response against bovine serum albumin and of the protection against infection with live influenza virus in mice (266). The hydrophobicity of the nanoparticles was altered by insertion of hydroxyl groups into the polymer or by the exchange of a methyl group of the acrylates against a cyano group and by variation in the length of the ester side chain. The hydrophobicity of the resulting polymers was investigated by the measurement of the water contact angles. Both the antibody determination and the mouse protection experiments led to the conclusion that the adjuvant effect increased with increasing hydrophobicity (226).

2. Adjuvant Effects with Nanoparticles Using Influenza or HIV-1 and HIV-2 Virus Vaccines

a. Influenza. The adjuvant effect of whole virus and of subunit influenza vaccines were tested in mice and guinea pigs. The antibody response using the chicken cell hemagglutination inhibition test (33,229,231,232) as well as the protection against infection by a challenge with live mouse-adapted influenza virus using a dose of 50 LD_{50} were determined (230,231). Both types of nanoparticle vaccine preparations, incorporation of the whole virus or of virus subunits into the nanoparticles in presence of the antigens and adsorption of the antigens on empty nanoparticles, were investigated.

The experiments showed that an optimal adjuvant effect was obtained 4 weeks after primary immunization in mice and guinea pigs at a nanoparticle concentration of 0.5% relative to the injection volume (33,231). This optimum in nanoparticle concentration was observable with the incorporated and with the adsorbed antigens. The optimum after adsorption, however, was much less pronounced than after incorporation. After incorporation, the antibody response with higher polymer concentrations of about 2% decreased significantly below that of the fluid vaccine, indicating that so much antigen surface was covered by the polymer that only a minimal immune response was exhibited. After prolonged time periods and boosting, the differences in antibody response caused by the variation of the polymer content decreased and high polymer content vaccines yielded equal or sometimes even better responses than the 0.5% concentration (33,231). This may be an indication that additional antigen surface becomes accessible to the immunocompetent cells, probably due to degradation of the nanoparticle polymer.

With whole virus vaccines, substantially higher antibody responses were obtained after incorporation of the virions into the nanoparticles used at an optimal concentration of 0.5% poly(methyl methacrylate) in comparison to the adsorbed vaccine and to aluminum hydroxide (33). The latter two vaccines yielded significantly higher antibody responses than the fluid vaccines. The protection obtained with these three adjuvants, however, was equal when the adjuvants were added in the above mentioned optimal concentration (230). The adjuvant effect disappeared when the adjuvants were diluted together with the antigen down to very low adjuvant concentrations. The protection experiments, on the other hand, indicated that both nanoparticle preparations were superior to aluminum hydroxide at low antigen concentrations and optimal adjuvant concentrations.

These observations were confirmed by the antibody determination in guinea pigs and mice after immunization with influenza subunits (232). Both nanoparticle preparations, incorporated as well as adsorbed product, were significantly superior to aluminum hydroxide and to the fluid vaccine. Again, by far the highest differences were observed at low antigen concentrations with the nanoparticles in comparison to the latter two vaccines (Fig. 27).

The antibody response and the protection with the nanoparticle adjuvants in comparison to the aluminum and the fluid vaccine were especially pronounced after longer time periods (231) (Fig. 28).

In order to determine the heat stability of nanoparticle influenza vaccines, vaccines with a low antigen content of 21 IU/mL were stored at 40°C for 0, 15, 30, 60, 120, and 240 hours and then administered intramuscularly to mice. After 4 weeks, the mice were challenged. The heat storage did not reduce significantly the protection obtained with both nanoparticle preparations.

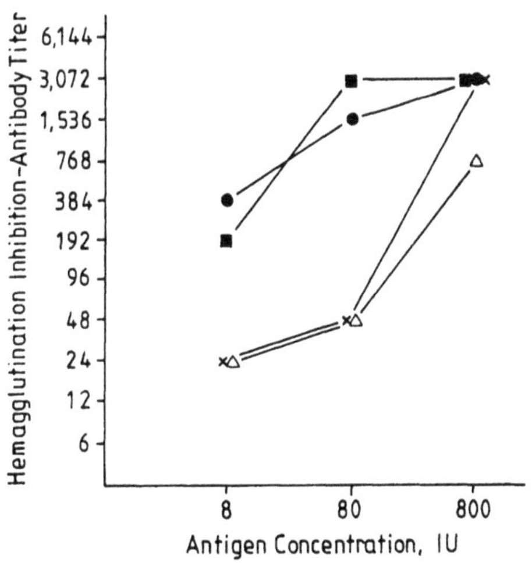

Figure 27 Dependence of the antibody response on the antigen concentration after 3 weeks. Sera were pooled before titration. (■) 0.5% PMMA + 0.5% PAA, polymerized in the presence of antigens; (●) 0.8% PMMA, polymerized in the absence of antigens; (×) 0.1 Al(OH); (△) fluid vaccine. Sera (n = 10) were determined individually. (Reproduced from Ref. (232) with permission of the copyright holder.)

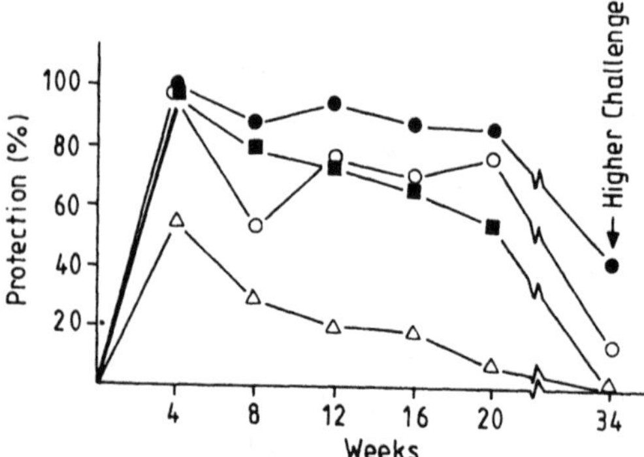

Figure 28 Protection of mice against infection with a dose of 50 LD_{50} of infectious mice-adapted influenza virus after s.c. immunization with 20 IU A PR 8 whole virus using the following adjuvants: (●) incorporation into 0.5% poly(methyl methacrylate); (○) adsorption onto 0.5% poly(methyl methacrylate); (■) adsorption onto 0.2% albumin hydroxide; (△) fluid vaccine without adjuvant. Higher challenge = 250 LD_{50}.

A significant drop of the protection was observed with aluminum hydroxide, and the protection disappeared totally after storage of the fluid vaccine for 60 hours (231).

b. HIV. Enormously higher adjuvant effects in comparison to the results obtained with influenza were observed with HIV-1 and especially with HIV-2 (227,422). The antibody responses against these antigens grown on Molt-4 clone 8 cell cultures were determined by ELISA after a single subcutaneous injection of 5 to 50 μg antigen/0.5 mL to mice. The virus was purified by sucrose gradient centrifugation, split with Tween® 80/ether, and inactivated with formalin 0.5% for 72 hours. Only nanoparticle preparations produced by adsorption were used in these experiments and compared to 0.2% aluminum hydroxide and to fluid antigen without adjuvant. The antibody response with HIV-2 was 10- to 200-fold higher than with aluminum hydroxide (Fig. 29). The latter adjuvant in turn yielded 10- to 100-fold higher responses than the fluid vaccine (227,422). With HIV-1 a delayed antibody response was observed. Significantly high antibody titers were reached after 10 weeks, whereas HIV-2 yielded high titers already after 4 weeks (Fig. 29). Although the HIV-1 antibody titers were about 5 to 10 times lower than with HIV-2, they still were about 10 times higher than those with aluminum hydroxide or with the fluid vaccine, even in comparison to HIV-2. Again, the titers obtained with aluminum hydroxide were generally statistically higher than those of the fluid vaccine.

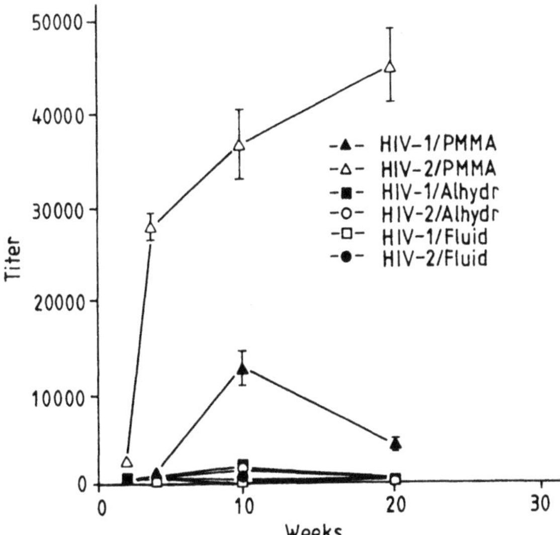

Figure 29 Antibody response (ELISA) of mice after immunization with HIV-1 and HIV-2 vaccines (10 μg antigen/mouse) containing different adjuvants. PMMA = poly(methyl methacrylate); Alhydr = aluminum hydroxide; Fluid = fluid vaccine without adjuvant. (Reproduced from Ref. (423) with permission of the copyright holder.)

In a final experiment using a single injection of 5 μg antigen and a single time point determination after 10 weeks, poly(methyl methacrylate) nanoparticles were compared to 24 different adjuvants including different aluminum compounds, Freund's complete and incomplete adjuvant, different liposomes, surfactants, Iscoms, and muramyl peptides. Again, the nanoparticles were by far the best adjuvant (423).

3. Conclusion

Poly(methyl methacrylate) nanoparticles are powerful adjuvants for a number of antigens including influenza whole virions and subunits, bovine serum albumin, rabies (unpublished preliminary observations), and especially HIV-1 and HIV-2 inactivated whole virion subunit antigens. This polymer material has been used in surgery for 50 years. Due to its biodegradability in nanoparticulate form and due to its hydrophobicity, poly(methyl methacrylate) seems to be the most promising polymer material for immunological adjuvants. Adjuvant effects with this material in comparison to other adjuvants were most pronounced at low antigen concentrations, with weaker antigens, and after prolonged time periods. In addition, poly(methyl methacrylate) nanoparticles seem to increase the heat stability of antigens.

The tissue reactions at the injection site 1 year after injection exhibited much less severe tissue reactions than aluminum hydroxide. These tissue reactions with whole influenza virions were not different from those with the same antigen in fluid form without adjuvant.

VIII. CONCLUSIONS

Nanoparticles represent interesting alternatives as drug delivery systems to liposomes. Due to their similar size they may be used for similar purposes. While liposomes have some advantages due to their composition, in that they consist of materials that are present as natural materials in the human body, the higher stability of nanoparticles yields longer shelf storage times as well as a better persistence of the particles in the body after their administration, and in addition it enables the administration by routes that are not practicable for liposomes, such as the peroral route.

Nevertheless, irrespective of stability considerations, nanoparticles and liposomes, as shown in Sec. VII.A, may be used alternatively for different diseases or for different types of tumors, as for instance in the specific ex ample reported by Beck et al. (132). In this example mitoxantrone bound to liposomes was more effective against P 388 leukemia, whereas the same drug was more effective against a solid B 16 melanoma after binding to nanoparticles.

Another aspect of the application of nanoparticles is the possibility of targeting them to specific organs or tissues by adsorbing and coating their surface with different substances. Here again nanoparticles might be superior to liposomes, because many coating materials such as surfactants can lead to the disintegration of liposomes.

Nanoparticles also seem to be useful for peroral administration and for this purpose seem to be superior to liposomes. Intestinal fluids contain bile salts, which have a high solubilizing power for liposomes, leading to their rapid disintegration. Nanoparticles are stable against these substances. Due to the possibility of monitoring their surface properties, their hydrophobicity, and their size, it may be possible to target them to different tissues in the gut and to influence their mode of interaction with and their uptake by the intestine. Since they are also able to stabilize peptides, they may enable the peroral delivery of peptides or of antigens for vaccination purposes.

In the same context, it should be remembered that the improved stability and slow release of at least some peptides after binding to nanoparticles may enable the controlled release and a substantially sustained delivery of these peptides also after parenteral administration.

For vaccination purposes the slowly degrading poly(methyl methacrylate) nanoparticles seem to have a number of advantages. The slow degradation enables a prolonged interaction of the antigen with the immunocompetent cells of the host body, leading to a prolonged immunostimulation and a prolonged protection of the host.

Last but not least, the recently observed tendency of nanoparticles and also of liposomes to accumulate in inflamed tissue may open up new possibilities of specific targeting of antiinflammatory drugs or of antibiotics to areas of inflammation. Especially antiinflammatory drugs possess considerable side effects. Targeting may reduce these side effects and could open up new possibilities for the therapy of diseases such as a number of rheumatic diseases.

In summary, colloidal carriers such as nanoparticles may enable new possibilities for therapy that presently have not been investigated. Compared to the efforts put into the discovery of new chemical entities, relatively little efforts have been put into the investigation of introducing nanoparticulate drug carriers into medical practice.

REFERENCES

1. Kreuter, J. (1992). Nanoparticles—preparation and applications. *Microcapsules and Nanoparticles in Medicine and Pharmacy* (M. Donbrow, ed.). Boca Raton, Fla.: CRC Press, p. 125.
2. Youssef, M., Fattal, E., Alonso, M.-J., Roblot-Treupel, L., Sauzières, J., Tancrède, C., Omnès, C., Couvreur, P., and Andremont, A. (1988). Effectiveness of nanoparticle-bound ampicillin in the treatment of *Listeria monocytogenes* infection in athymic nude mice. *Antimicrob. Agents Chemother. 32*: 1204.
3. Thews, G., Mutschler, E., and Vaupel, P. (1980). *Anatomie, Physiologie, Pathophysiologie des Menschen.* Stuttgart: Wissenschaftliche Verlagsgesellschaft, p. 229.
4. Little, K., and Parkhouse, J. (1962). Tissue reactions to polymers. *Lancet 2*: 857.
5. Nothdurft, H. (1955). Über die Sarkomauslösung durch Fremdkörperimplantationen bei Ratten in Abhängigkeit von der Form der Implantate. *Naturwiss. 42*: 106.
6. Nothdurft, H., and Mohr, H.-J. (1958). Sarkomerzeugung mit Fensterglas. *Naturwiss. 45*: 549.
7. Stinson, N. E. (1964). The tissue reaction induced in rats and guinea pigs by polymethylmethacrylate (acrylic) and stainless steel (18/8/mo). *Brit. J. Exp. Pathol. 45*: 21.
8. Stinson, N. E. (1965). Tissue reaction induced in guinea pigs by particulate polymethylmethacrylate, polythene and nylon of the same size range. *Brit. J. Exp. Pathol. 46*: 135.
9. Kreuter, J. (1983). Evaluation of nanoparticles as drug-delivery systems. I. Preparation methods. *Pharm. Acta Helv. 58*: 196.

10. Birrenbach, G., and Speiser, P. P. (1976). Polymerized micelles and their use as adjuvants in immunology. *J. Pharm. Sci. 65*: 1763.
11. Kramer, P. A. (1974). Albumin microspheres as vehicles for achieving specifity in drug delivery. *J. Pharm. Sci. 63*: 1646.
12. Widder, K. J., Senyei, A. E., and Ranney, D. F. (1979). Magnetically responsive microspheres and other carriers for the biophysical targeting of antitumor agents. *Adv. Pharmacol. Chemother. 16*: 213.
13. Kaufmann, U., Franke, F., and Lampert, F. (1979). Adherence of polymethylmethacrylate particles for characterization of B-lymphocytes. *Eur. J. Pediatr. 130*: 238.
14. Gallo, J. M., Hung, C. T., and Perrier, D. G. (1984). Analysis of albumin microsphere preparation. *Int. J. Pharm. 22*: 63.
15. Morimoto, Y., and Fujimoto, S. (1986). Albumin microspheres as drug carriers. *CRC Critical Rev. Therap. Drug Carrier Syst. 2*: 19.
16. Goodman, H., and Banker, G. S. (1970). Molecular entrapment as a precise method of controlled drug release I. Entrapment of cationic drugs by polymer flocculation. *J. Pharm. Sci. 59*: 1131.
17. Rhodes, C. T., Wai, K., and Banker, G. S. (1970). Molecular scale drug entrapment as a precise method of controlled drug release II. Facilitated drug entrapment to polymeric colloidal dispersions. *J. Pharm. Sci. 59*: 1578.
18. Rhodes, C. T., Wai, K., and Banker, G. S. (1970). Molecular scale drug entrapment as a precise method of controlled drug release III. In vitro and in vivo studies of release. *J. Pharm. Sci. 59*: 1581.
19. Boylan, J. C., and Banker, G. S. (1973). Molecular scale drug entrapment as a precise method of controlled drug release IV. Entrapment of anionic drugs by polymeric gelation. *J. Pharm. Sci. 62*: 1177.
20. Larson, A. B., and Banker, G. S. (1976). Attainment of highly uniform solid drug dispersions employing molecular scale drug entrapment in polymer latices. *J. Pharm. Sci. 65*: 838.
21. Fikentscher, H., Gerrens, H., and Schuller, H. (1960). Emulsionspolymerisation und Kunststoff-Latices. *Angew. Chem. 72*: 856.
22. Fitch, R. M., Prenosil, M. B., and Sprick, K. J. (1966). The mechanism of particle formation in polymer-hydrosols. I. Kinetics of aqueous polymerization of methyl methacrylate. *Amer. Chem. Soc., Polym. Preprints 7*: 707.
23. Fitch, R. M., Prenosil, M. B., and Sprick, K. J. (1969). The mechanism of particle formation in polymer-hydrosols. I. Kinetics of aqueous polymerization of methyl methacrylate. *J. Polymer Sci. C 27*: 95.
24. Fitch, R. M., and Tsai, C. (1970). Polymer colloids: Particle formation in nonmicellar systems. *J. Polymer Sci., Polym. Lett. 8*: 703.
25. Fitch, R. M. (1973). The homogenous nucleation of polymer colloids. *Brit. Polymer J. 5*: 467.
26. Roe, C. P. (1968). Surface chemistry aspects of emulsion polymerization. *Ind. Eng. Chem. 60*: 20.
27. Robb, I. D. (1969). Determination of the number of particles/unit volume of latex during the emulsion polymerization of styrene. *J. Polymer Sci., Part A-1 7*: 417.

28. Gardon, J. L. (1977). Interfacial, colloidal, and kinetic aspects of emulsion polymerization. *Interfacial Synthesis, Vol. 1* (F. Millich and C. E. Carraher, eds.). New York: Marcel Dekker, p. 205.
29. Berg, U. E., Kreuter, J., Speiser, P. P., and Soliva, M. (1986). Herstellung und in vitro-Prüfung von polymeren Adjuvantien für Impfstoffe. *Pharm. Ind. 48*: 75.
30. Kreuter, J. (1982). On the mechanism of termination in heterogeneous polymerization. *J. Polymer Sci., Polym. Lett. 20*: 543.
31. El-Egakey, M. A., Bentele, V., and Kreuter, J. (1983). Molecular weights of polycyanoacrylate nanoparticles. *Int. J. Pharm. 13*: 349.
32. Bentele, V., Berg, U. E., and Kreuter, J. (1983). Molecular weights of poly(methyl methacrylate) nanoparticles. *Int. J. Pharm. 13*: 109.
33. Kreuter, J., and Speiser, P. P. (1976). In vitro studies of poly(methyl methacrylate) adjuvants. *J. Pharm. Sci. 65*: 1624.
34. Kreuter, J., and Zehnder, H. J. (1978). The use of ^{60}Co-γ-irradiation for the production of vaccines. *Radiat. Effects 35*: 161.
35. Grislain, L., Couvreur, P., Lenaerts, V., Roland, M., Deprez-Decampeneere, D., and Speiser, P. (1983). Pharmacokinetics and distribution of a biodegradable drug-carrier. *Int. J. Pharm. 15*: 335.
36. Couvreur, P., Kante, B., Roland, M., Guiot, P., Baudhuin, P., and Speiser, P. (1979). Polyalkylcyanoacrylate nanocapsules as potential lysosomotropic carriers: Preparation, morphological and sorptive properties. *J. Pharm. Pharmacol. 31*: 331.
37. Douglas, S. J., Davis, S. S., and Holding, S. R. (1985). Molecular weights of poly(butyl-2-cyanoacrylate) produced during nanoparticle formation. *Brit. Polym. J. 17*: 339.
38. Vansnick, L., Couvreur, P., Christiaens-Leyh, D., and Roland, M. (1985). Molecular weights of free and drug-loaded nanoparticles. *Pharm. Res. 1*: 36.
39. Douglas, S. J., Illum, L., Davis, S. S., and Kreuter, J. (1984). Particle size and size distribution of poly(butyl-2-cyanoacrylate) nanoparticles. I. Influence of physicochemical factors. *J. Colloid Interface Sci. 101*: 149.
40. Douglas, S. J., Illum, L., and Davis, S. S. (1985). Particle size and size distribution of poly(butyl-2-cyanoacrylate) nanoparticles. II. Influence of the stabilizers. *J. Colloid Interface Sci. 103*: 154.
41. Seijo, B., Fattal, E., Roblot-Treupel, L., and Couvreur, P. (1990). Design of nanoparticles of less than 50 nm diameter: Preparation, characterization and drug loading. *Int. J. Pharm. 62*: 1.
42. Tsukiyama, S., and Takamura, A. (1974). Activation energy for the deformation and breakup of droplet on mechanical agitation. *Chem. Pharm. Bull. 22*: 2538.
43. Tsukiyama, S., and Takamura, A. (1975). Model for the deformation of droplet in agitation flow. *Chem. Pharm. Bull. 22*: 2565.
44. Lenaerts, V., Raymond, P., Juhasz, J., Simard, M. A., and Jolicoer, C. (1989). New method for the preparation of cyanoacrylic nanoparticles with improved colloidal properties. *J. Pharm. Sci. 78*: 1051.

45. Kreuter, J. (1990). Large-scale production problems and manufacturing of nanoparticles. *Specialized Drug Delivery Systems: Manufacturing and Production Technology* (P. Tyle, ed.). New York: Marcel Dekker, p. 257.
46. Rembaum, A., Yen, S. P. S., Cheong, E., Wallace, S., Molday, R. S., Gordon, I. L., and Dreyer, W. J. (1976). Functional microspheres based on 2-hydroxyethyl methacrylate for immunochemical studies. *Macromolecules 9*: 328.
47. Rembaum, A., Yen, S. P. S., and Molday, R. S. (1979). Synthesis and reactions of hydrophilic functional microspheres for immunological studies. *J. Macromol. Sci. Chem. A 13*: 603.
48. Rembaum, A. (1980). Synthesis, properties and biomedical applications of hydrophilic, functional, polymeric immunomicrospheres. *Pure Appl. Chem. 52*: 1275.
49. Molday, R. S., Dreyer, W. J., Rembaum, A., and Yen, S. P. S. (1975). New immunolatex spheres: Visual markers of antigens on lymphocytes for scanning electron microscopy. *J. Cell Biol. 64*: 75.
50. Yen, S. P. S., Rembaum, A., Molday, R. W., and Dreyer, W. (1976). Functional colloidal particles for immunoresearch. *Emulsion Polymerization* (I. Piirma and J. L. Gardon, eds.), ACS Symposium Series, Vol. 24. Washington, D.C.: American Chemical Society, p. 236.
51. Kreuter, J., Liehl, E., Berg, U., Soliva, M., and Speiser, P. P. (1988). Influence of hydrophobicity on the adjuvant effect of particulate polymeric adjuvants. *Vaccine 6*: 253.
52. Kreuter, J. (1974). Neue Adjuvantien auf Polymethylmethacrylatbasis. Diss. ETH 5417, Zürich, p. 43.
53. Lukowski, G., Müller, R. H., Müller, B. W., and Dittgen, M. (1992). Acrylic acid copolymer nanoparticles for drug delivery. I. Characterization of the surface properties relevant for in vivo organ distribution. *Int. J. Pharm. 84*: 23.
54. Scholsky, K. M., and Fitch, R. M. (1986). Controlled release of pendant bioactive materials from acrylic polymer colloids. *J. Controlled Rel. 3*: 87.
55. Ugelstad, J., Rembaum, J. T., Kemshead, K., Nustad, S., Funderud, S., and Schmid, R. (1984). Preparation and biomedical applications of monodisperse polymer particles. *Microspheres and Drug Therapy* (S. S. Davis, L. Illum, J. G. McVie, and E. Tomlinson, eds.). Amsterdam: Elsevier, p. 365.
56. Shahar, M., Meshulam, H., and Margel, S. (1986). Synthesis and characterization of microspheres of polystyrene derivation. *J. Polym. Sci. Chem. Ed. 24*: 203.
57. Schwartz, A., and Rembaum, A. (1985). Poly(vinylpyridine) microspheres. *Methods in Enzymology, Vol. 112* (K. J. Widder and R. Green, eds.). Orlando, Fla.: Academic Press, p. 175.
58. Chang, M., Richards, G., and Rembaum, A. (1985). Polyacrolein microspheres: Preparation and characteristics. *Methods in Enzymology, Vol. 112* (K. J. Widder and R. Green, eds.). Orlando, Fla.: Academic Press, p. 150.
59. Margel, S. (1985). Polyacrolein microspheres. *Methods in Enzymology, Vol. 112* (K. J. Widder and R. Green, eds.). Orlando, Fla.: Academic Press, p. 165.
60. Margel, S., Zisblatt, S., and Rembaum, A. (1979). Polyglutaraldehyde: A new reagent for coupling proteins to microspheres and for labeling cell-surface re-

ceptors. II. Simplified labeling method by means of nonmagnetic and magnetic polyglutaraldehyde microspheres. *J. Immunol. Meth. 28*: 341.
61. Mukherij, G., Murthy, R. S. R., and Miglani, B. D. (1989). Preparation and evaluation of polyglutaraldehyde nanoparticles containing 5-fluorouracil. *Int. J. Pharm. 50*: 15.
62. Margel, S. (1984). Characterization and chemistry of polyaldehyde microspheres. *J. Polymer Sci., Polym. Chem. Ed. 22*: 3521.
63. McLeod, A. D., Lam, F. C., Gupta, P. K., and Hung, C. T. (1988). Optimized synthesis of polyglutaraldehyde nanoparticles using central composite design. *J. Pharm. Sci. 77*: 704.
64. De Keyser, J.-L., Poupaert, J. H., and Dumont, P. (1991). Poly(diethyl methylidenemaloneate) nanoparticles as a potential drug carrier: Preparation, distribution and elimination after intravenous and peroral administration to mice. *J. Pharm. Sci. 80*: 67.
65. Birrenbach, G., and Speiser, P. P. (1976). Polymerized micelles and their use as adjuvants in immunology. *J. Pharm. Sci. 65*: 1763.
66. Ekman, B., Lofter, C., and Sjöholm, I. (1976). Incorporation of macromolecules in microparticles: Preparation and characteristics. *Biochem. 15*: 5115.
67. Ekman, B., and Sjöholm, I. (1978). Improved stability of proteins immobilized in microparticles prepared by a modified emulsion polymerization technique. *J. Pharm. Sci. 67*: 693.
68. Ljungstedt, I., Ekman, B., and Sjöholm, I. (1978). Detection and separation of lymphocytes with specific surface receptors by using microparticles. *Biochem. J. 170*: 161.
69. Edman, P., Ekman, B., and Sjöholm, I. (1980). Immobilization of proteins in microspheres of biodegradable polyacryldextran. *J. Pharm. Sci. 69*: 838.
70. Krause, H.-J., Schwarz, A., and Rohdewald, P. (1986). Interfacial polymerization, a useful method for the preparation of polymethylcyanoacrylate nanoparticles. *Drug Devel. Ind. Pharm. 12*: 527.
71. El-Samaligy, M. S., Rohdewald, P., and Mahmoud, H. A. (1986). Polyalkyl cyanoacrylate nanocapsules. *J. Pharm. Pharmacol. 38*: 216.
72. Gasco, M. R., and Trotta, M. (1986). Nanoparticles from microemulsions. *Int. J. Pharm. 29*: 267.
73. Campignano, R., Gasco, M. R., and Morel, S. (1991). Optimization of doxorubicin incorporation and of the yield of polybutylcyanoacrylate nanoparticles. *Pharm. Acta Helv. 66*: 28.
74. Gasco, M. R., Morel, S., Trotta, M., and Viano, I. (1991). Doxorubicine englobed in polybutylcyanoacrylate nanocapsules: Behaviour in vitro and in vivo. *Pharm. Acta Helv. 66*: 47.
75. Arakawa, M., and Kondo, T. (1980). Preparation and properties of poly(N^{α},N^{ε}-L-lysinediylterephthaloyl) microcapsules containing hemolysate in the nanometer range. *Can. J. Physiol. Pharmacol. 58*: 183.
76. Watanabe, A., Higashitsuji, K., and Nishizawa, K. (1978). Studies on electrocapillary emulsification. *J. Colloid Interface Sci. 64*: 278.
77. Al Khouri, N., Fessi, H., Roblot-Treupel, L., Devissaguet, J.-P., and Puisieux, F. (1986). Étude et mise au point d'un procédé original de préparation de nano-

capsules de polyalkylcyanoacrylate par polymérisation interfaciale. *Pharm. Acta Helv. 61*: 274.

78. Al Khouri Fallouh, N., Roblot-Treupel, L., Fessi, H., Devissaguet, J. P., and Puisieux, F. (1986). Development of a new process for the manufacture of polyisobutylcyanoacrylate nanocapsules. *Int. J. Pharm. 28*: 125.
79. Chouinard, F., Kan, F. W. K., Leroux, J.-C., Foucher, C., and Lenaerts, V. (1991). Preparation and purification of polyisohexylcyanoacrylate nanocapsules. *Int. J. Pharm. 72*: 211.
80. Rollot, J. M., Couvreur, P., Roblot-Treupel, L., and Puisieux, F. (1986). Physicochemical and morphological characterization of polyisobutyl cyanoacrylate nanocapsules. *J. Pharm. Sci. 75*: 361.
81. Fessi, H., Puisieux, F., Devissaguet, J. Ph., Ammoury, N., and Benita, S. (1989). Nanocapsule formation by interfacial polymer deposition following solvent displacement. *Int. J. Pharm. 55*: R 1.
82. Ammoury, N., Fessi, H., Devissaguet, J. P., Puisieux, F., and Benita, S. (1990). In vitro release pattern of indomethacin from poly(D,L-lactide) nanocapsules. *J. Pharm. Sci. 79*: 763.
83. Marchal-Heussler, L., Fessi, H., Devissaguet, J. P., Hoffman, M., and Maincent, P. (1992). Colloidal drug delivery systems for the eye. A comparison of the efficacy of three different polymers: polyisobutylcyanoacrylate, polylactic-co-glycolic acid, poly-epsiloncaprolacton. *S.T.P. Pharma Sci. 2*: 98.
84. Tice, T. R., and Gilley, R. M. (1985). Preparation of injectable controlled-release microcapsules by a solvent-evaporation process. *J. Controlled Rel. 2*: 343.
85. Gurny, R., Peppas, N. A., Harrington, D. D., and Banker, G. S. (1981). Development of biodegradable and injectable latices for controlled release of potent drugs. *Drug. Develop. Ind. Pharm. 7*: 1.
86. Krause, H. J., Schwarz, A., and Rohdewald, P. (1985). Polylactic acid nanoparticles, a colloidal drug delivery system for lipophilic drugs. *Int. J. Pharm. 27*: 145.
87. Tabata, Y., and Ikada, Y. (1989). Protein precoating of polylactide microspheres containing a lipophilic immunopotentiator for enhancement of macrophage phagocytosis and activation. *Pharm. Res. 6*: 296.
88. Coffin, M. D., and McGinity, J. W. (1992). Biodegradable pseudolatexes: Chemical stability of poly(D,L-lactide) and poly(ϵ-caprolactone) nanoparticles in aqueous media. *Pharm. Res. 9*: 200.
89. Jeffery, H., Davis, S. S., and O'Hagan, D. T. (1991). The preparation and characterization of poly(lactide-co-glycolide) microparticles. I: Oil-in-water solvent evaporation. *Int. J. Pharm. 77*: 169.
90. Koosha, F., Muller, R. H., and Washington, C. (1987). Production of polyhydroxybutyrate (PHB) nanoparticles for drug targeting. *J. Pharm. Pharmacol. 39*: 136P.
91. Koosha, F., Muller, R. H., Davis, S. S., and Davies, M. C. (1989). The surface chemical structure of poly(β-hydroxybutyrate) microparticles produced by solvent evaporation process. *J. Controlled Rel. 9*: 149.
92. Bodmeier, R., and Chen, H. (1990). Indomethacin polymeric nanosuspensions prepared by microfluidization. *J. Controlled Rel. 12*: 223.

93. Gallo, J. M., and Hassan, E. E. (1988). Receptor-mediated magnetic carriers: Basis for targeting. *Pharm. Res. 5*: 300.
94. Hassan, E. E., Parish, R. C., and Gallo, J. M. (1992). Optimized formulation of magnetic microspheres containing the anticancer agent, oxantrazole. *Pharm. Res. 9*: 390.
95. Bodmeier, R., Chen, H., Tyle, P., and Jarosz, P. (1991). Spontaneous formation of drug-containing acrylic nanoparticles. *J. Microencapsul. 8*: 161.
96. Zolle, I., Rhodes, B. A., and Wagner, H. N., Jr. (1970). Preparation of metabolizable radioactive human serum albumin microspheres for studies of the circulation. *Int. J. Appl. Radiat. Isotopes 21*: 155.
97. Zolle, I., Hosain, F., Rhodes, B. A., and Wagner, H. N., Jr. (1970). Human serum albumin millimicrospheres for studies of the reticuloendothelial system. *J. Nucl. Med. 11*: 379.
98. Scheffel, U., Rhodes, B. A., Natarajan, T. K., and Wagner, H. N., Jr. (1972). Albumin microspheres for study of the retinuloendothelial system. *J. Nucl. Med. 13*: 498.
99. Widder, K., Flouret, G., and Senyei, A. (1979). Magnetic microspheres: Synthesis of a novel parenteral drug carrier. *J. Pharm. Sci. 68*: 79.
100. Gupta, P. K., Hung, C. T., and Perrier, D. G. (1986). Albumin microspheres. II. Effect of stabilisation temperature on the release of adriamycin. *Int. J. Pharm. 33*: 147.
101. Yoshioka, T., Hashida, M., Muranishi, S., and Sezaki, H. (1981). Specific delivery of mitomycin C to the liver, spleen, and lung: Nano- and microspherical carriers of gelatin. *Int. J. Pharm. 8*: 131.
102. Tabata, Y., and Ikada, Y. (1987). Macrophage activation through phagocytosis of muramyl dipeptide encapsulated in gelatin microspheres. *J. Pharm. Pharmacol. 39*: 698.
103. Tabata, Y., and Ikada, Y. (1989). Synthesis of gelatin microspheres containing interferon. *Pharm. Res. 6*: 422.
104. Bungenberg de Jong, H. G., and Kruyt, H. R. (1929). Coazervation (partial miscibility in colloidal systems). *Proc. Koninkl. Akad. Wetensch. 32*: 849.
105. Derchivan, D. G. (1954). A comparative study of the flocculation and coazervation of different systems. *Discuss. Faraday Soc. 18*: 231.
106. Flory, J. (1953). *Principles of Polymer Chemistry*. Ithaca, N.Y.: Cornell University Press, p. 595.
107. Marty, J. J., Oppenheim, R. C., and Speiser, P. (1978). Nanoparticles—A new colloidal drug delivery system. *Pharm. Acta Helv. 53*: 17.
108. Marty, J. J. (1977). *The Preparation, Purification, and Properties of Nanoparticles*. D.Pharm. thesis, Victorian College of Pharmacy, Parkville, Australia, p. 30.
109. Juhlin, L. (1956). The spreading of spherical particles in dermal connective tissue. *Acta Dermato-Vener. 36*: 131.
110. El-Samaligy, M., and Rohdewald, P. (1982). Triamcinolone diacetate nanoparticles, a sustained release drug delivery system suitable for parenteral administration. *Pharm. Acta Helv. 57*: 201.

111. El-Samaligy, M. S., and Rohdewald, P. (1983). Reconstituted collagen nanoparticles, a novel drug carrier delivery system. *J. Pharm. Pharmacol. 35*: 537.

112. Mukherji, G., Murthy, R. S. R., and Miglani, B. D. (1990). Preparation and evaluation of cellulose nanospheres containing 5-fluorouracil. *Int. J. Pharm. 65*: 1.

113. Artursson, P., Edman, P., Laakso, T., and Sjöholm, I. (1984). Characterization of polyacryl starch microparticles as carriers for proteins and drugs. *J. Pharm. Sci. 73*: 1507.

114. Artursson, P., Edman, P., and Sjöholm, I. (1984). Biodegradable microspheres I. Duration of action of dextranase entrapped in polyacrylstarch microparticles in vivo. *J. Pharmacol. Exp. Therap. 231*: 705.

115. Artursson, P., Edman, P., and Sjöholm, I. (1985). Biodegradable microspheres II. Immune response to a heterologous and an autologous protein entrapped in polyacryl starch microparticles. *J. Pharmacol. Exp. Therap. 234*: 255.

116. Artursson, P., Martensson, I.-L., and Sjöholm, I. (1986). Biodegradable microspheres III. Some immunological properties of polyacryl starch microparticles. *J. Pharm. Sci. 75*: 697.

117. Artursson, P., Arro, E., Edman, P., Ericsson, J. L. E., and Sjöholm, I. (1987). Biodegradable microspheres V. Stimulation of macrophages with microparticles made of various polysaccharides. *J. Pharm. Sci. 76*: 127.

118. Laakso, T., Stjärnkvist, P., and Sjöholm, I. (1987). Biodegradable microspheres VI. Lysosomal release of covalently bound antiparasitic drugs from starch microparticles. *J. Pharm. Sci. 76*: 134.

119. Laakso, T., Edman, P., and Brunk, U. (1988). Biodegradable microspheres VII. Alterations in mouse liver morphology after intravenous administration of polyacryl starch microparticles with different biodegradability. *J. Pharm. Sci. 77*: 138.

120. Stjärnkvist, P., Artursson, P., Brunmark, A., Laakso, T., and Sjöholm, I. (1987). Biodegradable microspheres VIII. Killing of Leishmania donovani in cultured macrophages by microparticle-bound primaquine. *Int. J. Pharm. 40*: 215.

121. Laakso, T., and Sjöholm, I. (1987). Biodegradable microspheres X. Some properties of polyacryl starch microparticles prepared from acrylic acid-esterified starch. *J. Pharm. Sci. 76*: 935.

122. Artursson, P., Edman, P., and Ericsson, J. L.-E. (1987). Macrophage stimulation with some structurally related polysaccharides. *Scand. J. Immunol. 25*: 245.

123. Artursson, P., Ericsson, J. L.-E., and Sjöholm, I. (1988). Inflammatory response to polyacryl starch microparticles, role of acharidonic acid metabolites. *Int. J. Pharm. 46*: 149.

124. Artursson, P., Johansson, D., and Sjöholm, I. (1988). Receptor-mediated uptake of starch and mannan microparticles by macrophages: Relative contribution of receptors for complement, immunoglobulin, and carbohydrates. *Biomater. 9*: 241.

125. Artursson, P., Berg, A., and Edman, P. (1989). Biochemical and cellular effects of degraded starch microspheres on macrophages. *Int. J. Pharm. 52*: 183.
126. Stjärnkvist, P., Laakso, T., and Sjöholm, I. (1989). Biodegradable microspheres XII. Properties of the crosslinking chains in polyacryl starch microparticles. *J. Pharm. Sci. 78*: 52.
127. Fahlvik, A. K., Artursson, P., and Edman, P. (1990). Magnetic starch microspheres: Interactions of a microsphere MR contrast medium with macrophages in vitro. *Int. J. Pharm. 65*: 249.
128. Stjärnkvist, P., Degling, L., and Sjöholm, I. (1991). Biodegradable microspheres XIII. Immune response to the DNP hapten conjugated to polyacryl starch microparticles. *J. Pharm. Sci. 80*: 436.
129. Schröder, U., Stahl, A., and Salford, L. G. (1984). Crystallized carbohydrate spheres for slow release, drug targeting and cell separation. *Microspheres and Drug Therapy* (S. S. Davis, L. Illum, J. G. Mc Vie, and E. Tomlinson, eds.). Amsterdam: Elsevier, p. 427.
130. Kopf, H., Joshi, R. K., Soliva, M., and Speiser, P. (1976). Studium der Mizellpolymerisation in Gegenwart niedermolekularer Arzneistoffe. 1. Herstellung und Isolierung der Nanopartikel, Restmonomerenbestimmung, physikalisch-chemische Daten. *Pharm. Ind. 38*: 281.
131. Kopf, H., Joshi, R. K., Soliva, M., and Speiser, P. (1977). Studium der Mizellpolymerisation in Gegenwart niedermolekularer Arzneistoffe. 2. Bindungsart von inkorporierten niedermolekularen Modellsubstanzen an Nanopartikel auf Polyacrylamid-Basis. *Pharm. Ind. 39*: 993.
132. Beck, P., Kreuter, J., Reszka, R., and Fichtner, I. (1993). Influence of polybutylcyanoacrylate nanoparticles and liposomes on the efficacy and toxicity of the anticancer drug mitoxantrone in murine tumour models. *J. Microencapsul. 10*: 101.
133. Yalabik-Kas, H. S., Kreuter, J., Hincal, A. A., and Speiser, P. P. (1986). The adsorption of 5-fluorouracil from aqueous solustions onto methylmethacrylate nanoparticles. *J. Microencapsul. 3*: 213.
134. Harmia, T., Speiser, P., and Kreuter, J. (1986). Optimization of pilocarpine loading on to nanoparticles by sorption procedures. *Int. J. Pharm. 33*: 45.
135. Rolland, A. (1988). Pharmakokinetics and tissue distribution of doxorubicin-loaded polymethacrylic nanoparticles in rabbits. *Int. J. Pharm. 42*: 145.
136. Illum, L., Khan, M. A., Mak, E., and Davis, S. S. (1986). Evaluation of carrier capacity and release characteristic for poly(butyl-2-cyanoacrylate) nanoparticles. *Int. J. Pharm. 30*: 17.
137. Couvreur, P., Kante, B., Roland, M., and Speiser, P. (1979). Adsorption of antineoplastic drugs to polyalkylcyanoacrylate nanoparticles and their release in calf serum. *J. Pharm. Sci. 68*: 1521.
138. Guise, V., Drouin, J.-Y., Benoit, J., Mahuteau, J., Dumont, P., and Couvreur, P. (1990). Vidarabine-loaded nanoparticles: A physicochemical study. *Pharm. Res. 7*: 736.
139. Losa, C., Calvo, P., Castro, E., Vila-Jato, J. L., and Alonso, M. J. (1991). Improvement of ocular penetration of amikacin sulfate by association to poly(butyl cyanoacrylate) nanoparticles. *J. Pharm. Pharmacol. 43*: 548.

140. Egea, M. A., Valls, O., Alsina, M. A., Garcia, M. L., Losa, C., Vila-Jato, J. L., and Alonso, M. J. (1991). Interaction of amikacin loaded nanoparticles with phosphatidylcholine monolayers as membrane models. *Int. J. Pharm. 67*: 103.
141. Alonso, M. J., Losa, C., Calvo, P., and Vila-Jato, J. L. (1991). Approaches to improve the association of amikacin sulfate to poly(alkylcyanoacrylate) nanoparticles. *Int. J. Pharm. 68*: 69.
142. Bajwa, P. S., Couvreur, P., and Volz, P. A. (1987). The use of poly(isobutyl cyanoacrylate) nanoparticles with selected antifungal drugs. *FEMS Microbiol. Lett. 44*: 413.
143. Baszkin, A., Couvreur, P., Deyme, M., Henry-Michelland, S., and Albrecht, G. (1987). Monolayer studies on poly(isobutylcyanoacrylate)-ampicillin association. *J. Pharm. Pharmacol. 39*: 973.
144. Henry-Michelland, S., Alonso, M. J., Andremont, A., Maincent, P., Sauzières, J., and Couvreur, P. (1987). Attachment of antibodies to nanoparticles: Preparation, drug-release, and antimicrobial activity in vitro. *Int. J. Pharm. 35*: 121.
145. Youssef, M., Fattal, E., Alonso, M.-J., Roblot-Treupel, L., Sauzières, J., Tancrède, C., Omnès, A., Couvreur, P., and Andremont, A. (1988). Effectiveness of nanoparticle-bound ampicillin in the treatment of Listeria monocytogenes infection in athymic nude mice. *Antimicrob. Agents Chemother. 32*: 1204.
146. Fattal, E., Youssef, M., Couvreur, P., and Andremont, A. (1989). Treatment of experimental salmonellosis in mice with ampicillin-bound nanoparticles. *Antimicrob. Agents Chemother. 33*: 1540.
147. Couvreur, P., Fattal, E., Alphandary, H., Puisieux, F., and Andremont, A. (1992). Intracellular targeting of antibiotics by means of biodegradable nanoparticles. *J. Controlled Rel. 19*: 259.
148. Kubiak, C., Manil, L., and Couvreur, P. (1988). Sorptive properties of antibodies onto cyanoacrylic nanoparticles. *Int. J. Pharm. 41*: 181.
149. Chavany, C., Le Doan, T., Couvreur, P., Puisieux, F., and Hélène, C. (1992). Polyalkylcyanoacrylate nanoparticles as polymeric carriers for antisense oligonucleotides. *Pharm. Res. 9*: 441.
150. Marchal-Heussler, L., Maincent, P., Hoffman, M., Spittler, J., and Couvreur, P. (1990). Antiglaumatous activity of betaxolol chlorhydrate sorbed onto different isobutylcyanoacrylate nanoparticle preparations. *Int. J. Pharm. 58*: 115.
151. Marchal-Heussler, L., Fessi, H., Devissaguet, J. P., Hoffman, M., and Maincent, P. (1992). Colloidal drug delivery systems for the eye. A comparison of the efficacy of three different polymers: polyisobutylcyanoacrylate, polylactic-co-glycolic acid, poly-epsiolon-caprolacton. *S.T.P. Pharma Sci. 2*: 98.
152. Bonduelle, S., Foucher, C., Leroux, J.-C., Chouinard, F., Cadieux, C., and Lenaerts, V. (1992). Association of cyclosporin to isohexylcyanoacrylate nanospheres and subsequent release in human plasma in vitro. *J. Microencapsul. 9*: 173.

153. Brasseur, F., Couvreur, P., Kante, B., Deckers-Passau, L., Roland, M., Deckers, C., and Speiser, P. (1980). Actinomycin D adsorbed on polymethylmethacrylate nanoparticles: An increased efficiency against an experimental tumor. *Eur. J. Cancer 16*: 1441.
154. Couvreur, P., Kante, B., Lenaerts, V., Schailteur, V., Roland, M., and Speiser, P. (1980). Tissue distribution of antitumor drug associated with polyalkylcyanoacrylate nanoparticles. *J. Pharm. Sci. 69*: 199.
155. Kante, B., Couvreur, P., Lenaerts, V., Guiot, P., Roland, M., Baudhuin, P., and Speiser, P. (1980). Tissue distribution of [^{3}H] actinomycin D adsorbed on polybutylcyanoacrylate nanoparticles. *Int. J. Pharm. 7*: 45.
156. Hubert, B., Atkinson, J., Guerret, M., Hoffman, M., Devissaguet, J. P., and Maincent, P. (1991). The preparation and acute antihypertensive effects of a nanocapsular form of darodipine, a dihydropyridine calcium entry blocker. *Pharm. Res. 8*: 734.
157. Fouarge, M., Dewulf, M., Couvreur, P., Roland, M., and Vranckx, H. (1989). Development of dehydroemetine nanoparticles for the treatment of visceral leishmaniasis. *J. Microencapsul. 6*: 29.
158. Bertling, W. M., Gareis, M., Paspaleeva, V., Zimmer, A., Kreuter, J., Nürnberg, E., and Harrer, P. (1991). Use of liposomes, viral capsids, and nanoparticles as DNA carriers. *Biotech. Appl. Biochem. 13*: 390.
159. Rolland, A., Collet, B., Le Verge, R., and Toujas, L. (1989). Blood clearance and organ distribution of intravenously administered polymethacrylic nanoparticles in mice. *J. Pharm. Sci. 78*: 481.
160. Rolland, A., Bégué, J.-M., Le Verge, R., and Guilouzo, A. (1989). Increase of doxorubicin penetration in cultured rat hepatocytes by its binding to polymethacrylic nanoparticles. *Int. J. Pharm. 53*: 67.
161. Rolland, A. (1989). Clinical pharmakokinetics of doxorubicin in hepatoma patients after a single intravenous injection of free or nanoparticle-bound anthracycline. *Int. J. Pharm. 54*: 113.
162. Couvreur, P., Kante, B., Grislain, L., Roland, M., and Speiser, P. (1982). Toxicity of polyalkylcyanoacrylate nanoparticles II. Doxorubicin-loaded nanoparticles. *J. Pharm. Sci. 71*: 790.
163. Couvreur, P., Grislain, L., Lenaerts, V., Brasseur, F., Guiot, P., and Biernacki, A. (1986). Biodegradable polymeric nanoparticles as drug carrier for antitumor agents. *Polymeric Nanoparticles and Microspheres* (P. Guiot and P. Courvreur, eds.). Boca Raton, Fla.: CRC Press, p. 27.
164. Verdun, C., Couvreur, P., Vranckx, H., Lenaerts, V., and Roland, M. (1986). Development of a nanoparticle controlled-release formulation for human use. *J. Controlled Rel. 3*: 205.
165. Chiannikulchai, N., Driouich, Z., Benoit, J. P., Parodi, A. L., and Couvreur, P. (1989). Doxorubicin-loaded nanoparticles: Increased efficiency in murine hepatic metastases. *Selective Cancer Ther. 5*: 1.
166. Verdun, C., Brasseur, F., Vranckx, H., Couvreur, P., and Roland, M. (1990). Tissue distribution of doxorubicin associated with polyisohexylcyanoacrylate nanoparticles. *Cancer Chemother. Pharmacol. 26*: 13.

167. Chiannikulchai, N., Ammouri, N., Caillou, B., Devissaguet, J. Ph., and Couvreur, P. (1990). Hepatic tissue distribution of doxorubicin-loaded nanoparticles after i.v. administration in reticulosarcoma M 5076 metastasis-bearing mice. *Cancer Chemother. Pharmacol. 26*: 122.
168. Widder, K. J., Senyei, A. E., and Scarpelli, D. G. (1978). Magnetic microspheres: A model system for site specific drug delivery in vivo. *Proc. Soc. Exp. Biol. Med. 58*: 141.
169. Senyei, A., Widder, K. J., and Czerlinski, G. (1978). Magnetic guidance of drug-carrying microspheres. *J. Appl. Phys. 49*: 3578.
170. Widder, K. J., Senyei, A. E., and Ranney, D. F. (1980). In vitro release of biologically active adriamycin by magnetically responsive microspheres. *Cancer Res. 40*: 3512.
171. Widder, K. J., Morris, R. M., Poore, G. A., Howards, D. P., and Senyei, A. E. (1983). Selective targeting of magnetic albumin microspheres containing low-dose doxorubicin: Total remission in Yoshida sarcoma-bearing rats. *Eur. J. Cancer Clin. Oncol. 19*: 135.
172. Rettenmaier, M. A., Senyei, A. E., Widder, K. J., Stratton, J. A., Berman, M. L., and Disaia, J. P. (1985). In vivo alteration of RES phagocytosis by magnetic albumin microspheres. *J. Clin. Lab. Immunol. 17*: 99.
173. Morimoto, Y., Sugibayashi, and Kato, Y. (1981). Drug-carrier property of albumin microspheres in chemotherapy. V. Antitumor effect of microsphere-entrapped adriamycin on liver metastasis of AH 7974 cells in rats. *Chem. Pharm. Bull. 29*: 1433.
174. Sugibayashi, K., Okumura, M., and Morimoto, Y. (1982). Biomedical applications of magnetic fluids. III. Antitumor effect of magnetic albumin microsphere-entrapped adriamycin on lung metastasis of AH 7974 in rats. *Biomater. 3*: 181.
175. Gupta, P. K., Hung, C. T., and Perrier, D. G. (1986). Albumin microspheres. I. Release characteristics of adriamycin. *Int. J. Pharm. 33*: 137.
176. Gupta, P. K., Hung, C. T., and Perrier, D. G. (1987). Quantitation of the release of doxorubicin from colloidal dosage forms using dynamic dialysis. *J. Pharm. Sci. 76*: 141.
177. Gupta, P. K., Hung, C. T., Lam, F. C., and Perrier, D. G. (1988). Albumin microspheres. III. Synthesis and characterization of microspheres containing adriamycin and magnetite. *Int. J. Pharm. 43*: 167.
178. Gupta, P. K., Lam, F. C., and Hung, C. T. (1989). Albumin microspheres. IV. Effect of protein concentration and stabilization time on the release rate of adriamycin. *Int. J. Pharm. 51*: 253.
179. Gupta, P. K., Hung, C. T., and Lam, F. C. (1989). Factoral design based optimization of the formulation of albumin microspheres containing adriamycin. *J. Microencapsul. 6*: 147.
180. Gallo, J. M., Hung, C. T., Gupta, P. K., and Perrier, D. G. (1989). Physiological pharmacokinetic model of adriamycin delivered via magnetic albumin microspheres in the rat. *J. Pharmacokin. Biopharm. 17*: 305.
181. Gallo, J. M., Gupta, P. K., Hung, C. T., and Perrier, D. G. (1989). Evaluation of drug delivery following the administration of magnetic albumin microspheres containing adriamycin to the rat. *J. Pharm. Sci. 78*: 190.

182. Gupta, P. K., Hung, C.-T., and Rao, N. S. (1989). Ultrastructural disposition of adriamycin-associated magnetic albumin microspheres in rats. *J. Pharm. Sci. 78*: 290.
183. Gupta, P. K., and Hung, C. T. (1989). Effect of carrier dose on the multiple tissue disposition of doxorubicin hydrochloride administered via magnetic albumin microspheres in rats. *J. Pharm. Sci. 78*: 745.
184. Gupta, P. K., and Hung, C.-T. (1990). Albumin microspheres. V. Evaluation of parameters controlling the efficacy of magnetic microspheres in the targeted delivery of adriamycin in rats. *Int. J. Pharm. 59*: 57.
185. Gupta, P. K., Hung, c. T., and Lam, F. C. (1990). Application of regression analysis in the evaluation of tumor response following the administration of adriamycin either as a solution or via albumin microspheres to the rat. *J. Pharm. Sci. 79*: 634.
186. Oppenheim, R. C., Gipps, E. M., Forbes, J. F., and Whitehead, R. H. (1984). Development and testing of proteinaceous nanoparticles containing cytotoxines. *Microspheres and Drug Therapy* (S. S. Davis, L. Illum, J. G. Mc Vie, and E. Tomlinson, eds.). Amsterdam: Elsevier, p. 117.
187. Shankland, K., Whateley, T. L., and Oppenheim, R. C. (1987). Desolvation profiles of albumin during nanoparticle production. *J. Pharm. Pharmacol. 39*: 37P.
188. Miyazaki, S., Hashiguchi, N., Yokouchi, C., Takada, M., and Hou, W.-M. (1986). Antitumor effect of fibrinogen microspheres containing doxorubicin on Ehrlich ascites carcinoma. *J. Pharm. Pharmacol. 38*: 618.
189. Sjöholm, I., and Edman, P. (1979). Acrylic microspheres in vivo. I. Distribution and elimination of polyacrylamide microparticles after intravenous and intraperitoneal injection in mouse and rat. *J. Pharmacol. Exp. Therap. 211*: 656.
190. Edman, P., and Sjöholm, I. (1981). Prolongation of effect of asparaginase by implantation in polyacrylamide in rats. *J. Pharm. Sci. 70*: 684.
191. Couvreur, P., Tulkens, P., Roland, M., Trouet, A., and Speiser, P. (1977). Nanocapsules: A new type of lysosomotropic carrier. *FEBS Lett. 84*: 323.
192. Sugibayashi, K., Morimoto, Y., Nadai, I., and Kato, Y. (1977). Drug-carrier property of albumin microspheres in chemotherapy. I. Tissue distribution of microsphere-entrapped 5-fluorouracil in mice. *Chem. Pharm. Bull. 25*: 3433.
193. Sugibayashi, K., Morimoto, Y., Nadai, I., Kato, Y., Hasegawa, A., and Arita, T. (1979). Drug-carrier property of albumin microspheres in chemotherapy. II. Preparation and tissue distribution in mice of microsphere-entrapped 5-fluorouracil. *Chem. Pharm. Bull. 27*: 204.
194. Sugibayashi, K., Akimoto, M., Morimoto, Y., Nadai, T., and Kato, Y. (1979). Drug-carrier property of albumin microspheres in chemotherapy. III. Effect of microsphere-entrapped 5-fluorouracil on Ehrlich ascites carcinoma in mice. *J. Pharm. Dyn. 2*: 350.
195. Morimoto, Y., Akimoto, M., Sugibayashi, K., Nadai, T., and Kato, Y. (1980). Drug-carrier property of albumin microspheres in chemotherapy. IV. Antitumor effect of single-shot administration of microsphere-entrapped 5-fluorouracil on Ehrlich ascites or solid tumor in mice. *Chem. Pharm. Bull. 28*: 3087.

196. Kreuter, J., and Hartmann, H. R. (1983). Comparative study on the cytostatic effects and the tissue distribution of 5-fluorouracil in a free form and bound to polybutylcyanoacrylate nanoparticles in Sarcoma 180-bearing mice. *Oncology 40*: 363.
197. Grangier, J. L., Puygrenier, M., Gautier, J. C., and Couvreur, P. (1991). Nanoparticles as carriers for growth hormone releasing factor. *J. Controlled Rel. 15*: 3.
198. Gautier, J. C., Grangier, J. L., Barbier, A., Dupont, P., Dussossoy, D., Pastor, G., and Couvreur, P. (1992). Biodegradable nanoparticles for subcutaneous administration of growth hormone releasing factor (h GRF). *J. Controlled Rel. 20*: 67.
199. Malaiya, A., and Vyas, S. P. (1988). Preparation and characterization of magnetic nanoparticles. *J. Microencapsul. 5*: 243.
200. Vyas, S. P., and Malaiya, A. (1989). In vivo characterization of indomethacin magnetic polymethyl methacrylate nanoparticles. *J. Microencapsul. 6*: 493.
201. Ammoury, N., Fessi, H., Devissaguet, J.-P., Allix, M., Plotkine, M., and Boulu, R. G. (1990). Effect of cerebral blood flow of orally administered indomethacin-loaded poly(isobutylcyanoacrylate) and poly(DL-lactide) nanocapsules. *J. Pharm. Pharmacol. 42*: 558.
202. Ammoury, N., Fessi, H., Devissaguet, J.-P., Dubrasquet, M., and Benita, S. (1991). Jejunal absorption, pharmacological activity, and pharmacokinetic evaluation of indomethacin-loaded poly(d,l-lactide) and poly(isobutyl-cyanoacrylate) nanocapsules in rats. *Pharm. Res. 8*: 101.
203. Gürsoy, A., Eroglu, L., Ulutin, S., Tasyürek, M., Fessi, H., Puisieux, F., and Devissaguet, J.-P. (1989). Evaluation of indomethacin nanocapsules for their physical stability and inhibitory activity on inflammation and platelet aggregation. *Int. J. Pharm. 52*: 101.
204. Couvreur, P., Lenaerts, V., Kante, B., Roland, M., and Speiser, P. P. (1980). Oral and parenteral administration of insulin associated to hydrolysable nanoparticles. *Acta Pharm. Technol. 26*: 220.
205. Damgé, C., Michel, C., Aprahamian, M., and Couvreur, P. (1987). Polyalkylcyanoacrylate nanocapsules as drug carriers: Advantage in insulin treatment of streptozotocin-induced diabetic rats. *Adv. Biomater. 7*: 643.
206. Damgé, C., Michel, C., Aprahamian, M., and Couvreur, P. (1988). New approach for oral administration of insulin with polyalkylcyanoacrylate nanocapsules as drug carrier. *Diabetes 37*: 246.
207. Damgé, C., Michel, C., Aprahamian, M., Couvreur, P., and Devissaguet, J. P. (1990). Nanocapsules as carriers for oral drug delivery. *J. Controlled Rel. 13*: 233.
208. Michel, C., Aprahamian, M., Defontaine, L., Couvreur, P., and Damgé, C. (1991). The effect of site of administration in the gastrointestinal tract on the absorption of insulin from nanocapsules in diabetic rats. *J. Pharm. Pharmacol. 43*: 1.
209. Damgé, C., Aprahamian, M., Balboni, G., Hoeltzel, A., Andrieu, V., and Devissaguet, J. P. (1987). Polyalkylcyanoacrylate nanocapsules increase the intestinal absorption of a lipophilic drug. *Int. J. Pharm. 36*: 121.

210. Aprahamian, M., Michel, C., Humbert, W., Devissaguet, J.-P., and Damgé, C. (1987). Transmuscosal passage of polyalkylcyanoacrylate nanocapsules as a new drug carrier in the small intestine. *Biol. Cell 61*: 69.
211. Lee, K. C., Lee, Y. J., Kim, W. B., and Cha, C. Y. (1990). Monoclonal antibody-based targeting of methotrexate-loaded microspheres. *Int. J. Pharm. 59*: 27.
212. Opitz, N., Lübbers, D. W., Speiser, P. P., and Bisson, H. J. (1977). Determination of hydrogen ion activities via nanoencapsulated fluorescing pH-indicator molecules. *Pflüg. Arch. 368* (Suppl.): p.R. 49, No. 194.
213. Labhasetwar, V. D., and Dorle, A. K. (1990). Nanoparticles—A colloidal drug delivery system for primaquine and metronidazole. *J. Controlled Rel. 12*: 113.
214. Harmia-Pulkkinen, T., Tuomi, A., and Kristofferson, E. (1989). Manufacture of polyalkylcyanoacrylate nanoparticles with pilocarpine and timolol by micelle polymerization: Factors influencing particle formation. *J. Microencapsul. 6*: 87.
215. Harmia, T., Speiser, P., and Kreuter, J. (1986). A solid colloidal drug delivery system for the eye: Encapsulation of pilocarpine in nanoparticles. *J. Microencapsul. 3*: 3.
216. Harmia, T., Kreuter, J., Speiser, P., Boye, T., Gurny, R., and Kubis, A. (1986). Enhancement of the myotic response of rabbits with pilocarpine-loaded polybutylcyanoacrylate nanoparticles. *Int. J. Pharm. 33*: 187.
217. Diepold, R., Kreuter, J., Himber, J., Gurny, R., Lee, V. H. L., Robinson, J. R., Saettone, M. F., and Schnaudigel, O. E. (1989). Comparison of different models for the testing of pilocarpine eyedrops using conventional eyedrops and a novel depot formulation (nanoparticles). *Graefe's Arch. Clin. Exp. Ophthalmol. 227*: 188.
218. Gaspar, R., Préat, V., and Roland, M. (1991). Nanoparticles of polyisohexylcyanoacrylate (PIHCA) as carriers of primaquine: Formulation, physicochemical characterization, and acute toxicity. *Int. J. Pharm. 68*: 111.
219. Mbela, T. K. M., Poupaert, J. H., and Dumont, P. (1992). Poly(diethylmethylidene malonate) nanoparticles as primaquine delivery system to liver. *Int. J. Pharm. 79*: 29.
220. Li, V. H. K., Wood, R. W., Kreuter, J., Harmia, T., and Robinson, J. R. (1986). Ocular drug delivery of progesterone using nanoparticles. *J. Microencapsul. 3*: 213.
221. Krause, H.-J., and Rohdewald, P. (1985). Preparation of gelatin nanocapsules and their pharmaceutical characterization. *Pharm. Res. 2*: 239.
222. El Egakey, M. A., and Speiser, P. (1982). Drug loading studies on ultrafine solid carriers by sorption procedures. *Pharm. Acta Helv. 57*: 236.
223. Maincent, P., Devissaguet, J. P., Le Verge, R., Sado, P. A., and Couvreur, P. (1984). Preparation and in vivo studies of a new drug delivery system. *Appl. Biochem. Biotechnol. 10*: 263.
224. Maincent, P., Le Verge, R., Sado, P., Couvreur, P., and Devissaguet, J. P. (1986). Disposition kinetics and oral bioavailability of vincamine-loaded polyalkyl cyanoacrylate nanoparticles. *J. Pharm. Sci. 75*: 955.

225. Kreuter, J., Berg, U., Liehl, E., Soliva, M., and Speiser, P. P. (1986). Influence of the particle size of particulate polymeric adjuvants. *Vaccine 4*: 125.
226. Kreuter, J., Liehl, E., Berg, U., Soliva, M., and Speiser, P. P. (1988). Influence of the hydrophobicity on the adjuvant effect of particulate polymeric adjuvants. *Vaccine 6*: 253.
227. Stieneker, F., Kreuter, J., and Löwer, J. (1991). High antibody titres in mice with polymethylmethacrylate nanoparticles as adjuvant for HIV vaccines. *AIDS 5*: 431.
228. Kreuter, J., and Speiser, P. P. (1976). New adjuvants on a polymethylmethacrylate base. *Infect. Immunity 13*: 204.
229. Kreuter, J., and Haenzel, I. (1978). Mode of action of immunological adjuvants: Some physicochemical factors influencing the effectivity of polyacrylic adjuvants. *Infect. Immunity 19*: 667.
230. Kreuter, J., and Liehl, E. (1978). Protection induced by inactivated influenza virus vaccines with polymethylmethacrylate adjuvants. *Med. Microbiol. Immunol. 165*: 111.
231. Kreuter, J., and Liehl, E. (1981). Long-term studies of microencapsulated and adsorbed influenza vaccine nanoparticles. *J. Pharm. Sci. 70*: 367.
232. Kreuter, J., Mauler, R., Gruschkau, H., and Speiser, P. P. (1976). The use of new polymethylmethacrylate adjuvants for split influenza vaccines. *Exp. Cell Biol. 44*: 12.
233. Kreuter, J. (1983). Physicochemical characterization of polyacrylic nanoparticles. *Int. J. Pharm. 14*: 43.
234. Müller, R. H. (1990). *Colloidal Carriers for Controlled Drug Delivery and Targeting*. Stuttgart: Wissenschaftliche Verlagsgesellschaft, p. 49.
235. Shankland, K., and Whateley, T. L. (1992). Determination of refractive indices of polyalkylcyanoacrylate particles. *J. Colloid Interface Sci. 154*: 160.
236. Zimmer, A., Kreuter, J., and Robinson, J. R. (1991). Studies on the transport pathway of PBCA nanoparticles in ocular tissues. *J. Microencapsul. 8*: 497.
237. Magenheim, B., and Benita, S. (1991). Nanoparticle characterization: A comprehensive physicochemical approach. *S.T.P. Pharma Sci. 1*: 221.
238. Edstrom, R. D., Yang, X., Lee, G., and Evans, D. F. (1990). Viewing molecules with scanning tunneling and atomic force microscopy. *FASEB J. 4*: 3144.
239. Gould, S. A. C., Drake, B., Prater, C. B., Weisenhorn, A. L., Manne, S., Kelderman, G. L., Butt, H. J., Hasima, M., and Hansma, P. K. (1990). The atomic force microscope: A tool for science and industry. *Ultramicroscopy 33*: 93.
240. Hansma, P. K., and Tersoff, J. (1987). Scanning tunneling microscopy. *J. Appl. Phys. 61*: R1.
241. Hansma, P. K., Elings, V. B., Marti, O., and Bracker, C. E. (1988). Scanning tunneling microscopy and atomic force microscopy: Application to biology and technology. *Science 242*: 209.
242. Zasadzinski, J. A. N. (1989). Scanning tunneling microscopy with applications to biological surfaces. *Bio Techniques 7*: 174.
243. Binning, G., Quate, C. F., and Gerber, C. (1986). Atomic force microscope. *Phys. Rev. Lett. 56*: 930.

244. Drake, B., Prater, C. B., Weisenhorn, A. L., Gould, S. A. C., Albrecht, T. R., Quate, C. F., Cannell, D. S., Hansma, H. G., and Hansma, P. K. (1989). Imaging crystals, polymers, and processes in water with the atomic force microscope. *Science 243*: 1586.
245. Albrecht, T. R., Dovek, M. M., Lang, C. A., Grutter, P., Quate, C. F., Kuans, S. W., Frank, C. W., and Pease, R. F. W. (1988). Imaging and modification of polymers by scanning tunneling and atomic force microscopy. *J. Appl. Phys. 64*: 1178.
246. Denley, D. R. (1990). Practical applications of scanning tunneling microscopy. *Ultramicroscopy 33*: 83.
247. Fuchs, H., and Laschinski, R. (1990). Surface investigations with a combined scanning electron-scanning tunneling microscope. *Scanning 12*: 126.
248. Anders, H., Muck, M., and Heiden, C. (1988). SEM/STM combination for STM tip guidance. *Ultramicroscopy 25*: 123.
249. Ichinokawa, T., Miyazaki, Y., and Koga, Y. (1987). Scanning tunneling microscope combined with a scanning electron microscope. *Ultramicroscopy 23*: 115.
250. Vazquez, L., Bartolome, A., Garcia, R., Buendia, A., and Baro, A. (1988). Combination of a scanning tunneling microscope with a scanning electron microscope. *Rev. Sci. Instrum. 59*: 1286.
251. Gedde, U. W. (1990). Thermal analysis of polymers. *Drug. Dev. Ind. Pharm. 16*: 2465.
252. Benoit, J. P., Benita, S., Puisieux, F., and Thies, C. (1984). Stability and release kinetics of drugs incorporated within microspheres. *Microspheres and Drug Therapy* (S. S. Davis, L. Illum, J. G. McVie, and E. Tomlinson, eds.). Amsterdam: Elsevier, p. 91.
253. Kreuter, J. (1983). Solid dispersion and solid solution. *Topics in Pharmaceutical Sciences 1983* (D. D. Breimer and P. Speiser, eds.). Amsterdam: Elsevier, p. 359.
254. Bissery, M. C., Puisieux, F., and Thies, C. (1983). A study of process parameters in the making of microspheres by the solvent evaporation process. *Proc. 3rd Int. Confer. Pharm. Technol., Paris 1983,* Vol. 3, p. 233.
255. Müller, R. H. (1990). *Colloidal Carriers for Controlled Drug Delivery and Targeting.* Stuttgart: Wissenschaftliche Verlagsgesellschaft, p. 59.
256. Wilkins, D. J., and Myers, P. A. (1966). Studies on the relationship between the electrophoretic properties of colloids and their blood clearance and organ distribution in the rat. *Brit. J. Exp. Pathol. 47*: 568.
257. Roser, M. (1990). *Herstellung und Charakterisierung von Albumin-Partikeln.* Diss. Univ. Basel, p. 136.
258. Tröster, S. D., and Kreuter, J. (1988). Contact angles of surfactants with a potential to alter the body distribution of colloidal drug carriers on poly(methyl methacrylate) surfaces. *Int. J. Pharm. 45*: 91.
259. Carstensen, H., Müller, B. W., and Müller, R. H. (1991). Adsorption of ethoxylated surfactants on nanoparticles. I. Characterization by hydrophobic interaction chromatography. *Int. J. Pharm. 67*: 29.
260. Müller, R. H. (1990). *Colloidal Carriers for Controlled Drug Delivery and Targeting.* Stuttgart: Wissenschaftliche Verlagsgesellschaft, p. 99, 277.

261. Müller, R. H., Wallis, K. H., Tröster, S. D., and Kreuter, J. (1992). In vitro characterization of poly(methyl-methacrylate) nanoparticles and correlation to their in vivo fate. *J. Controlled Rel. 20*: 237.
262. Tröster, S. D., Wallis, K. H., Müller, R. H., and Kreuter, J. (1992). Correlation of the surface hydrophobicity of ^{14}C-poly(methyl methacrylate) nanoparticles to their body distribution. *J. Controlled Rel. 20*: 247.
263. Beck, P., Scherer, D., and Kreuter, J. (1990). Separation of drug-loaded nanoparticles from free drug by gel filtration. *J. Microencapsul. 7*: 491.
264. Benoit, E., Maincent, P., and Bessière, J. (1992). Applicability of dielectric measurements to the adsorption of drugs onto nanoparticles. *Int. J. Pharm. 84*: 283.
265. Illum, L., Khan, M. A., Mak, E., and Davis, S. S. (1986). Evaluation of carrier capacity and release characteristics for poly(butyl 2-cyanoacrylate) nanoparticles. *Int. J. Pharm. 30*: 17.
266. Benoit, J. P., Puisieux, F., Thies, C., and Benita, S. (1983). Characterization of drug-loaded poly(D,L-lactide) microspheres. *Proc. 3rd Int. Confer. Pharm. Technol., Paris 1983*, Vol. 3, p. 240.
267. Douglas, S. J., Illum, L., and Davis, S. S. (1986). Poly(butyl-2-cyanoacrylate) nanoparticles with differing surface charges. *J. Controlled Rel. 3*: 15.
268. Müller, R. H. (1990). *Colloidal Drug Carriers for Controlled Drug Delivery and Targeting*. Stuttgart: Wissenschaftliche Verlagsgesellschaft, p. 133.
269. Ratner, B. D. (1980). Characterization of graft polymers for biomedical applications. *J. Biomed. Mat. Res. 14*: 665.
270. Ratner, B. D., Johnston, A. B., and Lenk, T. J. (1987). Biomaterial surfaces. *J. Biomed. Mat. Res. Appl. Biomat. 21*: 59.
271. Leonard, F., Kulkarni, R. K., Brandes, G., Nelson, J., and Cameron, J. J. (1966). Synthesis and degradation of poly(alkylcyanoacrylates). *J. Appl. Polym. Sci. 10*: 259.
272. Kulkarni, R. K., Hanks, G. A., Pani, L. K., and Leonhard, F. (1967). The in vivo metabolic degradation of poly(methyl cyanoacrylate) via thiocyanate. *J. Biomed. Mater. Res. 1*: 11.
273. Mori, S., Ota, K., Takada, M., and Inou, T. (1967). Comparative studies of cyanoacrylate derivatives in vivo. *J. Biomed. Mater. Res. 1*: 55.
274. Vezin, W. R., and Florence, A. T. (1978). In vitro degradation rates of biodegradable poly-*n*-alkyl cyanoacrylates. *J. Pharm. Pharmacol. 30*: 5P.
275. Lenaerts, V., Couvreur, P., Christiaens-Leyh, D., Joiris, E., Roland, M., Rollmann, B., and Speiser, P. (1984). Degradation of poly(isobutyl cyanoacrylate) nanoparticles. *Biomater. 5*: 65.
276. Stein, M., and Hamacher, E. (1992). Degradation of polybutyl 2-cyanoacrylate microparticles. *Int. J. Pharm. 80*: R11.
277. Couvreur, P., and Vauthier, C. (1991). Polyalkylcyanoacrylate nanoparticles as drug carrier: Present state and perspectives. *J. Controlled Rel. 17*: 187.
278. Müller, R. H., Lherm, C., Herbot, J., and Couvreur, P. (1990). In vitro model for the degradation of alkylcyanoacrylate nanoparticles. *Biomater. 11*: 590.
279. Müller, R. H., Lherm, C., Herbot, J., Blunk, T., and Couvreur, P. (1992). Alkylcyanoacrylate drug carriers. I. Physicochemical characterization of nanoparticles with different alkyl chain length. *Int. J. Pharm. 84*: 1.

280. Vert, M., Li, S., and Garreau, H. (1991). More about the degradation of LA/GA-derived matrices in aqueous media. *J. Controlled Rel. 16*: 15.
281. Schäfer, V., v. Briesen, H., Rübsamen-Waigmann, H., Steffan, A. M., Royer, C., and Kreuter, J. (1994). Phagocytosis and degradation of human serum albumin microspheres and nanoparticles in human microphages. *J. Microencapsul.*, in press.
282. Kreuter, J., Mills, S. N., Davis, S. S., and Wilson, C. G. (1983). Polybutylcyanoacrylate nanoparticles for the delivery of (^{75}Se) norcholestenol. *Int. J. Pharm. 16*: 105.
283. Kreuter, J. (1983). Evaluation of nanoparticles as drug-delivery systems III: Materials, stability, toxicity, possibilities of targeting, and use. *Pharm. Acta Helv. 58*: 242.
284. Washington, C. (1990). Drug release from microdisperse systems: A critical review. *Int. J. Pharm. 58*: 1.
285. Cappel, M. J., and Kreuter, J. (1991). Effect of nanoparticles on transdermal drug delivery. *J. Microencapsul. 8*: 369.
286. Levy, M. Y., and Benita, S. (1990). Drug release from submicron O/W emulsion: A new in vitro kinetic evaluation model. *Int. J. Pharm. 66*: 29.
287. Kreuter, J., Täuber, U., and Illi, V. (1979). Distribution and elimination of poly(methyl-2-^{14}C-methacrylate) nanoparticle radioactivity after injection in rats and mice. *J. Pharm. Sci. 68*: 1443.
288. Kreuter, J. (1983). Evaluation of nanoparticles as drug-delivery systems II: Comparison of the body distribution of nanoparticles with the body distribution of microspheres (diameter $> 1\ \mu$m), liposomes, and emulsions. *Pharm. Acta Helv. 58*: 217.
289. Kreuter, J. (1985). Factors influencing the body distribution of polyacrylic nanoparticles. *Drug Targeting* (P. Buri and A. Gumma, eds.). Amsterdam: Elsevier, p. 51.
290. Lenaerts, V., Nagelkerke, J. F., van Berkel, T. J. C., Couvreur, P., Grislain, L., Roland, M., and Speiser, P. (1984). In vivo uptake of polyisobutyl cyanoacrylate nanoparticles by rat liver, Kupffer, endothelial and parenchymal cells. *J. Pharm. Sci. 73*: 980.
291. Kreuter, J., Wilson, C. G., Fry, J. R., Paterson, P., and Ratcliffe, J. H. (1984). Toxicity and association of polycyanoacrylate nanoparticles with hepatocytes. *J. Microencapsul. 1*: 253.
292. Waser, P. G., Müller, U., Kreuter, J., Berger, S., Munz, K., Kaiser, E., and Pfluger, B. (1987). Localization of colloidal particles (liposomes, hexylcyanoacrylate nanoparticles and albumin nanoparticles) by histology and autoradiography in mice. *Int. J. Pharm. 39*: 213.
293. Gipps, E. M., Groscurth, P., Kreuter, J., and Speiser, P. P. (1987). The effects of polyalkylcyanoacrylate nanoparticles on human normal and malignant mesenchymal cells in vitro. *Int. J. Pharm. 40*: 23.
294. Guise, V., Jaffray, P., Delattre, J., Puisieux, F., Adolphe, M., and Couvreur, P. (1987). Comparative cell uptake of propidium iodide associated with liposomes and nanoparticles. *Cell. Mol. Biol. 33*: 397.

295. Lherm, C., Müller, R. H., Puisieux, F., and Couvreur, P. (1992). Alkylcyanoacrylate drug carriers. II. Cytotoxicity of cyanoacrylate nanoparticles with different alkyl chain length. *Int. J. Pharm. 84*: 13.
296. Kante, B., Couvreur, B., Dubois-Krack, G., De Meester, C., Guiot, P., Roland, M., Mercier, M., and Speiser, P. (1982). Toxicity of polyalkylcyanoacrylate nanoparticles. I. Free nanoparticles. *J. Pharm. Sci. 71*: 786.
297. Cohn, Z. A., and Fedorko, M. E. (1969). The fate and formation of lysosomes. *Lysosomes in Biology and Pathology* (J. Dingle and F. Fell, eds.). Amsterdam: North Holland, p. 43.
298. Singer, J. M., Adlersberg, L., Hoenig, E. M., Ende, E., and Tchorsch, Y. (1969). Radiolabelled latex particles in the investigations of phagocytosis in vivo: Clearance curves and histological observations. *J. Reticuloendothel. Soc. 6*: 561.
299. Van Oss, C. J., Gillman, C. F., and Neumann, A. W. (1975). *Phagocytic Engulfment and Cell Adhesiveness as Cellular Surface Phenomena.* New York: Marcel Dekker.
300. Van Oss, C. J. (1978). Phagocytosis as a surface phenomenon. *Ann. Rev. Microbiol. 32*: 19.
301. Van Oss, C. J., Absolom, D. R., and Neumann, A. W. (1984). Interaction of phagocytes with other blood cells and with pathogenic and nonpathogenic microbes. *Ann. N.Y. Acad. Sci. 416*: 332.
302. Borchard, G., and Kreuter J. (1993). Interaction of serum components with poly(methylmethacrylate) nanoparticles and the resulting body distribution after intravenous injection in rats. *J. Drug. Target. 1*: 15.
303. Blunk, T., Hochstrasser, D. F., Rudt, S., and Müller, R. H. (1991). Two-dimensional electrophoresis in the concept of differential opsonication—An approach to drug targeting. *Arch. Pharm. 324*: 706.
304. Scieszka, J. F., and Cho, M. J. (1988). Cellular uptake of a fluid-phase marker by human neutrophils from solutions and liposomes. *Pharm. Res. 5*: 352.
305. Wassef, N. M., Matyas, G. R., and Alving, C. R. (1991). Complement-dependent phagocytosis of liposomes by macrophages: Suppressive effects of "stealth" lipids. *Biochem. Biophys. Res. Comm. 176*: 866.
306. Scieszka, J. F., Maggiora, L. L., Wright, S. D., and Cho, M. J. (1991). Role of complements C3 and C5 in the phagocytosis of liposomes by human neutrophils. *Pharm. Res. 8*: 65.
307. Kim, S. W., Lee, R. G., Oster, H., Coleman, D., Andrade, J. D., and Olsen, D. (1974). Platelet adhesion to polymer surfaces. *Trans. Amer. Soc. Artif. Intern. Organs 20*: 449.
308. Soderquist, M. E., and Walton, A. G. (1980). Structural changes in proteins adsorbed on polymer surfaces. *J. Colloid Interf. Sci. 75*: 386.
309. Ihlenfeld, J. V., and Cooper, S. L. (1979). Transient in vivo protein adsorption onto polymeric biomaterials. *J. Biomed. Mater. Res. 13*: 577.
310. Baszkin, A., and Lyman, D. J. (1980). The interaction of plasma proteins with polymers. I. Relationship between polymer surface energy and protein adsorption/desorption. *J. Biomed. Mater. Res. 14*: 393.

311. Chuang, H. Y. K., Mohammad, S. F., Sharma, N. C., and Mason, R. G. (1980). Interaction of human α-thrombin with artificial surfaces and reactivity of adsorbed α-thrombin. *J. Biomed. Mater. Res. 14*: 467.
312. Coleman, D. L. (1980). In vitro blood-materials interactions: A multi-test approach. Dissertation, University of Utah, Salt Lake City, p. 1.
313. Wilkins, D. J. (1967). The biological recognition of foreign from native particle as a problem in surface chemistry. *J. Colloid Interf. Sci. 25*: 84.
314. Wilkins, D. J. (1967). Interaction of charged colloids with the RES. *The Reticuloendothelial System and Arteriosclerosis* (N. R. Di Luzio and R. Paoletti, eds.). New York: Plenum Press, p. 25.
315. Illum, L., and Davis, S. S. (1983). Effect of the nonionic surfactant poloxamer 338 on the fate and deposition of polystyrene microspheres following intravenous administration. *J. Pharm. Sci. 72*: 1086.
316. Illum, L., and Davis, S. S. (1984). The organ uptake of intravenously administered colloidal particles can be altered using a non-ionic surfactant (poloxamer 338). *FEBS Letters 167*: 79.
317. Illum, L., Hunneyball, I. M., and Davis, S. S. (1986). The effect of hydrophilic coatings on the uptake of colloidal particles by the liver and by peritoneal macrophages. *Int. J. Pharm. 29*: 53.
318. Douglas, S. J., Davis, S. S., and Illum, L. (1986). Biodistribution of poly(butyl-2-cyanoacrylate) nanoparticles in rabbits. *Int. J. Pharm. 34*: 145.
319. Illum, L., Davis, S. S., Müller, R. H., Mak, E., and West, P. (1987). The organ distribution and circulation time of intravenously injected colloidal carriers sterically stabilized with a blockcopolymer—poloxamine 908. *Life Sci. 40*: 367.
320. Illum, L., and Davis, S. S. (1987). Targeting of colloidal particles to the bone marrow. *Life Sci. 40*: 1553.
321. Leu, D., Manthey, B., Kreuter, J., Speiser, P., and DeLuca, P. P. (1984). Distribution and elimination of coated polymethyl [2-^{14}C] methacrylate nanoparticles after intravenous injection in rats. *J. Pharm. Sci. 73*: 1433.
322. Tröster, S. D., Müller, U., and Kreuter, J. (1990). Modification of the body distribution of poly(methyl methacrylate) nanoparticles in rats by coating with surfactants. *Int. J. Pharm. 61*: 85.
323. Tröster, S. D., and Kreuter, J. (1992). Influence of the surface properties of low contact angle surfactants on the body distribution of ^{14}C-poly(methyl methacrylate) nanoparticles. *J. Microencapsul. 9*: 19.
324. Kreuter, J., Nefzger, M., Liehl, E., Czok, R., and Voges, R. (1983). Distribution and elimination of poly(methyl methacrylate) nanoparticles after subcutaneous administration to rats. *J. Pharm. Sci. 72*: 1146.
325. Adlersberg, L., Singer, J. M., and Ende, E. (1969). Redistribution and elimination of intravenously injected latex particles in mice. *J. Reticuloendothel. Soc. 6*: 536.
326. Tröster, S. (1990). Veränderung der Körperverteilung kolloidaler Arzneistoffträger, speziell von Nanopartikeln, nach intravenöser Applikation. Dissertation, University of Frankfurt, p. 178.

327. Müller, R. H. (1990). *Colloidal Carriers for Controlled Drug Delivery and Targeting*. Stuttgart: Wissenschaftliche Verlagsgesellschaft, p. 211.
328. Carstensen, H., Müller, B. W., and Müller, R. H. (1991). Adsorption of ethoxylated surfactants on nanoparticles. I. Characterization by hydrophobic interaction chromatography. *Int. J. Pharm. 67*: 29.
329. Wesemeyer, H., Müller, B. W., and Müller, R. H. (1993). Adsorption of ethoxylated surfactants on nanoparticles. II. Determination of adsorption enthalpy by microcalorimetry. *Int. J. Pharm. 89*: 33.
330. Johnson, B. A., Kreuter, J., and Zografi, G. (1986). Effects of surfactants and polymers on advancing and receding contact angles. *Colloids Surfaces 17*: 325.
331. Müller, R. H., and Wallis, K. H. (1993). Surface modification of i.v. injectable biodegradable nanoparticles with poloxamer polymers and poloxamine 908. *Int. J. Pharm. 89*: 25.
332. Ibrahim, A., Couvreur, P., Roland, M., and Speiser, P. (1983). New magnetic drug carrier. *J. Pharm. Pharmacol. 35*: 59.
333. Kharkevich, D. A., Alyautdin, R. N., and Filippov, V. I. (1989). Employment of magnet-susceptible microparticles for the targeting of drugs. *J. Pharm. Pharmacol. 41*: 286.
334. Widder, K. J., Marino, P. A., Morris, R. M., Howard, D. P., Poore, G. A., and Senyei, A. D. (1983). Selective targeting of magnetic albumin microspheres to the Yoshida sarcoma: Ultrastructural evaluation of microsphere disposition. *Eur. J. Cancer Clin. Oncol. 19*: 141.
335. Papisov, M. I., Savelyev, V. Y., Sergienko, V. B., and Torchillin, V. P. (1987). Magnetic drug targeting. I. In vivo kinetics of radiolabelled magnetic drug carriers. *Int. J. Pharm. 40*: 201.
336. Papisov, M. I., and Torchillin, V. P. (1987). Magnetic drug targeting. II. Targeted drug transport by magnetic microparticles: Factors influencing therapeutic effect. *Int. J. Pharm. 40*: 207.
337. Illum, L., Jones, P. D. E., Kreuter, J., Baldwin, R. W., and Davis, S. S. (1983). Adsorption of monoclonal antibodies to polyhexylcyanoacrylate nanoparticles and subsequent immunospecific binding to tumor cells in vitro. *Int. J. Pharm. 17*: 65.
338. Illum, L., Jones, P. D. E., Baldwin, R. W., and Davis, S. S. (1984). Tissue distribution of poly(hexyl 2-cyanoacrylate) nanoparticles coated with monoclonal antibodies in mice bearing human tumor xenografts. *J. Pharmacol. Exp. Therap. 230*: 733.
339. Manil, L., Roblot-Treupel, L., and Couvreur, P. (1986). Isobutyl cyanoacrylate nanoparticles as a solid phase for an efficient immunoradiometric assay. *Biomater. 7*: 212.
340. Widder, K. J., Senyei, A. E., Ovida, H., and Paterson, P. Y. (1981). Specific cell binding using staphylococcal protein A magnetic microspheres. *J. Pharm. Sci. 70*: 387.
341. Couvreur, P., and Aubry, J. (1983). Monoclonal antibodies for the targeting of drugs: Application to nanoparticles. *Topics in Pharmaceutical Sciences 1983* (D. D. Breimer and P. Speiser, eds.). Amsterdam: Elsevier, p. 305.

342. Illum, L., and Jones, P. D. E. (1985). Attachment of monoclonal antibodies to microspheres. *Methods in Enzymology, Vol. 112* (K. J. Widder and R. Green, eds.). Orlando, Fla.: Academic Press, p. 67.
343. Roland, A., Bourel, D., Genetet, B., and Le Verge, R. (1987). Monoclonal antibodies covalently coupled to polymethacrylic nanoparticles: In vitro specific targeting to human T lymphocytes. *Int. J. Pharm. 39*: 179.
344. Gipps, E. M., Arshady, R., Kreuter, J., Groscurth, P., and Speiser, P. P. (1986). Distribution of polyhexyl cyanoacrylate nanoparticles in nude mice bearing human osteosarcoma. *J. Pharm. Sci. 75*: 256.
345. Groscurth, P., Gipps, E., and Kreuter, J. (1985). Distribution of polyhexyl-cyanoacrylate nanoparticles in nude mice bearing human osteosarcoma. *Immune-Deficient Animals in Biomedical Research* (J. Rygaard, N. Brünner, N. Graem, and M. Spang-Thomsen, eds.). Basel: Karger, p. 401.
346. Illum, L., Wright, J., and Davis, S. S. (1989). Targeting of microspheres to sites of inflammation. *Int. J. Pharm. 52*: 221.
347. Diepold, R., Kreuter, J., Guggenbuhl, P., and Robinson, J. R. (1989). Distribution of poly-hexyl-2-cyano-[3-^{14}C]acrylate nanoparticles in healthy and chronically inflamed rabbit eyes. *Int. J. Pharm. 54*: 149.
348. Alpar, H. O., Field, W. N., Hyde, R., and Lewis, D. A. (1989). The transport of microspheres from the gastrointestinal tract to inflammatory air pouches in the rat. *J. Pharm. Pharmacol. 41*: 194.
349. Gupta, P. (1990). Drug targeting in cancer chemotherapy: A clinical perspective. *J. Pharm. Sci. 79*: 949.
350. Kreuter, J., Nefzger, M., Liehl, E., Czok, R., and Voges, R. (1983). Distribution and elimination of poly(methyl methacrylate) nanoparticles after subcutaneous administration to rats. *J. Pharm. Sci. 72*: 1146.
351. Schindler, A., Jeffcoat, R., Kimmel, G. L., Pitt, C. G., Wall, M. E., and Zweidinger, R. (1977). Biodegradable polymers for sustained drug delivery. *Contemporary Topics in Polymer Science, Vol. 2* (E. M. Pearce and J. F. Schaefgen, eds.). New York: Plenum Press, p. 251.
352. Oppenheimer, B. S., Oppenheimer, E. T., Danishefsky, I., Stout, A. P., and Eirich, F. R. (1955). Further studies of polymers as carcinogenic agents in animals. *Cancer Res. 15*: 333.
353. Kreuter, J. (1974). Neue Adjuvantien auf Polymethylacrylatbasis, Dissertation Nr. 5417, ETH Zürich, p. 100.
354. Kreuter, J., Müller, U., and Munz, K. (1989). Quantitative and microautoradiographic study on mouse intestinal distribution of polycyanoacrylate nanoparticles. *Int. J. Pharm. 55*: 39.
355. Kreuter, J. (1991). Peroral administration of nanoparticles. *Adv. Drug Deliv. Rev. 7*: 71.
356. Jani, P., Halbert, G. W., Langridge, J., and Florence, A. T. (1989). The uptake and translocation of latex nanospheres and microspheres after oral administration to rats. *J. Pharm. Pharmacol. 41*: 809.
357. Payne, J. M., Sansom, B. F., Garner, R. J., Thomson, A. R., and Miles, B. J. (1960). Uptake of small particles (1–5 μm diameter) by the alimentary canal of the calf. *Nature 188*: 586.

358. Sanders, E., and Ashworth, C. T. (1961). A study of particulate intestinal absorption and hepatocellular uptake. Use of polystyrene latex particles. *Exp. Cell Res. 22*: 137.
359. Scherer, D., Mooren, F. C., Kinne, R. H. K., and Kreuter, J. (1993). In vitro permeability of PBCA nanoparticles through porcine small intestine. *J. Drug Target 1*: 21.
360. Eldridge, J. H., Hammond, C. J., Meulbroek, J. A., Staas, J. K., Gilley, R. M., and Tice, T. R. (1990). Controlled vaccine release in the gut-associated lymphoid tissues. I. Orally administered biodegradable microspheres target the Peyer's patches. *J. Controlled Rel. 11*: 205.
361. Jani, P., Halbert, G. W., Langridge, J., and Florence, A. T. (1990). Nanoparticle uptake by rat gastrointestinal mucosa: Quantitation and particle size dependency. *J. Pharm. Pharmacol. 42*: 821.
362. Jani, P. U., Florence, A. T., and McCarthy, D. E. (1992). Further histological evidence of the gastrointestinal absorption of polystyrene nanospheres in the rat. *Int. J. Pharm. 84*: 245.
363. Alpar, H. O., Field, W. N., Hyde, R., and Lewis, D. A. (1989). The transport of microspheres from the gastrointestinal tract to inflammatory air pouches in the rat. *J. Pharm. Pharmacol. 41*: 194.
364. Nefzger, M., Kreuter, J., Voges, R., Liehl, E., and Czok, R. (1984). Distribution and elimination of polymethyl methacrylate nanoparticles after peroral administration to rats. *J. Pharm. Sci. 73*: 1309.
365. Sjöholm, I., and Edman, P. (1979). Acrylic microspheres in vivo. I. Distribution and elimination of polyacrylamide microparticles after intravenous and intraperitoneal injection in mouse and rat. *J. Pharmacol. Exp. Therap. 211*: 656.
366. Pimienta, C., Chouinard, F., Labib, A., and Lenaerts, V. (1992). Effect of various poloxamer coatings on in vitro adhesion of isohexylcyanoacrylate nanospheres to rat ileal segments under liquid flow. *Int. J. Pharm. 80*: 1.
367. Pimienta, C., Lenaerts, V., Cadieux, C., Raymond, P., Juhasz, J., Simard, M.-A., and Jolicoeur, C. (1990). Mucoadhesion of hydroxypropylmethacrylate nanoparticles to rat intestinal ileal segments in vitro. *Pharm. Res. 7*: 49.
368. Lee, V. H. K., and Robinson, J. R. (1979). Mechanistic and quantitative evaluation of precorneal pilocarpine disposition in albino rabbits. *J. Pharm. Sci. 68*: 673.
369. Fitzgerald, P., Hadgraft, J., and Wilson, C. G. (1987). A gamma scintigraphic evaluation of the precorneal residence of liposomal formulations in the rabbit. *J. Pharm. Pharmacol. 39*: 487.
370. Patton, T. F., and Robinson, J. R. (1976). Quantitative precorneal disposition of topically applied pilocarpine nitrate in rabbit eyes. *J. Pharm. Sci. 65*: 1295.
371. Wood, R. W., Li, V. H. K., Kreuter, J., and Robinson, J. R. (1985). Ocular disposition of poly-hexyl-2-cyano 3-[^{14}C] acrylate nanoparticles in the albino rabbit. *Int. J. Pharm. 23*: 175.
372. Fitzgerald, P., Hadgraft, J., Kreuter, J., and Wilson, C. G. (1987). A γ-scintigraphic evaluation of microparticulate ophthalmic delivery systems: Liposomes and nanoparticles. *Int. J. Pharm. 40*: 81.

373. Kreuter, J. (1990). Nanoparticles as bioadhesive ocular drug delivery systems. *Bioadhesive Drug Delivery Systems* (V. Lenaerts and R. Gurny, eds.). Boca Raton, Fla.: CRC Press, p. 203.
374. Lemperle, G., and Reichelt, M. (1973). Der Lipofundin® Clearance Test. *Med. Klin.* (München) *68*: 48.
375. Rolland, A., Étienne, P. L., Kerbrat, P., and Le Verge, R. (1988). Développement et perspectives thérapeutiques d'un nouveau système vecteur médicamenteux injectable, constitué d'un agent cytostatique (doxorubicine) fixé sur des nanosphères à base de copolymerès méthacryliques. *Cancer Commun.* *2*: 7.
376. Gipps, E. M., Groscurth, P., Kreuter, J., and Speiser, P. P. (1988). Distribution of polyhexylcyanoacrylate nanoparticles in nude mice over extended times and after repeated injection. *J. Pharm. Sci.* *77*: 208.
377. Trouet, A., and Tulkens, P. (1981). Intracellular penetration and distribution of antibiotics: The basis for an improved chemotherapy of intracellular infections. *The Future of Antibiotherapy and Antibiotic Research* (L. Ninet, P. E. Bost, D. H. Bouanchand, and J. Florent, eds.). London: Academic Press, pp. 337–349.
378. Brajtburg, J., Powderly, W. G., Kobayashi, G. S., and Medoff, G. (1990). Amphotericin B: Delivery systems. *Antimicrob. Agents Chemother.* *34*: 381.
379. Johnson, J. D., Hand, W. L., Francis, J. B., King-Thompson, N., and Corwin, R. W. (1980). Antibiotic uptake by alveolar macrophages. *J. Lab. Clin. Med.* *95*: 429.
380. Barza, M. (1981). Principles of tissue penetration of antibiotics. *J. Antimicrob. Chemother.* *8* (Suppl. C): 7.
381. Eltahawy, A. T. (1983). The penetration of mammalian cells by antibiotics. *J. Antimicrob. Chemother.* *11*: 293.
382. Lam, C., and Mathison, G. E. (1982). Intraphagocytic protection of staphylococci from extracellular penicillin. *J. Med. Microbiol.* *15*: 373.
383. Fattal, E., Rojas, J., Youssef, M., Couvreur, P., and Andremont, A. (1991). Liposome-entrapped ampicillin in the treatment of experimental murine listerosis and salmonellosis. *Antimicrob. Agents Chemother.* *35*: 770.
384. Fattal, E., Rojas, J., Andremont, A., and Couvreur, P. (1989). Efficacité comparée de nanoparticules et de liposomes chargés en ampicilline dans la traitement de la listériose et de la salmonellose expérimentales chez la souris. *Proc. 5th Int. Conf. Pharm. Technol., Paris 1989, Vol. 3*, APGI, Chatenay Malabry, p. 72.
385. Couvreur, P., Fattal, H., Alphandary, H., Puisieux, F., and Andremont, A. (1992). Intracellular targeting of antibiotics by means of biodegradable nanoparticles. *J. Controlled Rel.* *19*: 259.
386. Al-Khateeb, G. H., and Nolan, A. L. (1981). Efficacy of some drugs on *Leishmania donovani* in the golden hamster, *Mesocricretus auratus. Chemotherapy* *27*: 117.
387. Lherm, C., Couvreur, P., Loiseau, P., Bories, C., and Gayral, P. (1987). Unloaded polyisobutylcyanoacrylate nanoparticles: Efficacy against bloodstream trypanosomes. *J. Pharm. Pharmacol.* *39*: 650.

388. Gaspar, R., Préat, V., Opperdoes, F., and Roland, M. (1992). Macrophage activation by polymeric nanoparticles of polyalkylcyanoacrylates: Activity against intracellular *Leishmania donovani* associated with hydrogen peroxide production. *Pharm. Res. 9*: 782.
389. Gendelmann, H. E., Orenstein, J. M., Martin, M. A., Ferrua, C., Mitra, R., Phipps, T., Wahl, L. A., Lane, H. C., Fauci, A. S., Burke, D. S., Skillman, D., and Meltzer, M. A. (1988). Efficient isolation and propagation of human immunodeficiency virus on recombinant colony-stimulating-factor-1-treated monocytes. *J. Exp. Med. 167*: 1428.
390. Kühnel, H., v. Breisen, H., Dietrich, U., Adamski, M., Mix, D., Biesert, L., Kreutz, R., Immelmann, A., Meichsner, C., Andreesen, R., Gelderblom, H., and Rübsamen-Waigmann, H. (1989). Molecular cloning of two West African human deficiency virus type 2 isolates that replicate well in macrophages: A Gambian isolate, from a patient with neurologic acquired immunodeficiency syndrome, and a highly divergent Ghanian isolate. *Proc. Natl. Acad. Sci.* (Wash.) *86*: 2383.
391. v. Briesen, H., Andreesen, R., Esser, R., Brugger, W., Meichsner, C., Becker, K., and Rübsamen-Waigmann, H. (1990). Infection of monocytes/macrophages by HIV in vitro. *Res. Virol. 141*: 225.
392. Roy, S., and Wainberg, M. A. (1988). Role of the mononuclear phagocyte system in the development of the acquired immunodeficiency syndrome (AIDS). *J. Leukocyte Biol. 43*: 91.
393. Meltzer, M. S., Skillman, D. R., and Gendelmann, H. E. (1990). Macrophages and the human immunodeficiency virus. *Immunol. Today 6*: 217.
394. Schäfer, V. M., v. Briesen, H., Tröster, S. D., Andreesen, R., Kreuter, J., and Rübsamen-Waigmann, H. (1991). How to get antiviral drugs into HIV-infected cells? Phagocytosis of nanoparticles by human macrophages. *Microbiologist 2*: 117.
395. Schäfer, V., v. Briesen, H., Andreesen, R., Steffan, A.-M., Royer, C., Tröster, S., Kreuter, J., and Rübsamen-Waigmann, H. (1992). Phagocytosis of nanoparticles by human immunodeficiency virus (HIV)-infected macrophages: A possibility for antiviral drug targeting. *Pharm. Res. 9*: 541.
396. Esser, R., v. Briesen, H., Ceska, M., Glienke, W., Müller, S., Rehm, A., Rübsamen-Waigmann, H., and Andreesen, R. (1990). Secretory repertoire of HIV-infected human monocytes/macrophages. *Pathobiology 59*: 219.
397. Chavany, C., Saison-Behmoaras, T., Le Doan, T., Puisieux, F., Couvreur, P., and Hélène, C. Adsorption of oligonucleotides onto polyisohexylcyanoacrylate nanoparticles protects them against nucleases and increases their cellular uptake. *Pharm. Res.*, in press.
398. Beck, P., Kreuter, J., Müller, W. E. G., and Schatton, W. Improved peroral delivery of avarol with polybutylcyanoacrylate-nanoparticles. *Eur. J. Pharm. Biopharm.*, in press.
399. Alpar, H. O., Field, W. N., Hayes, K., and Lewis, D. A. (1990). A possible use of orally administered microspheres in the treatment of inflammation. *J. Pharm. Pharmacol. 41* (Suppl.): 50P.

400. Gurny, R. (1981). Preliminary study of prolonged acting drug delivery system for the treatment of glaucoma. *Pharm. Acta Helv. 56*: 130.
401. Gurny, R. (1986). Ocular therapy with nanoparticles. *Polymeric Nanoparticles and Microspheres* (P. Guiot and P. Couvreur, eds.). Boca Raton, Fla.: CRC Press, p. 127.
402. Gurny, R., Boye, T., and Ibrahim, H. (1985). Ocular therapy with nanoparticulate systems for controlled drug delivery. *J. Controlled Rel. 2*: 353.
403. Gurny, R., Ibrahim, H., Aebi, A., Buri, P., Wilson, C. G., Washington, N., Edman, P., and Camber, O. (1987). Design and evaluation of controlled release systems for the eye. *J. Controlled Rel. 6*: 367.
404. Ibrahim, H., Bindschaedler, C., Doelker, E., Buri, P., and Gurny, R. (1991). Concept and development of ophthalmic pseudo-latexes triggered by pH. *Int. J. Pharm. 77*: 211.
405. Losa, C., Marchal-Heussler, L., Orallo, F., Vila Jato, J. L., and Alonso, M. J. (1993). Design of a new formulations for topical ocular administration: Polymeric nanocapsules containing metipranolol. *Pharm. Res. 10*: 80.
406. Seale, L., Love, W. G., Amos, N., Williams, B. D., and Kellaway, I. W. (1992). Accumulation of polyvinylpyrrolidone within the inflamed paw of adjuvant-induced arthritic rats. *J. Pharm. Pharmacol. 44*: 10.
407. De Schrijver, M., Streule, K., Senekowitsch, R., and Friedrich, R. (1987). Scintigraphy of inflammation with nanometer-sized colloidal tracers. *Nucl. Med. Commun. 8*: 895.
408. Williams, B. D., O'Sullivan, M. M., Saggu, G. S., Williams, K. E., Williams, L. A., and Morgan, J. R. (1986). Imaging in rheumatoid arthritis using liposomes labelled with technetium. *Br. Med. J. 293*: 1143.
409. Love, W. G., Amos, N., Kellaway, I. W., and Williams, B. D. (1989). Specific accumulation of technetium-99m radiolabelled, negative liposomes in the inflamed paws of rats with adjuvant induced arthritis: Effect of liposome size. *Ann. Rheum. Dis. 48*: 143.
410. Mizushima, Y., Wada, Y., Etoh, Y., and Watanabe, K. (1983). Antiinflammatory effects of indomethacin incorporated in a lipid microsphere. *J. Pharm. Pharmacol. 35*: 398.
411. Mizushima, Y. (1985). Lipid microspheres as novel drug carriers. *Drugs Exptl. Clin. Res. 11*: 595.
412. Grafe, A. (1971). Hochdisperses δ-Al_2O_2-Aerosol als Adjuvans bei der Herstellung inaktiver Virus- oder Impfstoffe. *Arzneim.-Forsch. 21*: 903.
413. Kerkhof, N. J., White, J. L., and Hem, S. L. (1975). Effect of dilution on reactivity and structure of aluminum hydroxide gel. *J. Pharm. Sci. 64*: 940.
414. Nail, S. L., White, J. L., and Hem, S. L. (1976). Structure of aluminum hydroxide gel. I. Initial precipitate. *J. Pharm. Sci. 65*: 1188.
415. Nail, S. L., White, J. L., and Hem, S. L. (1976). Structure of aluminum hydroxide gel. II. Aging mechanism. *J. Pharm. Sci. 65*: 1192.
416. Nail, S. L., White, J. L., and Hem, S. L. (1976). Structure of aluminum hydroxide gel. III. Mechanism of stabilization by sorbitol. *J. Pharm. Sci. 65*: 1195.

417. Wunderli, H. K. (1950). Untersuchungen über die Adsorption des Maul- und Klauenseuchevirus an Aluminiumhydroxid. *Z. Hyg. Infektionskr. 132*: 1.
418. Pyl, G. (1953). Die Prüfung von Aluminiumhydroxid auf seine Eignung für die Maul- und Klauenseuchevakzine. *Arch. Exp. Veterinärmed. 7*: 9.
419. Muggleton, P. W., and Hilton, M. L. (1967). Some studies on a range of adjuvant systems for bacterial vaccines. *International Symposium on Adjuvants of Immunity, Utrecht 1966* (R. E. Regamey, W. Hennessen, D. Ikic, and J. Ungar, eds.). *Symposia Series Immunobiological Standardization, Vol. 6.* Basel: Karger, p. 29.
420. Haas, R., and Thommssen, R. (1960). Über den Entwicklungsstand der in der Immunbiologie gebräuchlichen Adjuvantien. *Ergeb. Mikrobiol. Immunitätsforsch. Exp. 34*: 27.
421. Jolles, P., and Paraf, A. (1973). Substances exhibiting an adjuvant effect: Preparation of 'crude material.' *Chemical Biological Basis of Adjuvants. Molecular Biology, Biochemistry, and Biophysics*, Vol. 13 (A. Kleinzeller, G. F. Springer, and H. G. Wittmann, eds.). Berlin: Springer, p. 5.
422. Stieneker, F., Kreuter, J., and Löwer, J. (1993). Different kinetics of the humoral immune response to inactivated HIV-1 and HIV-2 in mice: Modulation by PMMA nanoparticles adjuvant. *Vaccine Res. 2*: 111.
423. Kreuter, J. (1992). Physicochemical characterization of nanoparticles and their potential for vaccine preparation. *Vaccine Res. 1*: 93.

Index

9 780824 792145